MW00782487

Optimization Heuristics in Econometrics

Optimization Heuristics in Econometrics
Applications of Threshold Accepting

Peter Winker
University of Mannheim, Germany

JOHN WILEY & SONS, LTD

Chichester • New York • Weinheim • Brisbane • Singapore • Toronto

Copyright ©2001   John Wiley & Sons, Ltd
Baffins Lane, Chichester,
West Sussex, PO19 1UD, England

National   01243 779777
International (+44) 1243 779777

e-mail (for orders and customer service enquiries): cs-books@wiley.co.uk

Visit our Home Page on http://www.wiley.co.uk or http://www.wiley.com

Other Wiley Editorial Offices

John Wiley & Sons, Inc., 605 Third Avenue,
New York, NY 10158-0012, USA

Wiley-VCH Verlag GmbH
Pappelallee 3, D-69469 Weinheim, Germany

Jacaranda Wiley Ltd, 33 Park Road, Milton,
Queensland 4064, Australia

John Wiley & Sons (Asia) Pte Ltd, 2 Clementi Loop #02-01,
Jin Xing Distripark, Singapore 129809

John Wiley & Sons (Canada) Ltd, 22 Worcester Road,
Rexdale, Ontario, M9W 1L1, Canada

British Library Cataloguing in Publication Data

A catalogue record for this book is available from the British Library

ISBN 0-471-85631-2

Produced from PostScript files supplied by the author.
Printed and bound in Great Britain by Antony Rowe, Chippenham, Wiltshire.
This book is printed on acid-free paper responsibly manufactured from sustainable forestry in which at least two trees are planted for each one used for paper production.

Contents

IV   Conclusion and Outlook   299

Preface

This book is intended for statisticians and econometricians with a general interest in new optimization paradigms. It offers a general introduction to optimization heuristics in statistical and econometrical applications as well as a deeper description of the threshold accepting heuristic and its application to highly complex problems in statistics and econometrics. The applications are not only chosen as suitable examples, but are also of their own economic interest. Thus, the book may serve as a research monograph for this specific method, as well as including some textbook material on optimization heuristics in statistics and econometrics in general. It may also be used for studying the results for some specific optimization problems in economics, which are obtained by using the discussed methods.

This book bridges the gap between the marked advances of the 1980s and 1990s in the field of optimization and applied research in economics. Most of the global optimization algorithms discussed in this book have become standard tools in operational research, but have remained almost unnoticed in the fields of econometrics and statistics. However, many applications and problems in econometrics and statistics offer themselves naturally to the use of such optimization tools, as the examples in this book demonstrate. Thus, it might be expected that the use of such techniques will become more frequent in the near future.

The overall framework of the book is as follows. In the first part, the heuristic optimization paradigm is introduced by using some examples from statistics and econometrics, without yet giving all the theoretical background for the problems and all the implementation details of the algorithms. Nevertheless, the discussion of these problems shows the limits of classical methodologies in the face of highly complex problems and the opportunities of optimization heuristics. The second part provides a general introduction to optimization heuristics and a detailed discussion of the threshold accepting algorithm. Together with an in–depth description of some applications in the third part, a guideline for the implementation of this and similar techniques to other problems in statistics and econometrics is then presented.

Since applied statistics and econometrics consist mainly of solving optimization problems, a complete discussion of the potential of the heuristic optimization paradigm is clearly beyond the scope of this present book. However, it may open the door for researchers to attempt yet unsolved problems. Thus, the best outcome for this book is that many other problems in statistics and econometrics will be tackled by using methods similar to the ones described here.

How to read this book

The scope of this book as outlined in this preface is a quite comprehensive one. Consequently, besides material of general interest, the book also covers some more specific themes which might only be of interest to a fraction of its potential readers. A guideline to the different chapters of this book might be helpful in this case. Much effort has been spent on the introduction. This should provide the reader with a first intuitive understanding as to why optimization heuristics might be a useful tool in statistics and econometrics and what the main ingredients for a good implementation of such heuristics are. Therefore, it should be the starting point for all readers.

Researchers, who have already gained some acquaintance with non-standard optimization tools and are mainly interested in their own implementations of the threshold accepting heuristic, might skip directly to Chapter 6. However, the remaining chapters of Part II should be of interest, since they provide some technical details for the implementation of threshold accepting. Finally, readers might pick up one or more of the examples discussed in Part III, which may exhibit some similarities with the optimization problems that they want to treat. The discussion of these specific implementations should provide a guideline for the set–up of threshold accepting to further optimization problems from statistics or econometrics.

If the book is to be used as a first access to optimization heuristics in the context of statistics and econometrics, then Chapters 3 and 4 will provide an intuitive background to different optimization methods and the complexity issue, which is a major argument in favour of using optimization heuristics. Furthermore, Chapter 5 gives an outline of several optimization methods, including standard and heuristic approaches, which can be used to classify some features of the threshold accepting heuristic. While Chapter 6 provides the exact description of threshold accepting, the remaining chapters of Part II describe implementation and tuning details, which become most relevant when readers plan to implement threshold accepting for their own purposes. Again, a study of some of the examples in Part III might convince the reader that using this heuristic can be fruitful in different areas of statistics and econometrics.

Finally, the results of applied work in econometrics should also be of benefit to researchers less familiar with the specificities of different methods in this area. For this group of potential readers, Chapters 2 to 4 are possibly the most important ones, since they provide a substantiation for a change in the optimization paradigm, which is much more general than the few applications of the specific alternative heuristic provided in the later parts of the book suggest. If these chapters achieve their goal, the reader might become interested in reading more about the details. Then, in particular, Chapters 6 and 7, as well as Chapter 9 from Part II are recommended for further study before turning to some of the applications presented in Part III.

Of course, these guidelines to the contents of the book provide only my own personal view. Any other approach has its own rights and is also welcome.

Acknowledgements

The preparation of this book would have been impossible without the experience gathered in the use of threshold accepting over many years, and the support of my colleagues at the Universities of Mannheim and Konstanz. I am particularly indebted to G. Dueck for introducing me to the world of optimization heuristics as early as 1990, when I visited the Scientific Research Centre of IBM Germany in Heidelberg. During this stage my first application of threshold accepting to an economic problem, namely portfolio optimization, was implemented. Since then, I have continued to study the use of heuristics in different areas of statistics and econometrics, whenever complex problems resisted solution by using standard tools. I am grateful to W. Franz for supporting my research in this area, to J. Chipman, K.-T. Fang and B. Fitzenberger for contributing their efforts to some of the applications of threshold accepting, to M. Gilli, W. Smolny and K. Winckler, who contributed helpful comments on most of the contents of this book, and to the Editors R. Calver and S. Clutton and the referees, who significantly increased my motivation to finish this book and also helped to improve its readability. Helpful comments on parts of the material presented in this book have also been contributed by D. Belsley, P. Chan, B. Dorsey, F. Holland and A. H. Pesaran. I am also indebted to the Production Editor S. Corney. Any remaining errors and omissions lie solely with the author.

Mannheim, June 2000   Peter Winker

1

Introduction

The aim of this book may be summarized in a few sentences, as follows. It should provide a motivation which comes in two parts. First, a large part of all research in economics, statistics and econometrics is based explicitly or implicitly on some optimization paradigm. Secondly, the standard tools applied to the resulting optimization problems are often not appropriate. Either they provide an approximate solution of unknown quality, or, what might seem even worse, one is forced to use heavily simplifying assumptions in order to obtain a model which can be tackled by standard methods. Thus, the motivation for the material presented in this book is the limitations of standard methods applied to relevant optimization problems in economics, statistics and econometrics.

After providing this kind of motivation, the book aims at fascinating the reader. This goal is pursued by using some examples where optimization heuristics provide very powerful results despite their ease of implementation. Furthermore, the ideas behind some of these optimization heuristics or analogies to nature might be fascinating in themselves.

More information about the methods and their implementation is provided in the second part. Finally, new applications of a specific optimization heuristic, namely threshold accepting, are presented in Part III. These cover areas in statistics and econometrics where so far optimization was either not even made explicit or where only standard approaches have been used, thus resulting in results of poor quality. This documentation of successful applications, which might be fascinating or surprising each on their own, also transmits an idea about the potential of the optimization heuristic paradigm in related areas.

Therefore, my eventual goal is to provide readers with enough motivation, fascination, and information to start their own implementation of threshold accepting or similar algorithms in economics, statistics or econometrics. This is just the beginning.

1.1 Optimization is all around

When we talk about economics, statistics or econometrics, we talk about optimization. Households derive their *optimal consumption plans* from *utility maximization*, firms *maximize profits*, investment funds perform *portfolio optimization* with regard to risk and return, and firms in strategic interaction on some markets have to find *optimal strategies*. Statisticians search for efficient parameter estimators, i.e. estimators with *minimum variance*. Often, such estimators are *maximum–likelihood* estimators. For example, the *least squares estimator* most commonly used in econometrics is, under some assumptions, a maximum–likelihood estimator, but is always the result of explicit optimization. Finally, model selection, both in economic theory and econometrics, may serve as an example of optimization which is not always made explicit.

Although optimization is ubiquitous in these areas, it is handled differently depending on the specific application. While in the utility maximization approach, optimization is at the very heart of the analysis, the choice of relevant variables within a consumption model, in general, is not even considered as being an optimization problem. Similarly, in econometrics least squares estimation is often undertaken without explicitly noting the optimization background, since a closed form solution is available. In contrast, maximum–likelihood estimation of more complicated models often requires an extensive analysis of the optimization problem, in particular with regard to its numerical solution. Then, a closer look at the optimization problem and its solution is crucial.

Obviously, there are inherent differences in the optimization problems we tackle in economics and econometrics. Some are easy to solve and optimization is always applied to these instances. Other problems require a more careful analysis and sometimes simplifying assumptions in order to obtain a tractable optimization problem. Finally, there are problems which either are not yet analysed in the optimization context or are regarded as being intractable due to their high complexity.

Highly complex optimization problems also arise in the daily life of economic subjects. Quite often some ad hoc approaches are used, e.g. for daily consumption or stock market investments of private households. The decision rules to 'solve' these optimization problems are commonly not the result of some explicit optimization over the whole range of potential solutions taking into consideration all side constraints. In fact, most of such decisions are undertaken without even realizing that a complex, intertemporal, and possibly integer optimization problem is being solved. Instead, economic subjects follow some heuristic guidelines or personal experience. Given the costs of an explicit solution of the complete optimization problem and the gathering of all necessary information, such a heuristic approach may be even rational.

However, if the costs or benefits at stake are larger, the handling of optimization problems also becomes more elaborate. Flight or production scheduling, professional portfolio management, or internet routing have been described as optimization problems and the most up–to–date algorithms have been applied to provide at least some good approximations to the optimum. Obviously, it is rational to spend more resources on these problems where small gains in terms of relative performance correspond to large absolute gains. Nevertheless, some of the above mentioned applications exhibit such a high complexity that it is not possible to provide exact optimal solutions by using standard algorithms.

During the last two decades, an increasing number of heuristic optimization approaches has been developed and successfully implemented for such instances. In contrast to classical algorithms, they do not provide an exact solution. In practice, however, economic agents are satisfied with any improvement on the status quo situation, which commonly is given by ad hoc procedures based on experience or obtained from models using heavily simplifying assumptions. The success of simulated annealing, neural networks, or genetic algorithms in operational research indicates the potential of such approaches for the optimization problems that economic agents face in reality.

In contrast, we can hardly see a household using such methods for its scheduling or consumption plan problems. Again, this can be rationalized by the costs associated with such approaches. In fact, households, in general, do not even use classical optimization routines, but rely on experience, i.e. decision heuristics which have proved to be successful in the past.

The situation is different in economic theory and econometrics where classical optimization methods are applied routinely. Obviously, the cost associated with the gathering of data and the calculations involved in the optimization routines do not impose a binding constraint in this area. Therefore, it is even more surprising that successful optimization tools from economic practice are not used to a much larger degree in these areas. Instead, classical optimization tools prevail. More complex optimization problems are either reduced by simplifying assumptions, which are not always justified, or even neglected. It is, in fact, only recently, that a few applications of optimization heuristics have been proposed in econometrics. Compared to 'best practice', they have exhibited marked improvements.

Three reasons may be responsible for this neglect of modern optimization paradigms. First, ignorance about the existence, performance and easy implementation of optimization heuristics hinder their use. This book tries to make this argument less important by supplying both a guideline for the application of optimization heuristics and a collection of typical applications as they might arise in applied economic research. Secondly, the theoretical properties of optimization heuristics are much less well known than those of classical approaches. However, even for a large class of standard approaches no sound mathematical theory exists. Finally, a lack of computing power may have hindered the wider use of optimization heuristics in the past. Although the high complexity of some optimization problems precludes an exact solution, even with a large amount of computing power, the constraint imposed from this side can be expected to become smaller every year.

The approach proposed in this book is as follows. First, the tasks in economic theory or econometrics need to be analysed as to whether they may be described as being optimization problems. Then, it has to be decided whether these problems can be solved by using simple or standard optimization tools. If they seem to be too complex for the usual approaches, the application of optimization heuristics is indicated. Even if the problem cannot be solved to optimality by using such approaches and even if only a mediocre approximation to an optimum can be supplied, the results may still serve as a bench–mark for the approaches used so far, i.e. simplifying assumptions or neglecting the optimization framework.

1.2   A Flavour of Optimization Heuristics

Before turning to a short layout of this book, a few sentences about optimization heuristics in general and threshold accepting in particular may be appropriate in order to supply the reader with a flavour of the main ingredients in this volume.

The neglect of optimization issues and the limits of classical optimization tools became a topic earlier in other scientific disciplines, thus leading to a constant process of innovation in mathematics and optimization approaches. For example, the development of differential calculus may be regarded as a basic innovation for this purpose. This allowed simple optimization schemes like the one used in linear ordinary least squares regression, but also more refined gradient methods such as Newton's method and its further developments. However, for more complex optimization problems these methods failed to work efficiently. With the growing availability of computational resources, a new optimization paradigm based on concepts found in nature spread through different disciplines in the late 1980s and during the 1990s. The concepts include analogies to genetics, neural networks or simulated annealing. Following the seminal paper by Kirkpatrick, Gelatt and Vecchi (1983), such approaches grew in importance in disciplines ranging from physics to operational research.[1]

Let us consider a continuous optimization problem with a large number of local minima, as shown in Figure 1.1 as an example. Such problems arise in econometrics, e.g. in disequilibrium modelling or censored quantile regressions. The goal is to find x such that $f(x)$ reaches its minimum value on the considered interval, i.e. point F in Figure 1.1.

---

[1] Osman and Laporte (1996) provide a bibliography of related methods in operational research applications.

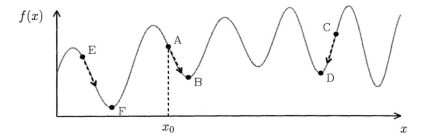

Figure 1.1 An optimization problem prototype.

A straightforward application of a standard gradient method will proceed as follows. An initial value x_0 is arbitrarily chosen in the interval. Starting from the corresponding point $(x_0, f(x_0))$, e.g. point A in the figure, the algorithm proceeds in the direction where the objective function values become smaller, i.e. along the dashed arrow from point A towards point B. If the parameters of the gradient method are suitably chosen, the algorithm will eventually reach point B, or a point very close to B. Unfortunately, this point represents only a local minimum of the objective function $f(x)$, i.e. close to point B no better solution can be found, but the value of $f(x)$ at this point is larger than at F. Thus, it does not represent the global minimum we are looking for. Running a standard gradient method from a point like A provides only information about objective function values along the path from A to B. Therefore, it is, in general, not possible to judge the quality of the result if many local minima are possible.

The result can be improved by starting the method several times from different initial values, e.g. points C and E. This heuristic extension in fact delivers the global minimum in this simple example, but it is unclear at the beginning how many restarts will be necessary to achieve this goal and which points should be chosen as starting points. Finally, there is no way of checking the result without knowledge about the whole function as provided in Figure 1.1.

The basic innovation of the kind of optimization heuristics which are used in this book, in particular, the threshold accepting heuristic, is their potential to move from points like B or D over the 'hills' of the functional landscape to better local minima such as F. It is this 'hill–climbing' behaviour which ensures that the algorithms do not get stuck in points like D, but eventually, i.e. with a large number of iterations, might reach the global minimum F, even if started from C.

Although Figure 1.1 might give a first rough idea about the working of different optimization approaches, it is not well suited to represent the complexity of the real optimization problems analysed in this volume. In particular, often the optimal solution **x** has to be found in a multivariate

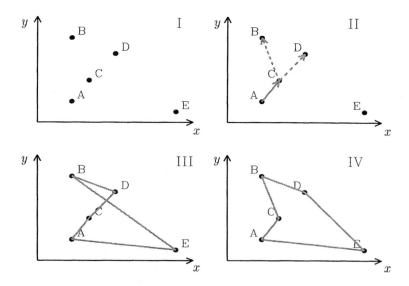

Figure 1.2   A simple travelling salesman problem.

set, with its components restricted to integer values, which, unfortunately, makes the task even more demanding. These problems provide hard bench–marks for standard and heuristic optimization heuristics. For example, the restart approach, which should work fine on a problem example like the one given in Figure 1.1, is much less efficient in a higher–dimensional framework.

Before providing additional motivation for the hill–climbing behaviour of modern optimization heuristics, a classical and complex optimization problem is first introduced, which shares some features with some of the econometric and statistical examples studied in Part III of this book. Figure 1.2 shows an instance of the so–called travelling salesman problem in the first quadrant. This is a classical integer optimization problem, which has been extensively studied in mathematical complexity theory and operational research.[2] Given a set of points (cities) {A,B,C,D,E} one has to find a circuit of shortest length connecting all these points. Obviously, problems like this have to be often solved by real economic subjects. Besides the travelling salesman we might think about general problems in logistics but as well of the layout of communication nets. Some of the problems arising in statistics and econometrics are also similar in their structure and complexity. However, this real life example may be better suited for introducing different concepts of optimization heuristics.

The first problem arises with the representation of the problem. While it is

---

[2] See also Cook, Cunningham, Pulleyblank and Schrijver (1998, pp. 241ff.) for a discussion of the problem and a recent overview on some solution methods.

intuitive to present it in the form of Figure 1.2, this representation is different
from the one used in Figure 1.1. In fact, the value of the objective function
'length of the circuit', cannot be easily derived from such a representation,
and the xs of Figure 1.1 are now given by complete circuits. A few other
presentation methods for such problems are presented below and in Part II of
this book, which are somewhat closer to the one given in Figure 1.1.

Let us move away from this problem of graphical presentation for the
moment and concentrate instead on the question, i.e. how the travelling
salesman could find his circuit. Although being a simple concept, it is
obviously not a good idea to follow the cities in lexicographical order. Another
more reasonable and constructive algorithm is described in quadrants II and
III of Figure 1.2. The starting point can be chosen arbitrarily, e.g. point A.
Then, the salesman is looking for the closest city to move to. This is city C.
There, he has to decide whether to continue via B or D. Obviously, the distance
to D is smaller, and thus he continues via D. The final circuit obtained by this
constructive approach is given in quadrant III.

Comparing the result of this sequential approach with the shortest circuit
given in quadrant IV, it becomes clear that the constructive approach does
not lead to an optimal solution. However, it is not evident how the approach
could be improved in order to result in the shortest circuit. A comparison of
the two circuits in the lower part of Figure 1.2 provides another starting point
towards finding the shortest tour. In fact, the suboptimal tour obtained by
the constructive approach can be transformed to the optimal one by simply
exchanging the position of cities B and D in the tour. Of course, for larger
problem examples, i.e. if the number of cities is much larger, it will not be
sufficient to exchange the position of just two cities. Then, repeating the
exchange steps of the kind just described might be a good idea. That is what
local search is all about! In fact, this idea allows us to describe the problem
similar to that in Figure 1.1 and to use hill–climbing optimization heuristics
for its (approximative) solution. We will come back to this aspect later in this
section.

Despite the intuition just given, the optimal circuit in quadrant IV has not
been obtained by such an exchange step. Unfortunately, even if it would have
been obtained that way, we would still not know whether it is the shortest tour.
In order to be sure of this, a third approach is followed, namely, all possible
circuits are enumerated and their lengths are calculated. Then, the shortest
tour is selected. With five cities, there are $4! = 24$ tours to be compared,[3] with
ten cities already $9! = 362\,880$ and for one hundred cities this number exceeds
10^{155}! Thus, this exact algorithm is not feasible for larger problem instances.

---

[3] As the tour length does not depend on the point of departure, one point can be selected
a priori. Furthermore, the tour length does not depend either on the direction of the
round trip. Consequently, by forming equivalence classes the number of tours to be
evaluated can be reduced slightly further.

Generalizing the findings for the travelling salesman problem, three different approaches towards optimization of complex functions can be distinguished: constructive algorithms, local search methods and enumeration algorithms. While the first two methods cannot guarantee to deliver the optimal solution to the problem (Aarts and Lenstra, 1997, p. 2), exact enumeration becomes infeasible for larger problem instances, since the number of potential solutions often grows at an exponential rate of the problem size.

While enumeration can be excluded for practical reasons for larger problems, the weighing up of the advantages and shortcomings of construction methods versus local search algorithms is less evident. Often, a suitably tailored construction method is more powerful for a specific problem instance than a more general local search heuristics approach. However, it is just the ease of implementation to new problems[4] which makes the use of the latter so appealing. Therefore, it is recommended to use local search heuristics at least for a first bench–mark implementation to optimization problems in statistics and econometrics. This choice can be rationalized by the following arguments:

1. Local search heuristics are easy to implement, and side constraints can be taken into consideration at low additional expense.
2. Local search heuristics share some familiarity with classical optimization paradigms such as gradient methods or deepest descent.
3. Local search heuristics have already been implemented to some real life problems as well as to applications in econometrics and statistics, where they exhibited a good or even very good performance. In fact, often it is difficult to better them even with highly problem–specific approaches.

Related to these advantages are reasons for the new or renewed interest in local search and other optimization heuristics. In particular, in research areas, such as statistics and econometrics, where they did not play an important role in the past, they have been used more frequently in recent years. This is attributed to four developments (Aarts and Lenstra, 1997, p. 2):

1. The appeal of analogies of local search heuristics with some real–world processes such as annealing, biological evolution or the Great Deluge.
2. The theory for some of the algorithms yielding results on their performance and some hints on successful implementations.
3. The large increase in computational resources has even enabled many applications to large problem instances. Consequently, a huge amount of practical experience was gathered.
4. The ease of implementation of some local search algorithms makes them a useful tool for tentative applications to real–world problems, where other methods have failed.

So far, we have emphasized the advantages of local search techniques and

---

[4] See the practical guide to the implementation of threshold accepting in Chapter 9.

mentioned that the optimization heuristics treated in this book, in particular threshold accepting, belong to the class of refined local search algorithms. The remaining parts of this section are devoted to the 'refinements' in local search added by these algorithms. In the example provided in Figure 1.2, a single exchange step leads to the optimal solution. However, as pointed out before, for real–world problems one single exchange step will not be sufficient and the algorithm should exhibit some hill–climbing potential to avoid those situations as depicted in Figure 1.1 for points B or D, where optimization stops in some local minimum, which is far away from the true global one. Therefore, in a local search heuristic we have to define which exchange steps to try and which exchange steps to accept. In fact, the differences between most of the local search algorithms mentioned here can be described in terms of these two choices. Obviously, for other optimization problems the meaning of 'exchange step' will be different and sometimes less intuitive than for the travelling salesman problem. Nevertheless, the following arguments apply to all other applications as well.

Let us have a look at the exchange steps first. Given the tour in quadrant III of Figure 1.2 it is a straightforward choice to exchange cities B and D and not, e.g. cities B and E. However, in more complex settings a graphical analysis is not possible. Therefore, this choice has to be made automatically by the algorithm. Maybe, there is a procedure describing our choice which can be coded in the algorithm? At least, it seems more natural to exchange cities, which are closer to each other than those far away. In fact, this idea is implemented in most local search heuristics, and not only for the travelling salesman problem. For any application, in each exchange step two potential solutions are compared, which are 'close' to one another under some definition of vicinity. We will come back to the general definition of such neighbourhoods later. In the case of the circuit problem, these neighbourhoods are given by requiring that the two cities to be exchanged in the tour are close to one another geographically. Thus, we could implement the algorithm in a way that it generates pairs of close cities and exchanges them in a given circuit. Of course, there is no reason to constrain such an exchange step to only two cities. In fact, in real applications the permutation of three and even four cities also resulted in highly performing algorithms. The ruin and recreate algorithm recently proposed by Schrimpf, Schneider, Stamm-Wilbrandt and Dueck (1998) even considers all cities situated in a quite large area which is randomly selected. Then, cities are not just exchanged, but a new sequence within the destroyed area is generated from scratch by following a simple constructive approach. The detailed discussion of several applications in Part II highlights the crucial importance of an adequate choice of such exchange steps, since by this definition the meaning of 'local' is pinned down. If the choice of exchange steps is not really local, then the resulting local search heuristic will be rather a random search heuristic. If on the other hand, the local neighbourhoods are too small, then the risk increases that a local search

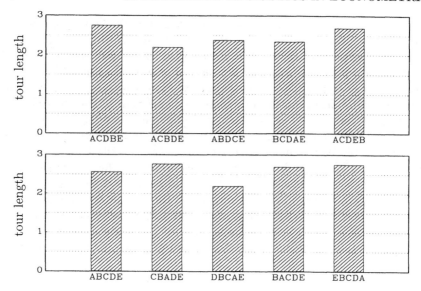

Figure 1.3   Local behaviour of tour length.

procedure will get stuck in some local minimum such as point B in Figure 1.1.

A rough idea about the local behaviour of the objective function, which is given by the tour length in the example, can be gained through the graphical presentations shown in Figure 1.3. In this figure, the tour lengths of different tours are plotted. In the upper panel, the first tour ACDBE is the tour from quadrant III in Figure 1.2. Starting with this tour, all possible exchanges of point B are analysed. These are sorted in increasing differences between B and the other points. Therefore, the second reported length is for tour ACBDE, since D is closest to B. It can be seen that exchanges with cities close to B reduce the tour length, while the exchange with the distant city E leaves the tour length more or less unchanged.

In the lower panel, the same experiment is repeated starting with the tour ABCDE, which is not the result of some constructive approach. Nevertheless, the shortest tour can also be obtained by simply exchanging points A and D. Notice that tour DBCAE is the same as tour ACBDE, just in reverse ordering, which, of course, does not change its length. However, this second example also shows that the local behaviour of the tour length, in general, is not monotonic in the distance of the points to be exchanged.

The second crucial ingredient to local search heuristics is the so–called acceptance criterion. After exchanging two cities in a tour, it will not always be the case that the new tour is shorter than the old one, as in the example of Figure 1.2. However, it is questionable whether it is a good idea to accept any such exchange and to hope that eventually, i.e. after a large number of exchanges, the algorithm will tend toward some short tour or even the shortest tour. To call such an approach questionable is a clear understatement. In fact, there is no chance that such an algorithm will tend towards good solutions. Even if by chance it reaches a short tour, the next exchange step will drive it away again.

A more promising acceptance criterion can be derived in analogy to the classical gradient methods depicted in Figure 1.1. There, all moves are directed towards smaller values of the objective function. For the travelling salesman, this analogy means that after each exchange step, the length of the resulting new tour is compared with the length of the previous tour. Only if the new tour is shorter, will it be accepted. Otherwise, a new exchange step is tried starting from the previous tour. Now, if the algorithm reaches a shortest tour at some iteration it will remain there since no further exchange step can reduce the objective function. This concept is headed 'classical' or 'standard local search' in the following. It represents a powerful heuristic for some integer optimization problems.

However, if a suitable graphical representation of the problem looks like Figure 1.1, where the different tours are mapped to the x–axis such that tours close to each other can be generated by exchanging two cities in the tours, and the y–axis shows tour lengths, any discrete analogy to standard gradient methods shares the same shortcomings. Starting with a tour corresponding to point A in Figure 1.1, a repeated application of exchange steps with the criterion of improving the objective function, i.e. finding shorter tours, will eventually lead to a tour corresponding to point B. Although this tour is much shorter than the original one, which might have been obtained by some simple constructive approach, it is not the shortest tour, which would correspond to point F.

How can this trap be avoided? For this purpose, a final ingredient has to be added, which makes the local search heuristic a 'meta–heuristic'. It is rather in this final ingredient that most of the refined local search algorithms differ than in the basic concepts described so far. Furthermore, it is in this final ingredient that analogies from nature, physics and other domains come into play, which provide some hill–climbing capacities or other features used to avoid local minima of bad quality.

Let us start with an analogy to human behaviour, which is more easily described in terms of a maximization problem. The assumption is that we are looking for the highest elevation in a given region of fixed size, e.g. the Himalayas. We abstract from any problem of measuring altitude and assume that it is easy to obtain an exact measure of altitude for any point in the landscape we reach. In Figure 1.4, an artificially generated landscape is provided which serves as an illustration for the argument.

What will our 'natural' algorithm for searching the highest elevation look like? First, we will tend to walk uphill. Starting from point A this will lead to B, starting from C to the lower of the two peaks at D, and only if we depart from a point like E will it lead to the highest elevation at F. Now, an alpinist starting

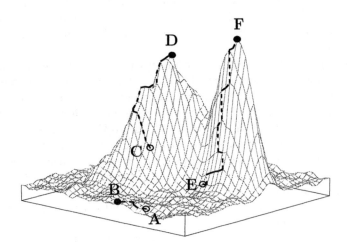

Figure 1.4   Looking for the tops of the mountains.

at A would probably not stop at B, but realize that he had reached only the top of some minor hill. Therefore, he retreats down for a while, until reaching the foot of the next hill or mountain. Obviously, this is a departure from the standard local search algorithm or standard gradient methods described above. In fact, this departure is the crucial one! However, we have not finished yet, since allowing any downhill path once we have reached a hill is not a good idea. It might lead us to the Gulf of Bengal instead of the summit of Mount Everest!

A more natural approach would be to move downhill only for a while and than look around to see whether there is some more attractive path maybe leading upwards again. Another idea is not to walk up to the top of the hill, if we are not sure that the latter is actually a high mountain. Instead, we may start going down again sometime during the ascent of the hill, i.e. before reaching the very top. However, when already climbing up the actual peak of Mount Everest, we are probably less inclined to go down a long distance than at the very beginning of our trip when we have to cross some river valleys, etc. Nevertheless, even on the slopes of Mount Everest, the alpinist sometimes has to go back some steps in order to find a feasible way.

This natural approach comes close to what several refined local search algorithms are simulating. We start with a rough description of what threshold accepting would look like on our mountain climbing problem. The algorithm starts somewhere in the region, e.g. at point A, looks in one randomly chosen direction and moves in this direction if the path is going uphill. It will also follow paths going downhill for a while, i.e. not descending too deep. After a

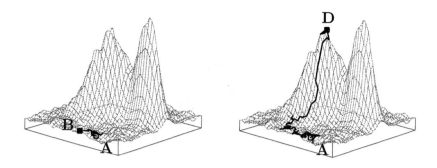

Figure 1.5 Climbing the mountains with threshold accepting.

few steps, the algorithm will stop and look around for some new direction. This process is repeated many times. As the algorithm proceeds, downhill moves become more and more restricted until at the very end only upward steps are accepted. Then, we might end up on top of Mount Everest (F) or — and that is a limitation of any optimization heuristic — with less chance of arriving on the top of K2 (D). Two sample paths followed by a threshold accepting implementation on the artificial landscape are provided in Figure 1.5.

Again, both paths depart from point A. While the path shown in the left panel allows only a total number of 100 steps, 400 steps are acceptable for the right panel. With a small number of steps (iterations), the threshold accepting algorithm climbs the same hill (B) as the standard local search approach, while for 400 steps, it steps down and up again several times, above eventually reaching 'K2' in point D. It should be noted that in contrast to the behaviour of the real alpinist described before, the algorithm acts only locally, i.e. when reaching point B, it cannot 'see' that there are some higher mountains behind. Nevertheless, even in a very foggy environment, an alpinist will probably follow some strategy similar to the one just described. Then, he will end up much higher than with a trivial local search algorithm, which would lead him up on the hill next to the airport in Katmandu, where he might stay in the belief of having climbed the Himalayas, when no further uphill movement is possible anymore.

The performance of threshold accepting in finding one of the highest elevations in Figure 1.5 is both astonishing and promising. Thus, one might try to find what a further increase in the number of allowed steps can lead to. Figure 1.6 provides such an answer. In fact, even after reaching a very marked peak at point D the algorithm moves on and tries several steps downward again, before eventually crossing the saddle and reaching the top of Mount Everest at F. The number of iterations used for this example is 4800, which is quite large compared to the total number of points in the landscape, which is only 1600. However, further tuning of the algorithm may considerably reduce

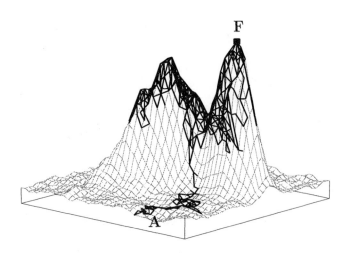

Figure 1.6   Finding Mount Everest with threshold accepting.

the number of iterations required to reach Mount Everest. Furthermore, in real applications the number of admissible points will be larger by several orders of magnitude. Then, also the relative advantages of refined local search algorithms become more apparent.

While moves downward become more and more restricted in size for the threshold accepting algorithm, they become less and less likely in the quite similar simulated annealing heuristic. Although the constraints on downhill moves are relative to the latest position in the threshold accepting and the simulated annealing algorithm, they are made absolute in the Great–Deluge algorithm. There, we can imagine the waters rising gradually in the valleys while we are looking for high elevations. Obviously, downhill moves become more and more restricted while the waters rise. Biblical analogy makes us believe that finally we will reach a high enough elevation to escape the Great Deluge. Of course, Noah is said to have done something else, but who knows?

Something more can be learned from the example of the alpinist. It may happen that on his trip over hills and through valleys the wanderer will come back to a point already visited before. Then, reasonably, he will decide not to pursue the same direction as for his first passage. Such a memory feature can be built into local search algorithms, and is given the title 'tabu search' (see Section 5.2.3).

Finally, if our aim is to find Mount Everest it may be a good idea to send out a complete team of researchers, and not just one. Depending on whether and how these researchers interact we may find different strategies which can also be used in heuristic optimization. If there is no exchange of knowledge between the researchers, the situation will be similar to a strategy in which one researcher is sent out after another from different places in the region. We find the restart version of search algorithms. The situation becomes more complex if interaction between agents is feasible, e.g. through satellite communication. Then, a strategy could be to concentrate forces in regions where high elevations have already been reached by someone previously. Although the analogy does not fit perfectly in this case, genetic algorithms do something similar. The use of analogies could be continued almost ad infinitum to motivate optimization heuristics already in use or to give hints for new approaches. I will, however, close such a list for now.

1.3 Outline of the Book

The outline of the book presented in this section aims at providing a guide for readers through the following chapters and indicates those parts which might be of the greatest interest for their own particular purposes.

The first part of this volume is dedicated to a broader discussion of the optimization background in economics, statistics and econometrics on the one hand, and to the potential of local search heuristics in these areas on the other hand. Therefore, Chapter 2 summarizes the importance of optimization paradigms in economics. In addition, it introduces a first ad hoc classification of optimization problems with regard to their complexity. While simple problems, i.e. optimization problems of low complexity, can be solved by standard approaches, they fail for highly complex problems. A few examples highlight the differences and show the potential of local search heuristics for such problems without going into the details of the algorithms. This exercise is repeated for problems stemming from statistics and econometrics, respectively, in Chapter 3. In addition, applied work in these areas is regarded through an optimizer's eye. Again by using some examples, it is highlighted that optimization has a larger scope in statistics and econometrics than its current use suggests. Finally, Chapter 4 contrasts the classical optimization paradigm with the heuristic optimization paradigm. The limits of both approaches are discussed and the use of local search algorithms, such as threshold accepting, is advocated at least as a bench-mark in cases where standard methods fail to provide high quality solutions.

Once motivation and general concepts are introduced through several examples in Part I, a deeper discussion of optimization heuristics is provided in Part II. The first chapter of this part, (Chapter 5) gives an overview on some classical and heuristic optimization tools. This list provides some references to the literature on other methods than threshold accepting, together with some references to applications, in particular within a context of statistics and econometrics. Furthermore, this overview serves as a setting for where to place the threshold accepting algorithm. The details of this algorithm are given in Chapter 6, while Chapter 7 discusses its relative performance, both in terms of asymptotic convergence and the application to real problems in competition with other optimization techniques. Aspects of the practical application of threshold accepting are the topic of the last two chapters in this part. First, Chapter 8 summarizes findings of some Monte Carlo evidence on the tuning of threshold accepting. Then, the concluding Chapter 9 provides a practical guide to the implementation of threshold accepting. By using the material in this book up to Chapter 9, the reader should be enabled to start with his first own implementation of threshold accepting in economics, statistics or econometrics.

Part III presents a collection of applications in statistics and econometrics covering some facets of the sequence of applied economic research, from the design of experiments, over model selection and the selection of databases, to the estimation of parameters in complex models. These applications are of their own interest in the respective areas. For those problems where other optimization approaches have been used before, the threshold accepting implementations exhibit, at least partially, large improvements over prior results. For other problems, which have not been treated explicitly as optimization problems before, they provide a first optimization bench–mark, which can be compared with the results of simple decision heuristics used previously. Furthermore, the diversity of the examples stresses the ubiquitousness of optimization problems in applied economic research. Hence, these applications might serve as a motivation for the implementation of threshold accepting or similar optimization heuristics to some more of the tremendous number of complex optimization problems in these areas which are suitable for an application.

Given this scope of the chapters in Part III, it becomes obvious that it is not possible to provide an extensive discussion and description of the fields of application covered. Instead, a general outline of the field is provided and some basic optimization problems within this field are introduced. Furthermore, some basic references are included for the reader who would like to gather a deeper understanding of a specific area of application. Next, the specific problem is introduced for which an implementation of threshold accepting is proposed. Other approaches to the solution of this problem and results on its complexity are reported. Then, some implementation details of the algorithm following the general guidelines outlined in Chapter 9 are provided. Finally, an analysis of the results obtained by this implementation highlights the gains and potential limitations of the threshold accepting algorithm. Each chapter within this part is concluded with an outlook to future research on the specific implementation and on related areas where threshold accepting could be a useful tool.

The first application about experimental design is presented in Chapter 11. Here, the main interest is not on the results obtained using some specific experimental design, but on the design itself. If we abstract from some a priori knowledge, the search for an optimal design corresponds to finding a

closest possible discrete approximation to the uniform distribution on some
d–dimensional unit cube. Thereby, discrepancy measures provide information
about the quality of this approximation. For this application, the threshold
accepting implementation is in direct competition with well developed number
theoretic methods obtained from algebraic geometry. Apart from this deep
mathematical background of the problem, there exist several applications in
more applied research, such as the design of experiments in the growing
field of experimental economics or response surface analysis for statistics
without closed–form solution, which have to be simulated for each parameter
constellation. Finally, quasi–Monte Carlo simulation methods are also based
on low–discrepancy point sets.

In Chapter 12, a central problem of economic and econometric modelling
is tackled, namely the problem of model selection. Of course, some natural
restrictions on model selection can be imposed a priori. However, in general a
large degree of subjective choice is left to the model builder. Implicitly, model
building is the same as solving some optimization problem, namely to find an
optimal presentation of interdependencies and causal interactions. In the case
of econometric model building, actual data provide a bench–mark for such
an optimized model. Again, a general discussion of this problem is beyond
the scope of this present book. Instead, the specific problem of lag–order
identification in multivariate vector autoregressive models is analysed. This
problem appears in the context of nonstructural macroeconometric models,
Granger causality tests or Johansen's test for cointegration, to name just a few
applications. In practice, often ad hoc solutions are used instead of solutions
obtained from an explicit optimization approach. Therefore, this application
serves also as an example of optimization problems in econometrics which
have to be made explicit first, before turning to optimization itself.

The problem of optimal aggregation of time series discussed in Chapter 13 is
also often treated in some ad hoc fashion. The finding of a very high complexity
for such a problem could serve as a substantiation for this observation.
Nevertheless, the use of threshold accepting provides marked improvements
over standard approaches, and the optimized results can be given some
meaningful economic interpretation in the context of modern trade theory,
where the latter seems more difficult from the results of the ad hoc approaches.

An estimation problem is considered in Chapter 14. Censored quantile
regressions gain importance in real economic applications due to the growing
availability of large individual data sets, e.g. from social security systems
which often exhibit severe censoring of some income variables. The use of
such models is already more common in other areas of applied statistics
such as quality control. Nevertheless, the standard algorithms used for this
purpose are not well suited for problem instances with a large degree of
censoring. Again, it is shown that the implementation of threshold accepting
is straightforward as soon as the estimation problem is described as some
integer optimization problem. A comparison of the results of the algorithm

based on threshold accepting with standard algorithms is also provided.

The last field of applications, outlined in Chapter 15, covers continuous global optimization problems as they appear in maximum likelihood estimation of complex econometric models. The censored quantile regression problem would fit into this class, if it could be described as an integer optimization problem due to some specific properties of potential solutions. Other potential applications include disequilibrium models, multivariate threshold autoregressive models or Markov switching models. Since a threshold accepting implementation for this kind of problems is not available yet, Chapter 15 outlines some relevant features of such an implementation. The potential of such an implementation is highlighted by some results obtained with simulated annealing and genetic algorithms for continuous global optimization problems, which exhibit several local optima and cannot be solved by standard methods.

Finally, Part IV summarizes the main arguments of this book in Chapter 16 before turning to the more important field of topics not covered here and, consequently left for future research in Chapter 17.

Part I

Optimization in Statistics
and Econometrics

2

Optimization in Economics

2.1 Introduction

Optimization is a central, if not *the* central ingredient to economic theory and practice.[1] However, economic theory mainly relies on standard optimization tools, in particular from differential calculus, while economic subjects use optimization heuristics routinely, successfully, and — given the complexity of problems and costs associated with exact solutions — rationally in quite different contexts. This behaviour is seldom taken into account or analysed for its own right in economic theory, nor do statistical procedures follow this successful approach. The latter aspect is discussed in more detail in Chapter 3. In this present chapter, the focus is on economic theory and practice.

The next section describes some optimization methods used in economic theory and economic practice, respectively. The observed gap between the methods used in theory and practice motivates a further analysis of the reasons which are responsible for these different approaches. Optimization concepts in economic theory are often selected for their mathematical tractability or clarity, while economic practice works on real life problem. Thus, it has to take into account features such as indivisibility constraints, nonlinearities, many side constraints, and the costs associated with solving optimization problems and gathering all necessary information for doing so. The latter may make it a rational decision not to solve the problem to optimality by some exact method, but to use optimization heuristics instead, either explicitly or in the disguise of solutions derived from experience or ad hoc guidelines, the so called 'rules of thumb' (McFadden, 1999, p. 79). Similar considerations also apply to the economics of computation or optimization itself, as will be discussed in Section 2.3.3.

Apparently, there exist optimization problems which allow for an exact solution with limited computational cost, while for other problems it is

---

[1] Beaumont and Bradshaw (1995, p. 159) even state that 'economics and optimization are nearly synonymous and, in recent years, the level of sophistication required to solve economic problems has increased dramatically'.

rational to renounce such an exact solution. The distinction between both classes of problems can be made based on their inherent complexity. A short discussion of the inherent complexity of economic optimization problems is undertaken in Section 2.3. Based on this analysis, the choice of optimization heuristics can be motivated. Since many real life economic problems are of much higher complexity than the problems typically considered in economic theory, the use of different optimization paradigms in theory and practice can be explained. The discrepancy between optimization approaches used in economic theory and practice may serve as point of departure for a general discussion of optimization paradigms in economics.[2]

2.2 Optimization in Economic Theory and Practice

It is beyond the scope of this section to provide a complete characterization of optimization approaches used in economic theory and practice. Instead, some examples taken from the behaviour of households, firms, and fiscal and monetary policy authorities are presented. On the one hand, these examples reflect the predominance of a particular optimization paradigm in economic theory. A few generalizations of the basic models, however, show that this paradigm is not very robust. On the other hand, real optimization behaviour of the same economic subjects is described by employing some further examples, which clearly do not fall under the standard optimization paradigm of differential calculus but, nevertheless, are obviously of relevance for real economic decisions.

2.2.1 Household behaviour

The standard model of household behaviour is based on utility maximization. It is assumed that households or individuals can appraise the utility of leisure time and consumption of different goods in the form of a utility function $\mathcal{U}(\cdot)$. The arguments of the utility function include the amount of leisure time l and the quantities q_i of consumption of n different goods $(i = 1, \ldots, n)$.

In order for this model to be operational, several additional assumptions are necessary. First, households must know all potential consumption goods. Secondly, there is no uncertainty about the function $\mathcal{U}$ for all possible arguments. Thirdly, the consumption of goods is not subject to any indivisibility constraints. Then, differential calculus can be applied to solve the utility maximization problem of the household subject to a budget constraint, which is imposed by available income, which itself is a negative function of leisure time. In order to guarantee a unique solution to this optimization problem, which can then be easily calculated by using Newton's methods or

---

[2] Judd (1998, p. 12) states: 'The normal approach of finding a model simple enough to come up with clean results can easily lead us away from the real truth'.

variants thereof,[3] the utility function has to fulfil additional requirements.

Assuming for the moment that all necessary conditions are fulfilled, the optimization problem of the household becomes

$$\max \mathcal{U}(l, q_1, \ldots, q_n) \quad \text{such that} \quad \sum_{i=1}^{n} p_i q_i = w(T - l), \qquad (2.1)$$

where the p_i are the unit prices of the different consumption goods, T denotes total available time for work and leisure, i.e. $(T - l)$ is the working time, and w is the wage rate.

This optimization problem is tractable from a mathematical or numerical point of view as long as n does not become too large. However, for the situation of a real household n is the number of all commodities available. Thereby, different qualities or time and locus of availability define different commodities. Consequently, it seems hard to imagine that households actually solve the above optimization problem explicitly. In fact, the results obtained from experimental economics cast some doubt on this optimization approach towards household behaviour.[4]

Since optimization procedures of households can rarely be observed directly, it is as difficult to assess non–standard techniques in real economic behaviour as it is for the theoretical approaches. In general, decisions on consumption and labour supply or leisure time, respectively, are subject to a much larger set of side constraints or side considerations in reality than usually modelled in economic theory. Furthermore, the assumption of complete information about all commodities is hardly realistic. The same holds true with regard to the utility function. This proved to be a successful tool to model household behaviour and to derive market demand, but real households do not know their utility function! Furthermore, habit persistence, e.g. in housing, or altruism, such as bequests and gifts, are departures from the standard approach. Of course, both the costs of information gathering and additional aspects like habit persistence and altruism can be incorporated in the utility framework. It is seldom done due to the increased complexity and, eventually, the utility framework would become tautological: what households do maximizes their utility under the given constraints! In addition, indivisibility constraints on consumption and working time decisions may hinder the solution of the optimization problem shown in equation (2.1) by standard methods.

After all, it seems questionable whether individuals or households really solve a simple optimization problem such as that represented by equation (2.1) by methods of differential calculus. They rather use rules of thumb and ad

---

[3] See Judd (1998, pp. 167ff.) and Section 5.1.2.

[4] For example, Rodepeter and Winter (1999) analyze life–cycle savings decisions. They find that simple rules of thumb and rules that have been found in experimental studies perform often almost as good as the solution of the intertemporal optimization problem, which is much harder to solve.

hoc decisions based on a limited set of commodities and information, which again is not the result of explicit optimization on information costs, but the result of experience.

More recent theoretical approaches to household behaviour try to take into account this discrepancy between optimization frameworks assumed in utility theory and observed in real life. A first approach under the heading of 'bounded rationality' (Selten, 1998) assumes that individuals may rather rely on rules of thumb which produce acceptable results on the average.[5] However, although this relaxation of the assumption of strict optimization seems natural, it is far less innocent than it might seem at first sight. In fact, Akerlof and Yellen (1985) demonstrate that second–order, i.e. small, deviations from strict optimization behaviour by a fraction of the population may result in first–order effects on the equilibrium of the economy. To put it differently, although the individuals have hardly an incentive to spend more resources on optimization, the resulting departure from strict optimization may lead to relevant changes in the global outcome.

The second approach considers aggregate market demand as a result of explicit aggregation of individual decisions. In order to derive the usual results at the aggregate level, it is not necessary to strait–jacket the microeconomic decisions to the same extent as in the usual utility maximization framework outlined above (Hildenbrand, 1994). Consequently, this approach avoids identifying a specific optimization paradigm at the household level, which makes it more robust to all the criticisms just mentioned.

Another decision of households, usually treated separately from consumption and labour supply decisions, concerns the allocation of financial assets. Standard portfolio theory is based on quadratic utility functions leading to the mean–variance approach introduced by Markowitz (1952), which can be solved by standard optimization techniques. For this example, the optimization behaviour of private households can be observed in some more detail. However, the results cast some doubt on the standard portfolio theory under a quadratic utility function. In fact, households may judge the positive deviations from some expected return differently from the negative deviations. Such an asymmetric utility function leads to highly complex optimization problems which cannot be solved analytically. In order to judge asset allocations recommended by financial advisors against some efficient bench–mark under such an utility, Canner, Mankiw and Weil (1997, p. 186) use a hill–climbing algorithm. Unfortunately, they do not provide further details about the algorithm or its performance. Nevertheless, they use the portfolios obtained by this algorithm as a bench–mark. The threshold accepting heuristic discussed in this book has also been applied to the problem of portfolio optimization under nonquadratic utility by Dueck and Winker (1992) and Gilli and Këllezi (2000).

---

[5] See McFadden (1999) for a recent survey. A direct link to complexity is provided by Velupillai (2000, pp. 136ff.).

2.2.2 *Firm behaviour*

While households' utility maximization is a theoretical concept, which cannot be observed directly, the standard assumption about firm behaviour, namely profit maximization, corresponds more closely to observed behaviour. It seems less contestable that firms maximize profits subject to constraints imposed by market demand, available technology, or strategic interaction with other firms, to mention just a few.

Nevertheless, a closer look at the assumptions of theoretical models exhibits potential shortcomings. First, a firm is not an economic subject, but ruled by individuals who try to maximize their own utility. A relevant fraction of the literature on industrial organization as well as contract theory discusses when and why firms may still behave as profit maximizing entities. Another foundation of the profit maximizing assumption is given by evolutionary arguments. It is argued that firms eventually have to behave as profit maximizing entities, since otherwise they would be swept out of the market. Then, however, it would seem more natural to model this evolutionary process instead of using the simplifying assumption of profit maximization. Obviously, a numerical treatment of such evolutionary approaches might require the use of evolutionary or genetic algorithms.[6]

Secondly, the standard assumption of continuity in input factors, output quantities, consumer demand and market shares is less harmless than it might seem at first sight. In fact, allowing for indivisibility leads to highly complex problems, which cannot be solved by standard optimization techniques. However, even theoretic results sometimes depend on this assumption. The example of Coase's conjecture that a durable–goods monopolist cannot earn supercompetitive profits in the continuous–time limit is discussed by Bagnoli, Salant and Swierzbinski (1989). They show that neither this conjecture nor Bulow's result, i.e. that for a monopolist renting a durable is always more profitable than selling it, hold when the number of consumers is finite!

Even if we take the profit maximization approach as a reasonable approximation to firms' decision problems, the usual optimization techniques may fail to provide results. This is the case, e.g. if strategic interactions between several firms are modelled. One example is the computation of the closed–loop Stackelberg solution for the repeated interaction of two firms. In fact, Basar and Vallée (1999) apply genetic algorithms to this problem of economic theory.

In contrast to household behaviour, the optimization approaches behind firms' decisions can be observed and are discussed to a much larger extent. In

---

[6] Arifovic (1994) studies the performance of a genetic algorithm for modelling learning in the cobweb model. It exhibits a rather fast convergence to the rational expectations equilibrium. This result provides a foundation for rational expectation models, which does not rely on individual rational behaviour. Birchenhall (1995) uses a similar approach to model technical change.

particular, the extensive research field of operational research covers mostly optimization approaches for decisions at the firm level. Firms use these techniques from the very basic ones, i.e. those based on differential calculus in a continuous setting or enumeration of scenarios in a discrete setting, to the most recent ones, including optimization heuristics. Nevertheless, there still exist firms and specific problems, where decisions are based on ad hoc rules or rules of thumb, sometimes even ignoring that the problem to be solved is an optimization one.

The optimization problems explicitly tackled using techniques from operational research are often integer problems. Many side constraints from available technology lead to highly nonconvex problems. In fact, problems such as the travelling salesman problem or its more realistic counterpart, the vehicle routing problem, belong to the most intensively studied hard optimization problems. Further tasks of this kind are chip layout, wood and cloth cutting, job shop scheduling, input shares for chemical reactions or steel production, and layout of telecommunication or computer networks, to mention just a few.[7]

Further optimization problems have to be solved based on less information, e.g. marketing decisions or strategic interactions with other firms. In these cases, the firms' actions help to gather information about the reaction of the market and other firms, which can then be used to improve the firms' actions in the future. In order to receive the maximum information out of such 'ballons d'essai', the 'experiments' should be well defined. This is another optimization problem which will be discussed in further detail in Chapter 11.

To sum up, optimization is omnipresent in firms' behaviour, although in different forms: explicit and implicit, using standard optimization techniques and refined methods from operational research, based on hard or fuzzy information, etc. It is easy to motivate the use of explicit optimization techniques in firms. As long as the expected cost reduction or improved market performance resulting from optimization exceeds the costs of their use, profit maximizing firms will use them.

2.2.3   Fiscal and monetary policy instances

The actions of fiscal and monetary policy instances complete this short overview on optimization in economic theory and practice. Obviously, the same caveat to treating institutions like individuals applies as in the case of firms. In order to highlight some additional aspects of fiscal and monetary policy in economic theory, we abstract from this problem.

Then, two broad approaches of modelling policy instances can be found. The first one takes some policy goals, e.g. low unemployment or price level stability,

---

[7] Some further examples are discussed in Biethahn and Nissen (1995).

as given and assumes that policy instances react to departures from these goals using their instruments. Although this approach of reaction function modelling is widespread in the empirical literature, it is difficult to base it on some theory unless the underlying optimization is made explicit. This is done in the second approach, which starts with some utility function for the policy instances. While it seems clear that departures from policy goals are negative, it is unclear a priori how to weigh different objectives and which functional form to use. Furthermore, in deciding about certain policy measures, e.g. short–term interest rates or public spending, institutions have to take into account both direct and indirect responses and effects on expectations about future actions. This interaction is fully mirrored in rational expectation models, which, however, still depend on specific assumptions about the utility function of policy institutions. A further assumption concerns the reactions of households and firms, which are assumed to base their expectations on the same model of the economy as the policy institutions do and to perform their optimization taking into account optimal policy responses. Although these assumptions are quite strong, the mathematics of solving these so–called rational expectation models can be quite demanding (Judd, 1998, Ch. 17).

As in the case of households, the objective or utility function of public institutions cannot be observed directly. As far as decisions are commented by the institutions, it seems doubtful that they follow a straightforward optimization approach for some underlying objective function.

While decisions of fiscal authorities appear to be rather discretionary and ad hoc, central banks often follow some rules. These rules might also be rather ad hoc than the result of some explicit optimization behaviour. For instance, prior to 1999 the Deutsche Bundesbank followed a money growth target derived from the growth of potential output and a 'tolerable' rate of inflation of about 2%. However, neither did the central bank achieve its growth target nor is there a widely accepted model of the transmission mechanism. Nevertheless, by using this rule of thumb inflation rates could be kept small over a considerable period of time.

Explicit optimization of policy reactions is also a difficult task. Neck and Karbuz (1995) provide a stochastic optimal control approach to this problem. This requires an explicit objective function and a model of the economy, both of which are certainly controversially discussed in real life.

2.3 Complexity and Heuristics

The previous section highlights that optimization problems faced by economic agents differ substantially in theory and practice. While a large fraction of optimization problems arising in economic theory can be solved by using standard differential calculus, this seems to be either impossible or irrational in economic practice. Furthermore, it is not clear a priori whether these differences in optimization behaviour are innocent small deviations from a reasonable conceptual framework or may have first–order effects in economic theory (Akerlof and Yellen, 1985). The use of standard optimization methods in theory has been motivated by their mathematical tractability which allows for analytical solutions. Some of the examples of real economic optimization behaviour given in Section 2.2 obviously did not fit into the standard optimization framework. Nevertheless, it has not been argued yet why economic subjects do not use other optimization techniques for these problems, but often rely on ad hoc decisions or rules of thumb. The following section on complexity supplies some arguments for this behaviour. At the same time, it motivates the use of optimization heuristics, which are discussed further in Section 2.3.2.

2.3.1 Complexity

In this section, only a first approach to complexity issues is provided. The discussion is continued in Section 3.3 by using examples from econometrics and statistics. Instead of having recourse to the mathematical complexity theory at this point,[8] we try to provide an intuitive understanding by using some of the examples mentioned before. For the purposes of this present book, it seems reasonable to summarize three categories of optimization problems.

Obviously, there exists a class of optimization problems for which the exact solution can be calculated at small computational costs. In particular, all problems which allow for an analytical solution belong to this class provided that the user is familiar with basic concepts from differential calculus. Thus, for suitable assumptions about functional forms, the standard utility optimization framework for households or profit maximization by firms in economic theory belong to this class of problems.

Secondly, analytical derivations for the same problems under different functional assumptions might lead to nonlinear systems of equations, which have to be solved in order to obtain the optimization result. In general, this kind of problem does not allow for an exact solution. However, under some additional assumptions about the functional form, there exist numerical algorithms giving reasonable close approximations to the optimum at relatively small computational costs. In fact, for many applications it is rather the floating–point arithmetic of the computer than the algorithm, which limits the approximation quality. These methods are well analysed (Dennis and Schnabel (1983) and Press, Teukolsky, Vetterling and Flannery (1992, Ch. 10)). In particular, most econometric or statistical software includes some of these algorithms to solve systems of nonlinear equations. The fact that these algorithms provide exact solutions only asymptotically, i.e. with the number of iterations going to infinity, is less relevant for practical purposes due to

---

[8] See Garey and Johnson (1979) and Section 13.9 for an introduction. Velupillai (2000, Ch. 8) discusses some aspects of computational complexity linked to economics.

their good approximation quality under the maintained assumptions of the problem. However, a major drawback of these methods is given by the fact that they may breakdown completely, i.e. fail to approximate the optimum, if the conditions on the objective function are not satisfied (McCullough and Vinod, 1999). The third class of optimization problem, i.e. not even allowing for a numerical solution by using standard methods, is outlined below.

All problems allowing for an exact analytical solution can also be tackled by using the above mentioned numerical methods. However, there also exist problems, which can be solved or approximated with reasonable quality by these numerical techniques, but which do not allow for an analytical solution. This difference in available solution methods provides the base of a general concept of complexity. Given two problems, different solution methods might be available for both problems. If we consider only the most efficient method for each problem, then the problem which requires less computational resources by using this most efficient method is considered as being less complex than the other. For example, if one optimization problem can be solved by differential calculus analytically, it is less complex than an optimization problem, which can be tackled solely by using a numerical method.

The difficulty for the practical use of this concept of complexity lies in the requirement of 'most efficient method', i.e. it is not relevant if the problem cannot be solved with some particular method, but it is required that no method can solve the problem, which requires less computational resources than a given amount, which defines its complexity. In Section 3.3, the example of matrix multiplication is discussed. While the usual algorithm for multiplying two 2×2 matrices requires 8 scalar multiplication, the most efficient method requires only 7. Thus, although most people would employ the standard method, the computational complexity of this problem is smaller! However, for practical purposes, i.e. for real economic behaviour, this strong concept of complexity might be less useful than a slightly weaker concept requiring that no method is available for the user solving the problem under the given limit with regard to computational resources.

Of course, for both concepts of complexity, computational resources have to be interpreted in relation to the size of the problem, the so-called instance size, e.g. to solve an optimization problem with regard to one variable is certainly easier than to solve one with regard to two variables. However, the complexity might be the same, if, e.g. the problem can be solved by analytical methods in both cases.

A further complication is introduced by the fact that most numerical algorithms deliver only approximating solutions for continuous problems. Thus, it is difficult to define the complexity of such algorithms. For the purpose of this section, a heuristic argument might be enough. In fact, for practical purposes, no economic subject would mind missing the global optimum by, let us say 0.0001%, except maybe in financial markets. Therefore, complexity is to be understood as the unavoidable computational input to obtain a satisfying approximation to the optimum.

Given this operational version of the complexity concept, most optimization problems still do not belong to either of the two classes already introduced. They are just not 'well behaved' enough. Continuous optimization problems, in general, are nonconvex, and exhibit several local optima. Then, the standard algorithms used for identifying optima may just provide any of those local optima. In general, an arbitrary local optimum is anything else than a good approximation to the global optimum, which the user is interested in. Additional complications arise, if the objective function exhibits discontinuities. Another subclass of these highly complex problems are many discrete optimization problems, as they arise in economic practice, for example, the travelling salesman problem or optimal chip layout. While for small problem instances, e.g. if the travelling salesman has to visit just four or five cities, the optimum can be easily obtained by enumeration of all alternatives, this approach quickly fails when the number of cities increases, as the number of alternative tours grows exponentially with the number of cities, i.e. the problem instance size. These highly complex integer or combinatorial optimization problems are the subject of mathematical complexity theory. They are also at the centre of interest of operational research approaches, since real economic life requires the solution or at least approximation of these kind of problems, as pointed out in Section 2.2.2.[9]

To summarize this first glance at complexity issues, one might state that optimization problems in conventional economic theory are often of low complexity, while many of the problems in real economic life belong to the highest complexity class. This provides the rationale for economic subjects not to rely on differential calculus or any other exact method for the solution of their problems, but on ad hoc rules, rules of thumb or more specific and refined heuristics.

2.3.2 Heuristics

This section provides a first intuitive approach to optimization heuristics, while a more formal introduction is provided in Chapter 4. By definition, highly complex problems cannot be solved by standard optimization approaches such as differential calculus, Newton's method or enumeration. Economic subjects take into account these constraints and employ 'heuristics'. In this broad understanding of heuristics, any method for obtaining approximate solutions to a complex problem will be summarized excluding

---

[9] It should be noted that some of the problems belonging to this last class of highly complex problems are not commonly treated as explicit optimization problems. In fact, Pirlot (1992) discusses the problem to convince managers that some of their real–world applications are optimization problems and should be tackled as such.

the standard methods just mentioned. Thus, the ad hoc decisions and rules of thumb frequently observed in this context could be considered as heuristics. If these approaches are based on a long experience or adapted from successful applications, the approximations to optimal solutions might be of high quality.

In this present book, the meaning of heuristics is a more focused one. Although it is not possible to provide an exact definition in this more restricted understanding, a few basic requirements can be listed (Barr, Golden, Kelly, Resende and Stewart, 1995, p. 12). If standard approaches like Newton's method can be applied to a problem, a heuristic should be used only if it can produce or identify high–quality solutions faster than these standard approaches. The examples discussed in this book, however, refer to problems when the standard approaches cannot provide a reasonable solution at all. For such problems, a heuristic procedure should contain elements which aim at producing high–quality results, i.e. good approximations to the global optimum. Secondly, they should be robust to changes in problem characteristics, tuning parameters or the introduction of additional constraints. Thirdly, they should be easy to implement to many problem instances, including new ones. Finally, a necessary requirement is that the solution approach consists of a procedure which does not depend on individual subjective elements. It is argued that threshold accepting is one heuristic which fulfills these requirements.

The last requirement seems operational and necessary in order to distinguish between heuristic procedures in our understanding and any heuristic approach in the broader meaning mentioned above, although some procedures depend on a large set of parameters, which open the door to some subjectivity. It is much harder to verify the other conditions on the quality of the results produced by the procedures. In fact, in a context of highly complex problems, heuristics can only produce approximations to optimal solutions. It is a matter of taste to define when a procedure is considered as aiming at high–quality approximations, since — as pointed out above — even ad hoc rules or rules of thumb may produce high–quality results for some instances. In the following, two criteria are used for verifying this condition. First, the procedures can be applied to small instances of the problem, i.e. to travelling salesman tours for only five to ten cities. Then, the procedure should be able to provide the global optimum. Secondly, it can be analysed whether the procedure is able to obtain the global optimum if it is allowed to use unlimited computer resources. This property will be discussed under the topic of global convergence in Section 7.3.

Some remarks on optimization heuristics developed during the last two decades may strengthen the intuitive idea of what an optimization heuristic should look like. Two main streams of optimization heuristics can be distinguished, namely constructive and local search methods (Aarts and Lenstra, 1997, p. 2). The difference is easily described by using the example of the travelling salesman problem. A constructive approach would consist in

starting in any of the cities and to continue to the closest city, which has not
been visited yet, until all cities are visited. Then the travelling salesman has to
return to his point of departure. This procedure does not operate on the whole
tour, but considers only small building blocks — the next city to be visited.
In contrast, a local search algorithm would start with any tour connecting
all cities. Then, it would try to improve this tour by suitable local changes
of the tour, e.g. by exchanging two cities in the sequence. Repeating this
exchange step several times, it might finally come up with a tour which is much
shorter than the initial one. The advantage of constructive approaches is their
velocity and close analogy to rules of thumb. However, a good performance of
constructive approaches requires a careful adapting to every specific problem.
Local search heuristicsare, typically, less problem dependent. The resulting
disadvantage is a weaker relation to the problem, i.e. it is more difficult to
obtain an intuitive understanding of their operation. Nevertheless, at least
three advantages make them a good choice for applications in economics as
well as in econometrics and statistics:

1. They are easy to implement for different problem classes, and side
   constraints can be taken into consideration at low additional expense.
2. They are similar in spirit to classical optimization techniques such as
   gradient methods or deepest descent.
3. They show a good or even very good performance for many practical
   problems, where it is difficult to better them even with highly problem–
   specific solutions.

In Section 5.2, some local search algorithms are presented, including
genetic algorithms, tabu search, simulated annealing, and the Great–Deluge
algorithm, which besides the advantages just mentioned have some intuitively
appealing analogy to natural or technical procedures. The discussion of
threshold accepting in Part II and of different applications in econometrics
and statistics in Part III stresses that threshold accepting also belongs to this
class of local search algorithms.

2.3.3   Economics of computation

High complexity prevents the successful use of standard and, consequently,
cheap optimization routines for many real–life economic problems. Optimization heuristics may improve on the results obtained by ad hoc decisions. However, the costs of using these heuristics can be high. Hence, when discussing
optimization tools in economic applications it seems natural to say at least a
few words on the economics of computing.

Optimization can be described as a production process using different
inputs. First, qualified human labour is required to analyse the problem and
to describe it as an optimization problem. Further labour input is necessary
to code the problem in some programming language. Given a computer code

for some optimization heuristics, the solution quality may depend on the run time on a computer and available computer storage space, which also come at a cost. All of these inputs have to be combined to obtain an approximation of the 'true optimum' of a certain quality (Judd, 1998, p. 33ff.). The huge set of classical optimization methods and algorithms derived from the heuristic optimization dogma provide the production possibility set. The user has to make an efficient choice based on his indifference curves describing the trade–off between higher costs for factor input and solution quality.

The most expensive factor in producing optimization results is certainly qualified labour input. Thus, the use of standard software is to be preferred whenever it might provide solutions of the required quality within a reasonable amount of computing time. However, as the above discussion on complexity pointed out, there are many problems where this condition is not met, even after taking into account the decreasing prices of raw computing power. For these instances, either an algorithm tailored to the specific problem can be used requiring, in general, a considerable amount of qualified labour input, or a general purpose heuristic can be used. The latter requires a much smaller amount of labour input, although at the cost of a potentially larger required computing time to obtain results of reasonable quality.

Finally, once the decision of the most efficient optimization technology is made, it has to be decided whether the expected gain in solution quality compared to standard methods or ad hoc approaches justifies the expense. This is more likely to be the case when the larger the expected gain is and the smaller the costs are. The gain depends positively on the economic problem size, i.e. the total turn around, and on the potential for improvements, i.e. the cost share which can be avoided by more efficient optimization routines. The costs depend on the complexity of the optimization problem and the prices for the input factors. Given the decreasing costs of computing, qualified labour will become more and more the limiting cost factor. Therefore, the approach advocated throughout this book and manifested in the threshold accepting heuristic is to use a rather simple multi–purpose heuristic. Since this algorithm can be easily implemented to a large number of highly complex optimization problems, the probability that its use is economically rational is large for many problems where the use of standard algorithms is not possible and the development of especially tailored algorithms is too expensive. Of course, these arguments also apply to the field of optimization problems in statistics and econometrics.

3

Optimization in Statistics and Econometrics

3.1 Statistics is Optimization

Applied statistics and econometrics is concerned with the analysis of data. Three sources of data can be distinguished. Observational data come from real processes. The researcher cannot influence the parameters of the process. However, it might be possible to select specific observational entities, e.g. in survey data sampling, or at least to define the methodology for gathering the data, e.g. for national accounts data. The second source of data is from experiments, where the researcher can set all conditions. Typically, this approach is used in science and engineering. It is only recently that experimental approaches have become more widely used in economics. Finally, instead of relying on observational data from real processes or experimental settings, which might be noisy, artificially generated data can be used. This last approach is mainly used for assessing the performance of statistical methods against a controlled bench–mark, which, in general, is not possible for real data.

The task of data analysis can be further categorized into description, inference and modelling. However, the borderline between these categories is not clear cut. Nevertheless, by using these categories it seems easier to describe the whole range of applied statistics and econometrics and the relevance of optimization methods over this whole range.

Under an optimization point of view, the description of data is synonymous with extraction of information from the available data. Then, those descriptive methods are preferable which are able to extract a maximum of information from the limited set of available data. This argument is followed in Section 3.1.1. Subsection 3.1.2 covers optimization in the modelling step, which itself consists of several tasks. All of these are undertaken under the aim of finding a model, which explains the given data best (Burnham and Anderson, 1995, p. 316), i.e. is congruent with the observations and has forecasting power. The next step of data analysis, i.e. inference, consists of deciding whether the data are congruent with some hypothesis about their generation or an underlying model. Statistical tests are used for this purpose. In this step, those tests are preferable which are most likely to detect departures from the assumptions under the constraint that they do not reject too often, when the data fit within the assumptions. The importance of this optimality criterion is discussed in Section 3.1.3. The final Section 3.1.4 exhibits some of the optimization methods typically used within this statistical context.

3.1.1 Data description

In this section, data description means descriptive statistics and graphical tools, while distributional properties and data description by means of models are covered in the following sections. For a given data set, there exists an almost unlimited number of descriptive statistics and graphical presentations. It is not the purpose of this section to provide an overview as this can be found in most textbooks. Instead, a few examples might highlight that behind most, if not all, descriptive statistics stands some, typically implicit optimization. This result motivates the interpretation of data description as a general optimization approach in order to extract some useful information.

The first example concerns the probably most frequently used descriptive statistic for any data set, namely its arithmetic mean. Let $x_1, \ldots, x_T$ denote the values of a variable for observations $i = 1, \ldots, T$, which might correspond to individuals or time periods. Then, the arithmetic mean $\bar{x}$ is defined by

$$\bar{x} = \frac{1}{T} \sum_{i=1}^{T} x_i \,.$$

Obviously, this definition is only reasonable for metric variables, although it is often also applied to ordinal data in practice.

It is not obvious why or in which context the calculation of the arithmetic mean corresponds to the solution of an optimization problem. In fact, if the x_i describe, e.g. the rates of change of employment for the firms in a specific sector, a straightforward interpretation of $\bar{x}$ is the mean or expected change, i.e. if we are interested in the employment change of some firm not included in the sample, we might assume that typically it should be around $\bar{x}$, if no other information is available on that firm. This assumption, however, is only reasonable under additional conditions. A sufficient condition is the assumption that the x_i are drawn from a normal distribution with mean μ and variance σ. It is well known that under this setting $\bar{x}$ is a maximum–likelihood estimator of μ, i.e. $\mu = \bar{x}$ is the value which makes it most likely that the x_i are random drawings from a normal distribution with mean μ. Then, the expectation for any further drawing, i.e. the employment change of a further firm, is equal to μ. Obviously, this interpretation of the arithmetic mean describes it as the solution to an explicit optimization problem, although the calculation itself is trivial.

A related descriptive statistic is the median of a sample. This provides the central observation. If T is uneven, and the observations are sorted in increasing order, i.e.

$$x_1 \leq x_2 \leq \ldots \leq x_{\frac{T+1}{2}} \leq \ldots x_T ,$$

this central observation is given by $x_{(T+1)/2}$. For T even the usual convention is to choose $1/2(x_{T/2} + x_{(T+1)/2})$. In both cases the relevant property of the median estimator is that there is an equal number of observations above and below the median. The advantage of this location parameter is that it is independent of some outliers at the upper or lower end of the empirical distribution given by the x_i. One disadvantage is given by the computational burden for calculating the median estimator. Usually, the observations are not sorted. Therefore, for a straightforward estimator of the median, a sorting of all observations is required.[1] Despite the use of fast sorting algorithms, this takes much more time than the calculation of a mean. In particular, if the number of observations becomes very large, e.g. several million people in social security data files, it may take several seconds on modern computers. This computational burden does not pose a real problem if only one such calculation is necessary. However, in a bootstrap or simulation setting, the calculation of the median might have to be repeated many times. In such a case, the computational costs become relevant.

The calculation of the median can also be described as the solution to an optimization problem. Let $u(x) = \#\{x_i \mid x_i \geq x\}$ denote the number of points in the sample greater than or equal to x, and $l(x) = \#\{x_i \mid x_i \leq x\}$ the number of points less than or equal to x. Then, the solution x^{opt} of the optimization problem

$$\min_x \mid u(x) - l(x) \mid$$

is the median, if T is uneven. For T even, any x in the interval $[x_{T/2}, x_{(T+1)/2}]$ is a solution to the optimization problem and, if T is large and the x_i are closely spread around the median, becomes a close approximation to the median. Using this description of the median, several simple heuristics can be thought of for its approximation. One idea is to solve the problem only on a subset of the x_i and use the solution as an approximation to the total set (subset 1). A second approach is to repeat this approximation on a subset several times and use the mean of the solutions (subset 2). A slightly more refined algorithm would proceed by an iterative refinement (iterative). Given the non-sorted observations x_i, this starts, e.g. with the mean $\bar{x}$. Then,

---

[1] In fact, the sorting of all observations is not necessary; see Koenker (1997, p. 17), and Floyd and Rivest (1975).

Table 3.1 Approximation of the median

Algorithm					
Subset 1		Subset 2		Iterative	
$\Delta^{a)}$	$time^{b)}$	$\Delta^{a)}$	$time^{b)}$	$\Delta^{a)}$	$time^{b)}$
Sample size 1000					
0.080 09	2.35%	0.024 41	13.11%	0.000 066	22.46%
Sample size 10 000					
−0.006 46	0.89%	0.002 72	5.39%	-4×10^{-7}	11.51%
Sample size 100 000					
0.000 67	0.84%	0.002 58	5.00%	1×10^{-6}	9.99%
Sample size 1 000 000					
−0.003 15	0.89%	−0.001 42	4.78%	-1×10^{-7}	8.24%

[a] Mean difference to exact median estimator.
[b] Computing time relative to exact median estimator by complete sorting.

$u(\bar{x})$ and $l(\bar{x})$ are calculated. If $u(\bar{x}) > l(\bar{x})$, the candidate solution has to be increased; otherwise it is decreased.

Table 3.1 shows some results of an ad hoc implementation of such approximation algorithms in GAUSS. For the subset algorithms, subsets of size 1% of the sample are used. For subset 2, five randomly selected subsets are considered. For the iterative algorithm (iterative), the change of the candidate solution is made dependent on the relative difference $\mid u(x) - l(x) \mid /T$ with a factor starting from 2 and decreasing by multiplication with 0.8, whenever the sign of $u(x) - l(x)$ changes. The algorithms were run on 100 sets of observations randomly generated from a standard normal distribution. The columns Δ give the mean difference to the exact median estimator[2] for the given data sets over all 100 replications. The columns *time* indicate the relative computing time of the approximating methods compared to the implementation of the median in GAUSS. A straightforward implementation of the algorithm provided by Floyd and Rivest (1975), which avoids a complete sorting of all observations, appeared not to be competitive for the sample sizes considered in this example.

Although all three methods are completely ad hoc, they provide high–

---

[2] The true median is equal to zero for the standard normal distribution. The difference between estimated medians and this true value is found to be usually larger than the difference between estimated median and its approximations.

quality approximations to the median. In particular, the error of the iterative approach would be considered negligible in most applications. In fact, the algorithm guarantees that exactly 50% of all observations are greater or equal to the approximation of the median. At the same time, the approximation methods are much faster. Since the iterative method approaches the median from some starting point by using only local information (how many points above or below), it represents a simple local search heuristic. Much more refined methods could be thought of, but are not the subject of this present chapter.[3] Here, the example of the median should highlight the importance of optimization in descriptive statistics and the potential for the use of heuristics in this context.

Another example is the calculation of a simple bivariate regression line. Given pairs of observations (x_i, y_i) on two variables, it is possible to generate a scatterplot. One possibility to concentrate the information content of such a plot consists in associating it with a linear relationship. In the graphical presentation, this corresponds to drawing a straight line in the plot. Of course, one is interested in a line representing a maximum of information, i.e. one which mimics the obvious relationship between the variables as good as possible. A useful measure of this fit consists in the distances of the points from the regression line. Depending on whether one minimizes the sum of the squares of these distances or rather their absolute sum, one obtains the classical least squares or the minimum absolute deviation estimator. While the first is closely related to the mean estimator, the second is a generalization of the median estimator discussed above. Both estimators are the result of an explicit optimization approach. The prevalence of the least squares regression in practice can be attributed, at least partially, to the fact that methods for solving this optimization problem became available earlier than for the minimum absolute deviation estimator.[4] Thus, not only are statistical methods often based implicitly or explicitly on optimization problems, but the tractability of these problems might even determine the choice of the methods.

The last example already indicates a clear borderline case to the modelling part of statistics. Hence, the few examples given in this present section must suffice for the motivation that even data description is an optimization task if interpreted under the aspect of extracting the most important information.

---

[3] Koenker (1997) discusses an interior point method for approximating the median and more general L_1–regression problems. An application of threshold accepting in a similar context, i.e. the calculation of censored quantile regressions, is discussed in Chapter 14.

[4] See Koenker (1997, pp. 15ff.), in particular the reference to Gauss. However, a comprehensive treatment of the numerics of least squares is still a book–filling undertaking (see, e.g. Gill, Murray and Wright (1991)).

3.1.2   Modelling

Estimation of model parameters has been viewed as an optimization problem
for at least eight decades. For example, the use of least squares estimators
became so common that users were not always aware of the optimization
problem behind these estimators.[5] The other aspects of modelling, however,
i.e. model selection and inference, have been discussed in an explicit
optimization context for a much shorter time.[6] The information–theoretic
paradigm introduced by Akaike (1969; 1973) might be considered as a point
of departure.[7]

A hypothetical example is used to exhibit implicit and explicit optimization
in modelling. It is assumed that an endogenous variable y depends on several
exogenous variables x_i, where $i = 1, \dots, N$. Either a limited number T of
observations is available from empirical data or can be generated through
experiments or simulations. The importance of optimization in the latter case
is discussed in Chapter 11. Here, the observations y_t, where $t = 1, \dots, T$, and
x_{it}, where $i = 1, \dots, N$ and $t = 1, \dots, T$, are assumed to be given as usual,
e.g. for macroeconomic data.

Although the selection of variables is usually discussed in the context of
econometric estimation, it is hardly ever made explicit that this first modelling
step consists of solving an optimization problem. The procedure consists of
selecting a subgroup of the variables x_i, which are likely to be relevant in a
parsimonious model for y. This selection is based on theoretical arguments,
a priori expectations or ease of access to the data. In addition, part of the a
priori modelling step consists of various assumptions about the functional form
of the relationship between endogenous and exogenous variables. Sometimes
linear or loglinear relationships can be derived from theory, but more often a
simple functional form is assumed, since otherwise it would not be possible
to use empirical data for the further steps. One alternative approach consists
in nonparametric modelling techniques, which nevertheless are also based on
optimization considerations.

---

[5] The examples provided in McCullough and Vinod (1999) indicate that users should
be aware of the underlying optimization procedures. For example, they present three
different full–information maximum–likelihood estimators of Klein's Model I which are
all based on the same original data, and which do not even share the same sign or a
single significant digit!

[6] Burnham and Anderson (1995) present a textbook on optimization–based model
selection and inference for empirical data. On p. 315 of this book, they state:
'Conceptually, there is information in the observed data, and we want to express this
information in a compact form via a "model". Such a model is then the basis for
making inferences about the process or system that generated the data. [...] The (not
achievable) goal of model selection is to attain a perfect 1–to–1 translation such that
no information is lost in going from the data to a model of the information. Models are
only approximations and we cannot hope to perfectly achieve this idealised goal'.

[7] For a deeper analysis of model selection based on information measures, see Chapter 12.

Let us presuppose for this example that it is reasonable to assume that y is mainly determined by the first three exogenous variables x_1, x_2 and x_3 and that a quadratic polynomial is a suitable functional relationship. This finishes the design step of modelling in the terminology of Ericsson and Marquez (1998, p. C230). The next step, model evaluation, consists of model selection and estimation within this constraint framework.

The most general model within the selected class is given by

$$y_t = \alpha_0 + \sum_{i=1}^{3} \alpha_i x_{it} + \sum_{i=1}^{3} \beta_i x_{it}^2 + \sum_{i=1}^{3} \sum_{j>i} \gamma_{ij} x_{it} x_{jt} + \varepsilon_t, \qquad (3.1)$$

where for the given parameters α_i, β_i and γ_{ij}, the ε_t are the residuals between y_t and the model's forecast. This most general model depends on ten parameters α_i, β_i and γ_{ij}. Given at least ten observations on the y_t and x_{it}, it is possible to assign the parameters optimal values with regard to some criterion such as least squares or least absolute deviations. However, despite the effort spent on estimation of these parameters, which might be considerable depending on the chosen criterion, it cannot be guaranteed that the resulting numerical model is in fact useful. In particular, some of the terms might show up as superfluous, thus making the model less parsimonious than possible.

In order to find a parsimonious model, which nevertheless mirrors the dependency of y on the x_i as close as possible, several approaches can be found in the literature. The most common ones are probably the top–down and the bottom–up approaches. The top–down approach starts with the general model (see equation (3.1)) and sequentially removes non–relevant elements. The decision on removal is based on statistical inference, e.g. t–statistics. The bottom–up method, on the contrary, starts with a simplistic submodel, e.g. only the linear part of equation (3.1) and sequentially adds relevant additional terms. A drawback of both approaches is that the resulting model depends on the chosen sequential strategy and on statistical inference which might be challenged when the 'true' model is not known.

A third approach, which does not rely on statistical inference, but uses a direct optimization approach, is model selection based on information criteria such as Akaike's final prediction error criterion. This approach provides a weighting of 'goodness of fit' of the model, which is larger for more general models, and parsimony, which makes the model more useful for practical use.[8] Optimal model selection based on such information–theoretic criteria is described easily for the model class of our example. In principle, it is possible to consider all submodels of equation (3.1) and to estimate optimal parameters for each of these submodels according to the estimation method chosen a priori. The next step consists in calculating the information criterion

---

[8] See Burnham and Anderson (1995, p. 319) and the discussion in Chapter 12.

for each estimated submodel. Finally, the model maximizing the information criterion is retained as the most useful approximation to the data. Of course, this approach depends on the specific model selection criterion.[9] Furthermore, it may result in a high computational load. The example above already allows for $2^{10} - 1 = 1023$ different submodels. If the number of potential important exogenous variables becomes larger, the model class becomes more general, or the estimation method for the parameters itself becomes computationally burdensome, a complete enumeration and comparison of all candidate models is no longer possible. Then, a heuristic approach combining ideas from the top–down and bottom–up approaches, i.e. leaving out or adding a few elements sequentially, and from the information–theoretic optimization framework can be useful. One such example, the identification of multivariate lag structures in vector autoregressive linear models, is treated in Chapter 12 by using the threshold accepting optimization heuristic. An example of this is provided in Section 3.2.2.

Once model selection and estimation is finished, a third important step of statistical modelling comes on the agenda, i.e. the post–evaluation analysis (Ericsson and Marquez, 1998, p. C230). This consists of statistical inference on estimated parameters, a point we will come back to in Section 3.1.3, and further tests of the model such as simulations and forecasts. This aspect of post–evaluation analysis also implies some optimization aspects, as highlighted later through the discussion of quasi–Monte Carlo simulation techniques in Section 11.6.3.

3.1.3 Statistical inference

The final aspect of optimization in statistics covered in this section is statistical inference. In general, statistical models do not provide a one–to–one description of empirical or simulated data. In fact, this is the reason why the residuals ε_t had to be introduced in equation (3.1) in the previous section. These departures are either due to missing controls, stochastic components or the parsimony requirement, which precludes the inclusion of all aspects within one model. This stochastic component in the modelling step implies that estimates of parameters are also subject to some uncertainty. If the stochastic component of the data changes, the parameter estimates, in general, will also change. Therefore, most estimation procedures do not solely provide parameter estimates, but also some information on the probability distribution of these parameters, which, of course, depends on the assumptions about the probability distribution of the stochastic components.

In the case of a straightforward linear model, it is often assumed that the residuals are drawn from a normal distribution, i.e. they can be considered

---

[9] See Section 12.3 for a discussion of some criteria and their asymptotic convergence properties.

as stochastic normally distributed shocks. Then, least squares parameter estimates are also stochastic variables. Under the given assumptions, they follow a t–distribution with degrees of freedom equal to the number of observations minus the number of estimated parameters. Asymptotically, i.e. with the number of observations going to infinity, they are normally distributed. Least squares estimation also provides an estimate of the variance of this distribution. This feature is used to derive a number of tests. The most widely used test is the so–called 't–test' on parameter significance. If a parameter estimate $\hat{\beta}$ has the estimated variance $\hat{\sigma}_\beta$, it is analysed whether it is likely that $\hat{\beta}$ is drawn from the 'true' distribution with mean zero and variance $\hat{\sigma}_\beta$. If the estimate is large relative to its variance, it is unlikely to stem from a distribution with mean zero. The likelihood is given by the mass of the tail of the normal distribution to the right of $\hat{\beta}$ in this case. A typical limit is imposed by a 5% significance level. Then, $\hat{\beta}$ is regarded as being significantly different from zero if the mass of the tail right of it is less than 0.025.

Nevertheless, despite which significance level is imposed, it may happen that $\hat{\beta}$ is large by pure change although the true distribution has a mean of zero. The probability of a test statistic to reject such a null hypothesis, when it is true, is called the 'size' of a test. This corresponds to the significance level. However, if the number of observations is small, the real size of the test may differ from its nominal size under the assumption of the asymptotic distribution. While the size of a test should be as small as possible, the main interest in testing is in rejecting the null hypothesis if it is false. Therefore, the 'power' of a test is also very important. This is defined as the probability to reject the null hypothesis when the latter is not true. In the example of the t–test, it becomes clear that a low significance level corresponds to a low power of the test. Thus, there is a trade–off between size and power (Davidson and MacKinnon, 1993, pp. 405ff.).

In this and more complicated test settings, a choice has to be made, which again is the result of an optimization under the constraints imposed by the trade–off between size and power. Optimal testing strategies are the result of this procedure.

3.1.4 Optimization methods in statistics

The previous sections substantiate the omnipresence of optimization in statistics. Some of these optimization problems are solved by statistical or econometric software almost routinely. The user does not even have to be aware of this fact, in contrast to the early years of applied econometrics, when due to constraints imposed by computational resources, people were well aware of the expense involved in performing such calculations. Nevertheless, the examples provided in McCullough and Vinod (1999) indicate that optimization routines may fail to produce reasonable results when applied to problems which do not fit into the standard framework. Furthermore, one might also think of performing explicit optimization for those problems which so far have been treated by ad hoc methods or have not been considered as being optimization problems at all.

Since the early years of computation, both computer technology and the approaches used for description, inference and modelling in statistics and econometrics have continued to evolve. Therefore, the question may be raised as to whether the classical optimization toolbox, which comprises analytical solutions and numerical methods based on differential calculus, is still adequate for all problems. While it seems natural to calculate the mean of N observations $x_1, \ldots, x_N$ by $\bar{x} \equiv 1/N \sum_{i=1}^{N} x_i$, standard methods may fail when applied to some complex maximum–likelihood problem with multiple local maxima, as they come up, e.g. when estimating disequilibrium models (Dorsey and Mayer, 1995) or Markov switching autoregressive models (Clements and Krolzig, 1998, p. C61). Another example is the estimation of linear threshold method, which requires a complete enumeration of all possible threshold values, which becomes infeasible for large sample sizes or multivariate settings (Hansen, 2000, p. 578). However, even if classical methods could solve the optimization problem, the examples provided by McCullough and Vinod (1999) show that implementations in statistical or econometric software packages often fail to do so. Finally, often a modelling problem like selecting the right variables for the model is not even described as an optimization problem. If it was, a complex combinatorial optimization problem results. Some further problems of this type are discussed in Part III.

The failure of standard optimization techniques to solve problems arising in statistics and econometrics makes it at least doubtful whether the optimization techniques of the last century are still adequate for all optimization problems related to the most recent methods and upcoming trends in statistics and econometrics. There are a large number of problems in econometrics and statistics, including numerical models in economics, e.g. for computable general equilibrium models or quantitative game theory (Judd, 1998, pp. 133ff and 187ff, respectively), which are documented, for which standard optimization approaches may fail to provide solutions at all or would require tremendous amounts of computing resources. Therefore, the question as to whether new optimization paradigms could be useful in statistics and econometrics can be answered by a clear–cut 'yes'.

3.2 Econometrics through an Optimizer's Eye

The previous section highlighted the importance of optimization in statistics. Since econometrics consists of particular statistical procedures applied to economic models and data, similar arguments apply to the methods applied in econometrics. However, the weight attributed to the steps of an econometric analysis may differ from those typically used in other statistical applications.

Therefore, this section provides a description of econometric analysis from an optimization point of view.

The steps of an econometric analysis include model selection, both at the level of theory and empirical model, data selection, model fit and model tests, interpretation of the results and applications for simulation or forecast purposes. Of course, this list is far from being comprehensive or typical for all applications in econometrics. Nevertheless, it covers some of the most prominent steps of a broad range of econometric applications.

In order to stress the importance of optimization in almost all links of this chain, a specific example is used throughout this section, namely the money demand function, which has been analysed extensively in the empirical literature.[10] It is not the aim of this section to provide a detailed discussion of money demand and its econometric estimation. Therefore, it is assumed that something like a money demand function exists, that observed monetary aggregates correspond to money demand, i.e. disequilibrium on money markets is excluded by assumption, and that money demand is driven by transaction, speculative and precautionary motives. The standard theory of money demand implies a relationship between money demand M, output Y as a measure of the transaction volume, prices P, and interest rate i as an indicator for the costs of holding money.

3.2.1 Money demand model selection

Starting with a general conception such as a money demand function, a first step of model selection takes place before applying any statistical techniques. The researcher has to answer various questions, i.e. which monetary aggregate comes closest to the theoretical concept, how output should be measured and which interest rates are relevant? In empirical studies, the first question is most frequently discussed as a choice between narrow monetary aggregates, which include currency and sight deposits, on the one hand and a wider concept such as M3, which also comprises short–term time deposits, on the other hand. Then, optimization with regard to the most suitable approximation to the theoretic concept boils down to a choice between a few alternatives. Even then, however, this choice can be crucial for the results, and, therefore, has to be made in an optimal way. Of course, this optimization could be performed over a much larger set of alternatives, including liquidity stemming from foreign monetary assets, credit cards or electronic cash. Still, the problem persists, that an optimal choice has to be made among several or even many alternatives.

The same holds true for the other variables of the money demand model.

---

[10] See Fase (1994) for an overview on more than a hundred econometric studies focusing on money demand. Some recent contributions can be found in the volume edited by Lütkepohl and Wolters (1999).

Usually, output is approximated by gross domestic product (GDP) or similar concepts. Nevertheless, it can be argued that money demand is much larger in sectors of the economy not mirrored in official gross domestic product such as trade in narcotics and other black market activities. Then, the choice of GDP is not optimal. While the GDP deflator provides a good approximation to the development of all prices, demand for money may depend more on household behaviour. Then, a consumer price index (CPI) with all problems of measurement might be more appropriate. Finally, the interest rates relevant for the costs of holding monetary assets may differ markedly between individuals due to differences in time preferences, to name just one example. Therefore, instead of one interest rate the inclusion of several rates mirroring the term structure of interest rates could be adequate. These few examples highlight that all these choices are based on some implicit optimization calculus in econometric research: 'Which observable variable(s) mirror some theoretic concept most closely?' Unfortunately, the answer to this question may sometimes be blurred by another effect: 'Which data can be observed and gathered at reasonable cost?'.

Besides the optimization steps involved in the selection of adequate variables, another element deserves attention, namely the choice of functional form. As pointed out in Section 3.1, theoretical models only rarely provide an explicit functional form of the relationship between variables of interest. If this is not the case, again some optimization is necessary in order to decide which functional form should be used in the econometric analysis. Arguments like parsimony, interpretation of estimated coefficients and model fit enter the objective function in this step of model selection. Very often, standard formulations in the form of a linear or loglinear model are chosen if the data do not clearly indicate some nonlinear relationship.

3.2.2   Estimation of money demand

Let us assume that the first optimization problems in modelling money demand are solved and result in a set of variables and the assumption that the relationship is loglinear in the money aggregate, output and prices, and linear in the interest rate. For example, Brüggemann and Wolters (1998) employ quarterly German data of the narrow money aggregate M1, the real gross national product, the deflator of gross national product and the average bond rate. Let $m1$, p, and y denote the logarithms of the first three variables, and R the average bond rate. Brüggemann and Wolters (1998) also include the import price index as a further explanatory variable.

Given a set of variables and the kind of functional relationship, the next modelling step consists of a specified functional relationship between the variables. While straightforward application of money demand theory would result in a model with $m1$ as the single endogenous variable, more recent estimation approaches take into account that the other variables may also depend on monetary shocks. Therefore, a univariate modelling like

$$m1 = f(y, p, R) + \varepsilon,$$

which implicitly assumes that y, p and R are exogenous with regard to $m1$, is replaced by multivariate approaches, which treat all variables as potentially endogenous. The simplest non–structural approach for this purpose is given by a vector autoregressive model:

$$X_t = \begin{pmatrix} m1_t \\ p_t \\ y_t \\ R_t \end{pmatrix} = \sum_{k=1}^{K} \Gamma_k X_{t-k} + \varepsilon_t, \tag{3.2}$$

where $\varepsilon_t = (\varepsilon_{1t}, \ldots, \varepsilon_{4t})$ are identically and independently distributed as $\mathcal{N}_4(0, \Sigma)$, where Σ denotes the 4×4 covariance matrix of the error distribution, and the matrices

$$\Gamma_k = \left(\gamma_{i,j}^k \right)_{1 \leq i,j \leq 4}$$

provide the autoregression coefficients at lag k. Now, the modelling problem is reduced to the selection of a maximum lag length K and the estimation of the parameters Γ_k.

As long as a specific choice of $K > 0$ implies that all parameters Γ_k for $k \leq K$ are estimated, this final step is not too complex. In fact, for a given K, the estimation can be carried out efficiently by using ordinary least squares for each equation separately. Then, the optimal choice of k can be based on information criteria (see Chapter 12). However, with a four–dimensional model as the money demand system, a lag length of $k = 4$ corresponding to four quarterly observations requires the estimation of 64 parameters. Consequently, the number of available observations limits the usefulness of this approach. Therefore, in recent approaches to modelling money demand, e.g. Brüggemann and Wolters (1998) and Lütkepohl and Wolters (1998), a more general model is considered. Thereby, most elements of the Γ_k are restricted to zero a priori, and only the remaining few are estimated freely. Again, the optimal choice can be based on information criteria. However, the number of possible alternatives becomes much larger than in the standard approach. For the money demand example with a maximum lag length of 4, there exist $2^{64} \sim 1.8 \times 10^{19}$ different lag structures. Consequently, calculating an information criterion for each lag structure and then selecting the best one is not feasible within a reasonable time period. Furthermore, given such a lag structure, which may exhibit some 'holes', an efficient estimation of equation (3.2) requires a system approach, e.g. seemingly unrelated regression (SURE) (Zellner, 1962).

Despite these problems, Brüggemann and Wolters (1998) present estimation results for the more complicated model based on a sequential selection procedure. Due to problems of nonstationarity of the data they consider a

Table 3.2   Money demand system

	$\Delta m1_t$	Δp_t	Δy_t	ΔR_t
$\Delta m1_{t-1}$				+ T
$\Delta m1_{t-2}$	T		+	
$\Delta m1_{t-3}$			+	
$\Delta m1_{t-4}$			T	
Δp_{t-1}		+ T		
Δp_{t-2}	T	+ T	+	
Δp_{t-3}				
Δp_{t-4}	+	+ T		
Δy_{t-1}				
Δy_{t-2}	+	+ T	+ T	
Δy_{t-3}	T	+ T		
Δy_{t-4}			+ T	
ΔR_{t-1}	+ T	+		+ T
ΔR_{t-2}				
ΔR_{t-3}				
ΔR_{t-4}				
ec_{t-1}	+ T			
Δpm_{t-4}		+		

vector error correction model and, as pointed out before, include lagged import prices as further explanatory variables. Table 3.2 summarizes the chosen lag structure for the dynamic part of the model for $(\Delta m1_t, \Delta p_t, \Delta y_t, \Delta R_t)'$ and indicates whether the error correction term (ec_{t-1}) and lagged changes in import prices (Δpm_{t-4}) are included. Thereby, a '+' marks an element of the Γ_k matrices not restricted to zero.

Given the high inherent complexity of the lag order selection problem, it seems rational to employ explicit optimization heuristics for this purpose. Chapter 12 presents an implementation of threshold accepting to this problem. This implementation has been applied to the vector error correction model of money demand by using the data from Brüggemann and Wolters (1998). It results in the lag structure marked by 'T' in Table 3.2. Furthermore, the error correction term, lagged changes in the import price index, seasonal dummies and a shift dummy corresponding to German unification have been taken into account as potential additional exogenous regressors.

The results of explicit optimization based on a threshold accepting implementation and the specification proposed by Brüggemann and Wolters (1998) share some similarities, both in the modelling of the dynamic part and with regard to the inclusion of exogenous variables. In particular, the error correction term is included exclusively in the equation for $\Delta m1_t$ by both approaches, while the lagged import price index appears to be less important in the specification obtained by threshold accepting. In fact, while this variable is highly significant in the specification proposed by Brüggemann and Wolters (1998) it is not significant at the 5% level if added to the specification obtained by the threshold accepting implementation. Furthermore, the coefficient for the error correction term is almost halved when using the optimized model structure.

The information criterion used for model selection by the threshold accepting implementation is smaller for the optimized model than for the model proposed by Brüggemann and Wolters (1998). However, the outcome might be different when using other selection criteria. Although it is not possible to test one model against the other, because they are not nested, the comparison of the results clearly indicates that the automatic lag order identification procedure based on threshold accepting results in a dynamic structure, which mimics most of the features of the ad hoc solution. In contrast, using the standard approach of including all lags up to a given maximum lag length would result in choosing 48 lags for the example, if the same information criterion was used.

It is not the aim of this example to discuss the results in more depth with regard to their meaning for money demand theory. However, the findings clearly indicate that the use of an optimization heuristic like threshold accepting can help to solve optimal model selection in econometric analysis. A more extensive analysis of this application of optimization heuristics in econometrics is presented in Chapter 12. However, other steps of the modelling process may also result in optimization problems of high complexity, which cannot be solved satisfyingly by using standard methods. For example, the analysis of the results may require further optimization steps, e.g. for detecting structural breaks.

3.2.3 Implications of optimization

An optimal, or at least optimized, lag structure like the one presented in Table 3.2 might be interesting on its own, if one is mainly interested in the dynamic adjustment processes. Then, the use of optimization techniques provides an immediate gain in result quality. However, the results of several intermediate steps of the modelling process often do not provide such intrinsically valuable information to the researcher. Nevertheless, these steps have to be performed carefully, i.e. in an optimal way, because the findings of further steps of the analysis may depend on previous steps. In fact, modelling the dynamic part of the money demand model is a typical example of such an instance. The central interest in studying this relationship is the question as to whether there exists a long–run relationship between the variables. Given that the variables are non–stationary, Johansen (1988; 1991; 1992) and Johansen and Juselius (1990) present a procedure for detecting the existence of such long–run relationships within a vector error correction framework. Meanwhile, it is well known that the maximum–likelihood–based test statistics of this procedure depend on a correct specification of the dynamic part of the model. Thus, in the case of the money demand model, a bad choice of the dynamic structure may lead to erroneous results for an application of Johansen's procedure.

It will be the subject of further research and further applications of optimization heuristics like threshold accepting to assess the impact of such explicit optimization methods within the econometric modelling process. Wherever ad hoc heuristics have been used so far in preliminary modelling steps, it can be expected that the use of explicit optimization methods will improve the results. Whether this has an impact on subsequent hypothesis tests, like Johansen's procedure, depends on the robustness of these statistics. Thus, this question has to be answered by using Monte Carlo evidence.

3.3 Complexity Issues

Obviously, the optimization problems described in the previous sections differ markedly. Some of them are rather trivial, and others are easy to solve by standard methods, but at least one of them requires tremendous computing resources or heuristic methods. Since such a mix of different optimization problems is typical in applied statistics and econometrics, it seems necessary to provide a classification of these problems in order to decide which kind of optimization approach is best suited for which problem.

Such a classification of problems is provided by the notion of complexity (Ausiello, Crescenzi, Gambosi, Kann, Marchetti-Spaccamela and Protasi, 1999, pp. 1–37). This concept is widely used in different areas of computational methods without necessarily referring, 'one–to–one', to the exact mathematical concept. In this section, a rather loose and intuitive concept of complexity is also introduced and used. A reference to more formal concepts is provided later in Section 13.9.

In a very broad sense, the complexity of a problem can be defined as the amount of computational resources which are necessary to obtain a solution. In order to make this definition operational, a few further qualifications have to be made.

First, the amount of computational resources spent on a problem has to be specified. In most cases, this refers to the number of elementary operations, such as additions or multiplications, which a specific routine, software or algorithm needs to solve the problem. On a given computer,[11] this number is closely related to the total time the computer needs to solve the problem. Thus, one might replace the amount of computational resources

---

[11] To simplify notation, only single–user machines are considered or it is assumed that multi–user machines are used by only one user at a time.

in the above definition by the time a specific computer needs to solve the problem with a given software or algorithm. An alternative approach consists in considering the computer memory required for the solution of a problem. The corresponding concept of complexity is called 'space' complexity, in contrast to the already introduced 'time' complexity. Problems with high space complexity also exhibit high time complexity, while the contrary case does not apply. Consequently, the approaches are not equivalent.

Secondly, it is not of much interest typically to know the computational resources required to solve one specific problem, e.g. an ordinary least squares regression with 3 explanatory variables and given data for 20 periods. Rather, one would like to know how much time it needs in general to solve an ordinary least squares regression with k regressors and T periods. Even more important is the relationship between the computational time t^c and the instance or input size,[12] which in the example is given by the number of regressors k times the number of periods T. Thus, the complexity of a specific algorithm for the solution of a problem is a function of the problem size. For the ordinary least squares example, a $k \times k$ linear equation system has to be solved, which requires computational time proportional to k^3 if the standard algorithm for matrix inversion is used.[13] In addition, matrix multiplications require a time which is proportional to kT^2 when using the standard procedures. Therefore, the total complexity of ordinary least squares estimation is given by

$$t^c_{OLS} = \mathcal{O}(\max\{k^3, kT^2\}) = \mathcal{O}(kT^2),$$

since the number of observations T has to be larger than the number of regressors k. Thereby, $\mathcal{O}(\cdot)$ means that the time requirements are of the order of the terms in parentheses. Even for the simple OLS problem, the actual amount of resources spent on its solution may differ for different instances of the same size depending on the data. Then, a distinction has to be made between worst–case complexity, i.e. the amount of resources necessary to solve the most complex problem instance of a given kind, and average–case complexity (Ausiello et al., 1999, pp. 4ff.).[14] Finally, one might consider measures which weigh the amount of computing resources against the achieved approximation quality, if the problem is not solved to optimality.

Thirdly, after having specified the time complexity of a specific algorithm for a specific class of problems, the intrinsic complexity of the problem can be defined. Intrinsic complexity is also a function of problem instance size, but in

---

[12] A formal definition of input size requires some encoding of any possible input, e.g. using the binary alphabet, and measures input size as the number of digits of such an encoding (Ausiello et al., 1999, p. 5).
[13] This holds true only under the assumption that one multiplication of real numbers can be carried out with sufficient accuracy within one basic computer operation.
[14] Closely related to a consideration of average–case complexity is the probabilistic analysis of approximation algorithms. References to this subject are given in Ausiello et al. (1999, pp. 287ff.).

contrast to the complexity of a specific algorithm it is defined as the minimum over all feasible algorithms, i.e. including all known and unknown procedures for the solution of the problem. Consequently, the time complexity of a specific algorithm, such as provided above for the ordinary least squares case, provides an upper bound for its intrinsic time complexity. However, there might be other algorithms working faster. To provide exact lower bounds is therefore not a trivial task and is not solved for most of the applications studied in this book. In fact, mathematical complexity theory is mainly concerned with the calculation of lower bounds for highly complex problems.[15]

Since the intrinsic complexity of a problem is sometimes very hard to obtain, a more heuristic approach is followed for the rest of this section. Given that a problem might be solved by different algorithms, which may also differ in their complexity, the problem is considered as being hard to solve, i.e. highly complex, if the best available algorithm exhibits high complexity. By using this concept, problems of relatively low complexity can be distinguished from those with high complexity based on the following classification. An algorithm is considered as being of low or modest time complexity if it operates in polynomial time in the problem instance size. For example, the instance size of the regression model with k explanatory variables and T observations is proportional to kT. Hence, the time complexity t^c_{OLS} of $\mathcal{O}kT^2$ is certainly polynomial in the input size kT. Such an algorithm is called a polynomial time algorithm. Problems, for which such a polynomial time algorithm is known, are considered as being of low complexity, while the complexity of problems, for which all known algorithms cannot be bounded by polynomials in the problem size, is considered as being high. This classification is rationalized by the calculations provided in Table 3.3. This shows computing times for polynomial and non–polynomial time complexity functions. The factor of proportionality is set equal to one for these calculations. The times in the table correspond to a computer facility with $1\,000\,000\,000$ arithmetic operations per second, i.e. 1000 MFLOPs or 1 GFLOP, which corresponds to the technical frontier at the time of writing this book.

Table 3.3 indicates that polynomial time complexity functions grow rather modestly in their time requirements depending on the instance size, whereas the time needed by non–polynomial algorithms grows explosively. From this table, it becomes clear that even an increase of computing speed by a factor of 1000 or $1\,000\,000$ would not make the non–polynomial problems more tractable as soon as the instance size reaches a certain level.

Regarding this table, one might agree with computer scientists about looking for polynomial time algorithms for a given problem. When there is possibly no polynomial time algorithm which can solve the problem it will be called intractable. Of course, there exist non–polynomial time algorithms

---

[15] See Section 13.9 for a specific application and the literature cited therein.

Table 3.3 Polynomial and non–polynomial time complexity functions

Time complexity function	Instance size T				
	10	20	40	100	1000
T	1×10^{-8} seconds	2×10^{-8} seconds	4×10^{-8} seconds	1×10^{-7} seconds	1×10^{-6} seconds
T^2	1×10^{-7} seconds	4×10^{-7} seconds	1.6×10^{-6} seconds	1×10^{-5} seconds	0.001 seconds
T^5	0.0001 seconds	0.0032 seconds	0.1024 seconds	10 seconds	11.574 days
$2^{T/\ln T}$	2×10^{-8} seconds	1×10^{-7} seconds	1.8×10^{-6} seconds	0.0034 seconds	1.2×10^{27} centuries
3^T	0.59×10^{-4} seconds	3.49 seconds	385.5 years	1.6×10^{31} centuries	4.2×10^{460} centuries

which work quite well in practice, such as the simplex algorithm for linear programming.[16] However, such examples of non–polynomial time algorithms which work well in practice are rare, not yet well understood and always risk having enormous time requirements for some specific instances.

It is certainly reasonable to look for polynomial time algorithms. However, for some applications arising in economics, econometrics and statistics, intrinsic complexity is non–polynomial even when using the exact mathematical definition. For many more problems, no polynomial time algorithm is known which provides an exact solution of the problem. Nevertheless, these problems are solved in practice by using ad hoc heuristics or problem–specific heuristics. A 'solution' in this context is not an exact solution of the original problem, but is either a solution to a simplified version of the problem or an approximate solution to the original problem. Therefore, a broader understanding of complexity has to take into account not only worst–case measures for exact solutions as the concept introduced above, but also solution quality for ad hoc and approximation algorithms (see also Judd (1998, pp. 48ff.)). This point will be reconsidered when introducing optimization heuristics in the next section.

After introducing the concept of complexity, it is now possible to provide

---

[16] Another example is the quite complex algorithm for the travelling salesman problem discussed in Holland (1987).

a classification of optimization problems in statistics and econometrics. We might distinguish problems which are easy to solve since they belong to the class of polynomial time problems and a polynomial time algorithm is known. Other problems are known to require a higher computational input, although, at least for particular situations, results are still obtainable with reasonable effort. Next, there are problems which are known to be hard to solve, i.e. of high computational complexity, and finally, problems for which it is not even common procedure to treat them as explicit optimization problems.[17]

3.3.1 *Low complexity problems*

Ordinary least squares estimation has already been used as an example of econometric problems where an exact algorithm requiring computational time which grows only polynomially in problem size is available at least as long numerical problems do not hinder an accurate enough calculation of the solution of a system of linear equations. Another, but completely different example is the selection of a maximal lag order in a univariate autoregressive process, i.e. to choose $k \in \mathbb{N}$, $k \leq K$, such that the explanatory content — as measured by some information criterion — of the regression

$$x_t = \sum_{i=1}^{k} \alpha_i x_{t-i} + \varepsilon_t$$

reaches its maximum. Of course, in this case it suffices to calculate the information criterion for all K regression models by varying k from 1 to K, and then to select the optimal model.

3.3.2 *More complex problems*

For a large number of optimization problems, there exist numerical algorithms giving reasonably close approximations to the optimum at relatively small computational costs. This is the case for continuous optimization problems on globally convex functions as they arise, e.g, in non–linear least squares estimators, probit models and some maximum–likelihood problems. The methods used for this kind of problem are well documented and much work has been devoted to the analysis of their convergence properties.[18] If the assumptions about the function properties are fulfilled, these methods provide high–quality approximations to the optimization problem at hand. Often, accuracy is limited rather by floating point errors of the software used than

---

[17] A similar problem seems to exist in real–world applications of optimization heuristics, where managers have to be convinced that some problem can and should be treated as an optimization problem (Pirlot, 1992).

[18] See Press et al. (1992, Ch. 10).

by the number of iterations which can be performed with a given amount of computing resources. However, as already mentioned, a major drawback of these methods does exist. If the objective function does not fulfil the necessary properties and we are not aware of this fact, these algorithms may break down completely, i.e. fail to approximate the optimum.

3.3.3 Highly complex problems

Most optimization problems arising in economics, econometrics and statistics are not 'well behaved' enough to fall into one of the previous categories. Such problems are highly complex. Continuous optimization problems with a large number of local minima as they come up in disequilibrium modelling, or censored quantile regressions are examples of such problems.[19] Large–scale discrete optimization problems, e.g. the above–mentioned lag order selection problem in a multivariate setting, when subsets of all K lags have to be chosen, is another. A number of such examples are introduced in Part III of this book in order to demonstrate the power of optimization heuristics. In fact, the intrinsic computational complexity of these problems is the basic reason for the study and application of optimization heuristics, which is motivated in the next chapter.

---

[19] See Chapters 14 and 15 for a discussion of threshold accepting in this context.

4

The Heuristic Optimization Paradigm

4.1 The Classical Optimization Paradigm

Chapters 2 and 3 presented a few examples of typical optimization problems in economics, econometrics and statistics. Most of them can be expressed in mathematical terms as

$$\max_{\mathbf{x} \in \mathcal{X}} f(\mathbf{x}), \tag{4.1}$$

where the set $\mathcal{X}$ is a — possibly discrete — subset of some n–dimensional real vector space, i.e. $\mathcal{X} \subset \mathbb{R}^n$.[1] This formal definition of the optimization problem is often used synonymously with a solution $\mathbf{x}^{\mathbf{opt}}$, which is assumed to exist and — often — to be unique. McCullough and Vinod (1999, p. 635) state: 'Many textbooks convey the impression that all one has to do is use a computer to solve the problem, the implicit and unwarranted assumptions being that the computer's solution is accurate and that one software package is as good as any other'. The examples provided in their article demonstrate that this is certainly not the case.

Unfortunately, the sole representation in mathematical formulas does neither answer the question about existence and uniqueness of a solution, nor does it provide a solution or a method to check whether a candidate solution in fact solves the optimization problem. As Polak (1997, p. 1) formulates it: 'It is impossible to tell whether a piece of yellow metal is gold without submitting it to physical and chemical tests. It is equally impossible to tell whether a vector is a solution of an optimization problem. This fact has profound consequences on the extent to which one can hope to "solve" an optimization problem'.

Although optimization problems are probably as old as human civilisation, and the roots of mathematics reach back several millenniums, standard techniques for the solution of some optimization problems have evolved only

---

[1] A minimization problem for g can be represented by using the transformation $f(\mathbf{x}) = -g(\mathbf{x})$.

during the last two to three hundred years. For a given finite set of alternatives it was always possible to decide which alternative is the preferred one and, consequently, optimal. However, this simple enumeration algorithm is neither operational for real valued problems, as they arise in modern economic theory or econometrics, nor for discrete optimization problems with a tremendous number of alternatives, as are typical in economic applications.

At least for the former class of problems, the introduction of differential calculus represented a major breakthrough. For problems accessible to this approach, optimality can be checked directly through the well known first– and second–order conditions. The first–order condition is a necessary condition, i.e. it must be satisfied by any local maximizer. In order to be a local maximizer, additional conditions have to be met, e.g. a second–order condition. Sufficient conditions are those implying that a point is a local maximizer. If $\mathbf{x} \in \mathcal{X}$ is a local maximizer, the values of f close to $\mathbf{x}$ are smaller or equal to $f(\mathbf{x})$. However, a solution to the above optimization problem is a point $\mathbf{x^{opt}}$, so that $f(\mathbf{x}) \leq f(\mathbf{x^{opt}})$ for all $\mathbf{x} \in \mathcal{X}$. Such a global maximizer can be obtained through standard differential calculus only under additional assumptions about f or $\mathcal{X}$ (see Section 4.2.).

Given that the optimization problem falls into the category which can be solved by differential calculus methods, at least two additional complications can show up. First, the resulting global optimizer does not need to be unique. For real life applications, where f might be, e.g. a profit function, such a case seems less problematic, while it certainly is in economic theory or for econometric estimates, in particular, if the method delivers one global maximizer without information about the potential existence of further solutions. Secondly, differential calculus reduces the solution of an optimization problem to the solution of a (system of) equation(s). However, except for a few cases, e.g. linear or quadratic functions, the solution of this (system of) equation(s) also requires numerical optimization methods such as those described in Section 5.1 with all related problems (see Section 4.2 and McCullough and Vinod (1999)). In particular, the obtained solution is not necessarily as exact as the mathematical formulation implies.

To sum up the reading of the 'classical optimization paradigm' in this chapter, it is understood as being the implicit identification of the mathematical formulation of an optimization problem as in equation (4.1) with its (approximate) solution by means of simple enumeration or differential calculus. In particular, the existence of a (unique) solution is presumed, as well as the convergence of classical optimization methods for the solution of the corresponding first–order conditions. Consequently, most of the standard optimization problems of economic theory considered in Section 2.2 fall within this category, and also the ordinary least squares estimator from Section 3.1.2 as well as some maximum–likelihood estimators. However, many other optimization problems outlined in the previous chapters obviously resist this standard approach. The reasons for this failure of the classical paradigm to fulfil all optimization needs in economics and econometrics are summarized in the following section.

4.2 Limits of the Classical Paradigm

The limits of the classical optimization paradigm have already appeared in the preceding chapters. These can be summarized under three headings. First, many optimization routines currently applied in real economic life do not fit into the rigid framework provided by this paradigm. Secondly, standard optimization routines are applied to problems which do not fulfil the requirements of these methods (Goffe, Ferrier and Rogers, 1994, p. 68). Consequently, claim and reality may fall apart for these applications. Thirdly, in cases where the standard optimization paradigm can be applied, problem sizes may hinder efficient calculation even for problems with polynomial time complexity.

In the classical paradigm, optimization problems can be solved exactly. However, except for a few special cases of f and $\mathcal{X}$, the local maximizers, which can solely be obtained by all standard techniques in nonlinear optimization (see Section 5.1.), do not have to be global maximizers (Horst and Tuy, 1996, p. 5). In general, there is no local criterion for deciding whether such a local solution is global or at least close to a global solution. Thus, e.g. the firm maximizing a profit function given complex production technologies, factor input functions and side constraints, by using standard nonlinear optimization techniques, might come up with a local maximum which is much smaller than the global optimum. In fact, this corresponds to the case depicted earlier in Figure 1.1 as going from point A to B. It is not only in real life economics that problems resist solution with the standard approach. The same holds true for problems in econometrics, e.g. maximum–likelihood estimation may exhibit multiple local optima (Dorsey and Mayer, 1995). A further class of optimization problems not fitting in the classical paradigm are integer optimization ones such as the travelling salesman problem, which are treated in operational research by using a quite different methodological approach. The reason why the classical paradigm fails for these instances is the intrinsic high complexity.[2] Finally, headings like 'fuzzy optimization' indicate that even the formulation of optimization problems such as those represented by equation (4.1) can be misleading for specific applications.

Despite the fact that the standard optimization approach requires strong assumptions about f and $\mathcal{X}$, it is routinely applied both in economic theory and in econometrics.[3] In economic theory, the requirements for the local

---

[2] As John Holland (1992) states: '... complexity makes discovery of the optimum a long, perhaps never–to–be completed task, so the best among tested options must be exploited at every step'. (Birchenhall, 1995, p. 237).

[3] Dennis and Schnabel (1989, p. 3) state: 'It should be noted that in most nonlinear

maximizers to be global ones are obtained by assumption. In some cases, these assumptions are made solely to obtain an analytical or at least unique solution to the optimization problems of economic agents. However, there is no reason to believe that these assumptions are congruent with the real behaviour of economic agents, nor is it possible to give a guarantee that the solutions obtained by using this assumptions are at least close to the ones which might have been obtained by using more realistic assumptions. Of course, the use of more realistic assumptions might imply that the optimization problem cannot be solved within the classical optimization paradigm. In econometrics it is more difficult to obtain the desirable features of f or $\mathcal{X}$ by assumption, since data may lead to a rejection of these assumptions. Therefore, often standard techniques are also applied to optimization problems with several local minima. Sometimes, a restart version of the optimization routine is used to improve its performance. Such a restart version is often based on a regular grid, which is laid over the search space.[4] Nevertheless, the fundamental limit of the classical paradigm to local optima cannot be resolved in this way. Although the resulting estimates are the — up to numerical problems — exact solution to a local optimization problem, they do not necessarily share any information with the global optimum. The use of standard techniques implies erroneously high quality, if not exact results (McCullough and Vinod, 1999, p. 651).

Since even the solution of optimization problems with the classical toolbox requires numerical approximations, if the linear or quadratic framework is left, it cannot be guaranteed to work efficiently for all problem instances. In particular, if the dimension of the optimization problem is large and the features of f do not correspond to the requirements of the standard optimization procedures, they may fail even in this case. Typical examples are the solution of an ill conditioned system of linear equations or the performance of gradient methods on a very flat part of the likelihood function.

The examples used in Part III of this book to demonstrate the versatility of the threshold accepting heuristic mainly refer to problems of intrinsic high complexity, or problems where the assumptions for standard optimization tools to work efficiently are not satisfied (several local minima). Nevertheless, one might also consider using it in cases where standard methods are adequate, but still too slow. However, in this case the development of parallel algorithms and more powerful computers may help to overcome the limits of the classical paradigm (Dennis and Schnabel, 1989, pp. 64ff.). Other problems cannot be solved solely by improving computational facilities, even at the tremendous rate we have experienced during the last two decades, owing

---

optimization problems solved today, the variables are continuous, $f(\mathbf{x})$ is differentiable, and a local minimizer provides a satisfactory solution. This probably reflects available software as well as the needs of practical applications'.

[4] See also Section 5.1.4.

to their high intrinsic computational complexity. This opens the door for a different approach, namely the heuristic optimization paradigm. Of course, optimization heuristics do not represent the sole cure to the deficiencies of the classical paradigm. Meanwhile, there also exist deterministic global optimization methods, which are successfully applied to many problems.[5] However, the discussion of complexity and heuristics in Sections 2.3 and 3.3 gives the rationale that such approaches may provide high–quality solutions for specific problems, but do not provide a new paradigm which can be used for all problems arising in economics and econometrics.

4.3 Heuristics

The lack of adequate optimization approaches was, and still is a limiting factor for the development of science. Other scientific disciplines have faced similar problems to those described in the previous section during the last centuries. In order to overcome these obstacles, a constant process of innovation in mathematics and optimization methods evolved. The development of differential calculus, represented as the core ingredient of the classical optimization paradigm, may be regarded as a basic innovation for this purpose. This allowed simple optimization schemes like the one used in linear ordinary least squares regression, but also more refined gradient methods such as Newton's method. In combination with powerful computers, these tools allow us to tackle a huge range of optimization problems in economics and econometrics. However, for more complex optimization problems these methods fail to work efficiently.[6]

Since the limits of the classical optimization paradigm also apply to other disciplines, it is not too surprising that alternative approaches have been considered. Besides deterministic global optimization approaches[7] a new optimization dogma based on concepts found in nature spread through different disciplines as a consequence of growing computational power in the late 1980s and 1990s. These concepts include analogies to genetics, neural networks or simulated annealing. They are particularly appealing for applications in economics, as they model optimization behaviour directly instead of relying on some optimality criterion. Following the seminal paper by Kirkpatrick et al. (1983), such approaches grew in importance in disciplines ranging from physics to operational research. It is only recently, however, that the first applications of this new basic innovation in optimization can also be

---

[5] See Horst and Tuy (1996) and Kan and Timmer (1989) and the references given in these contributions.

[6] Judd (1998, p. 99) states: 'In any case, global optimization problems are difficult outside of special cases, and one must develop careful heuristics to reduce the chance of gross error'.

[7] See Horst and Tuy (1996) for a recent overview.

found in economics, econometrics and statistics. It is one purpose of this book to advocate a wider dissemination of such approaches in these areas.

From the two reasons for the recourse to alternative optimization paradigms — to mirror the optimization behaviour of economic agents more realistically and to provide approximate solutions to highly complex problems, which cannot be tackled within the classical paradigm — this book focuses on the latter. However, if the intrinsic high complexity of real problems does not allow for their exact solution as assumed in the classical paradigm, a reconsideration of the latter might be also useful for the modelling of the behaviour of economic agents, who have to solve those real optimization problems (McFadden, 1999). In fact, some approaches followed in evolutionary economics proceed along these lines.

After this pleading for a change in optimization paradigms, which hopefully is well founded on the examples provided in the previous chapters, it is time to introduce the alternative paradigm, namely optimization heuristics. Like the classical methods, optimization heuristics aim at providing high–quality solutions to optimization problems. However, given the intrinsic complexity of those problems which withstand the classical approach and are therefore tackled by heuristics, they cannot pretend to produce the exact solution in every case with any certainty.[8] Furthermore, many of the powerful techniques exhibit some stochastic elements. Consequently, the results provided by these algorithms may differ depending on the random numbers used for the simulation of stochastic components.[9] While these characteristics of optimization heuristics may seem disadvantageous at first sight, it should be kept in mind that optimization heuristics are applied solely in cases when the classical paradigm fails to provide solutions at all or cannot guarantee to provide global optima. Then, a stochastic high–quality approximation of a global optimum is better than a deterministic poor–quality local minimum provided by a classical method or no solution at all.

During the last two decades, many powerful techniques have been developed for highly complex optimization problems. A complete overview is definitely not only beyond the scope of this present chapter, but also of the whole book.[10] Nevertheless, some widespread methods will be outlined as a bench–mark for threshold accepting in the next chapter. A few of these are also well suited to provide a more vivid understanding of optimization heuristics

---

[8] Farley and Jones (1994, p. 167) state on this aspect: 'One way to deal with the complexity problem is to accept non–optimal results. Though always nice to be able to find "the best" solution, we will be concerned with the more practical goal of finding "a good" solution; that is, the best one we can find in a reasonable amount of time'.

[9] See also Section 11.6.3 for a deterministic approach for simulating stochastic components, which can also be applied to optimization heuristics; see, e.g. Fang, Hickernell and Winker (1996).

[10] Osman and Laporte (1996) provide a recent and extensive bibliography of related methods in operational research applications.

through analogies to real–world behaviour and phenomena.

The methods provided in the literature can be divided into two subclasses, namely constructive and local search methods (Aarts and Lenstra, 1997, p. 2). The former are more closely related to optimality conditions in the classical paradigm and have advantages for specific problem instances, e.g. in operational research.[11] The latter are more commonly used in statistics and econometrics. This can be attributed to at least three causes:

1. Most of them are easy to implement for different problems. Thus, a new complex optimization problem can be tackled almost immediately by using these techniques, while problem–specific construction methods require a deeper understanding of the mathematical structure of the problem. Furthermore, side constraints on the solution of almost any kind can be taken into consideration at low additional coding expense. To a certain extent, local search heuristics replace manpower by computer power.[12]
2. Local search heuristics proceed quite similarly to Newton's method or other gradient methods. Thus, knowledge acquired through the use of these methods can be partially carried over to the heuristic framework.
3. Even simple local search heuristics often show a good or even excellent performance on many problem instances arising in economic or econometric practice. It appears to be difficult to obtain better results even with highly problem–specific methods.

The examples of optimization heuristics presented in Section 5.2 as a background for the introduction of threshold accepting are chosen from this second class of local search algorithms. The appeal of these methods lies in the intuitive correspondence between the iteration steps of the algorithms and the optimization steps which can be observed in human behaviour or natural processes. Using the example of climbing the Himalayas given in Chapter 1, it is much more intuitive to look for a path leading uphill than to test for a given point of the landscape as to whether it fulfils some local optimality property. Although heuristic methods cannot avoid referring to more formal mathematical concepts, their basic concepts are often linked more closely to an intuitive understanding of optimization. Examples of such algorithms include genetic algorithms, which imitate Darwinism, tabu search, which forbids the way back to bad parts of the landscape, simulated annealing, which copies properties of metal melting, or the Great–Deluge algorithm, which tries to find high elevations in the functional landscape following Noah's traces by avoiding already flooded areas.

Besides this intuitively appealing properties of some optimization heuristics, at least three further developments have contributed to the increasing interest

---

[11] A few applications, including the travelling salesman problem, are described in Ausiello et al. (1999, pp. 40ff.).

[12] See also Section 2.3.3 for a discussion of the economics of computation.

in these techniques, even in research areas where they have not played an important role in the past. Statistics and econometrics certainly belong to this group. According to Aarts and Lenstra (1997, p. 2), the other developments can be summarized as follows:

1. While the application of most heuristic concepts started with real–world analogies in the beginning and was motivated by their — sometimes surprising — success, at least for some of the algorithms, a deeper analysis of their stochastic properties followed. Results from this analysis indicate nice convergence properties or delivers additional hints for successful implementations.[13]

2. Optimization heuristics are used if either the problem cannot be tackled at all by other approaches due to its intrinsic complexity or if the development of a specially tailored algorithm would require a large input of human resources. Consequently, the tremendous decrease in costs of computation resulting from a parallel increase in computational resources over the last decade has made the use of optimization heuristics relatively more attractive.[14]

3. The ease of implementation of some local search algorithms suggest their routine use for real–world problems when other optimization methods have failed to provide results of appropriate quality. In this context, the application of optimization heuristics provides a bench–mark for other methods at comparatively low costs.

The arguments put forward in this section provide several good reasons for a careful look at optimization heuristics whenever a non–trivial optimization problem has to be tackled. Furthermore, they allow the formulation of problems in an optimization context, which so far have only been considered by using ad hoc approaches. Nevertheless, the application of heuristic methods also requires some general criteria for deciding when a heuristic can be considered as a valuable tool or a contribution for the solution of a class of optimization problems (Barr et al., 1995). First, they should produce or identify high–quality solutions faster than standard approaches. Secondly, they should be more robust to changes in problem characteristics, tuning parameters or the introduction of additional constraints. Thirdly, they should be easy to implement to many problem instances, including new ones.

Before turning to the description of some classical and heuristic optimization methods with a focus on threshold accepting in Part II of this text, it should be stated, that the claim of this book is to demonstrate that the chosen heuristic fulfils these requirements for the selected problems — and potentially for many other optimization problems in statistics and econometrics. Thus, the bench–mark for evaluating the threshold accepting

---

[13] These aspects are treated for threshold accepting in some detail in Chapters 6 and 7.
[14] See also Section 2.3.3.

heuristic is high, namely to beat classical approaches in quality and speed or to solve problems which cannot be tackled within the classical paradigm due to, e.g. their computational complexity.

4.4 Conclusions

To sum up the arguments put forward in this introductory part of the book, the following might be stated. First, optimization is omnipresent in economics, econometrics and statistics. The decisions of some economic agents, such as firms or monetary authorities, is based on explicit optimization approaches. The behaviour of other economic agents, in particular private households, is modelled as if households solve optimization problems. In fact, at least if the understanding of solving optimization problems is extended beyond the classical paradigm, this seems to be a reasonable approach. In econometrics and statistics, many of the optimization problems appear in explicit form like maximum–likelihood estimation, optimal model selection or least squares estimation. However, the influence of optimization approaches extends beyond these obvious cases. In fact, any descriptive statistic, any decision about functional forms or variables to be included in a model, and any interpretation of results from statistical inference enclose the solution of some explicit or — more often — implicit optimization problem. The share of optimization-related tasks and decisions within these disciplines is much larger than the explicit mentioning of optimization topics suggests.

Secondly, a relevant portion of these problems cannot be solved by standard optimization routines within the classical optimization paradigm. This share increases heavily if the often unrealistic assumptions are relaxed, which are imposed in order to make the model or estimation problem fit into the Procrustean bed of problems accessible through standard (non) linear optimization tools. The basic reason for this failure of the standard methods to cope with this large fraction of optimization problem is their high intrinsic complexity. Even if complexity does not hinder solution by using standard methods, numerical problems will sometimes do so.

Thirdly, for some problems, which are not accessible by standard tools, there exist more refined global optimization methods. However, these methods work well only for specific problem groups and may imply huge computational costs for some instances. Given the complexity argument, it cannot be hoped to obtain similar deterministic and high–quality solutions to all optimization problems arising in economics, econometrics and statistics, including the examples cited in the previous chapters. For this reason, heuristic methods are considered as valuable alternatives. They have already been successfully applied in many disciplines, including operational research. Furthermore, some recent theoretical results provide substantiation for this success. In addition, some of these heuristics, including the threshold accepting algorithm considered in this volume, are quite easy to implement on many different problems. Therefore, the results obtained by this method may serve as a bench–mark for any other more refined, more problem–specific solution method, or for solutions obtained by relying on some unrealistic additional assumptions.

If the given optimization aspects of the problems appearing in economics, econometrics and statistics are to be considered as they arise, i.e. without recourse to unrealistic assumptions or ad hoc solutions, a shift to a new, heuristic optimization paradigm is indicated.

Part II

Heuristic Optimization: Threshold Accepting

5

Optimization Methods

This part of the book is devoted to the optimization heuristic threshold accepting, and its properties, performance and tuning. Finally, Chapter 9 provides a practical guide to the implementation of threshold accepting for optimization problems appearing in statistics and econometrics, such as those described in the first part of this book.

Before turning to the threshold accepting heuristic, this present chapter discusses some basic classical and heuristic optimization methods.[1] The classical methods are included for reference purposes.[2] First, optimization heuristics mimic some features of classical methods. Secondly, the results obtained by classical methods may serve as a bench–mark for optimization heuristics. When classical methods are adequately applied, a global optimum is found which should be reproduced or at least closely approximated by heuristics. If classical methods are applied, and although some of the requirements may not be fulfilled, the results in this case represent a bench–mark which can be improved by the use of optimization heuristics. The extent to which the results can be improved in these cases provides information on the quality of the heuristic. Finally, classical methods also provide a bench–mark for the input of computational resources.

The aim of this book is to demonstrate that optimization heuristics can be successfully applied to many problems in econometrics and statistics when standard approaches fail. The threshold accepting heuristic is chosen as a very efficient representative of optimization heuristics. Nevertheless, it is not stated that this heuristic is the best candidate for any optimization problem. In fact, although the results presented later in Section 7.4indicate a slight advantage in some real implementations, the motivation for the choice of threshold accepting is also a heuristic one. Most authors dealing with optimization heuristics emphasize that the performance of such heuristics still depends — and following Aarts, Korst and van Laarhoven (1997, p. 117)

---

[1] For a general classification of optimization approaches, see also Pham and Karaboga (2000, pp. 244ff.).

[2] A more detailed overview can be found in Judd (1998, pp. 93ff.).

always will — to a large extent on theoretical knowledge, taste and practical experience. Hence, the acquaintance gained over the last few years with several implementations of threshold accepting are a strong argument in favour of using this heuristic. Nevertheless, Section 5.2 also gives a short overview on some other optimization heuristics which are used or could be considered in this context. Some results of the performance of threshold accepting relative to some of these other optimization heuristics are provided in Section 7.4. Thus, they also serve as a bench–mark for a successful implementation of threshold accepting.

5.1 Classical Optimization Methods

Since a significant percentage of optimization problems arising in statistics and econometrics are data–fitting problems (Dennis and Schnabel, 1989, p. 2), this section starts with a description of least squares estimation, which it the most often used optimization–based method in this context. If data fitting or other optimization problems lead to nonlinear, though still well behaved, e.g. at least twice continuously differentiable, optimization problems, gradient methods are applied to solve the nonlinear system of equations resulting from the first–order conditions. These methods are outlined in Section 5.1.2. Corresponding to least squares for the case of a continuous parameter set, enumeration is a strikingly simple method for solving optimization problems over a discrete set, while grid search is used both for integer problems and continuous problems if several local optima are to be expected. These classical methods for integer problems are considered in Subsections 5.1.3 and 5.1.4, respectively. Of course, grid search could also be considered as an optimization heuristics, since it does not necessarily deliver a high–quality solution. The same holds true for the classical local search algorithm discussed in Section 5.1.5, which is considered as a link between the classical optimization paradigm and the universe of optimization heuristics introduced in Section 5.2.

5.1.1   Least squares

Given the data $(\mathbf{x}_t, y_t)$, $t = 1, \ldots, T$ and a model $g(\mathbf{x}, \beta)$, where β denotes the parameter vector of the model, the difference between the model's prediction conditional on $\mathbf{x}_t$ and the y_t, which might be observed with some error, is given by

$$\varepsilon_t = y_t - g(\mathbf{x}_t, \beta). \tag{5.1}$$

The residuals ε_t can also be considered as the part of the variation of y_t which cannot be 'explained' by the model given the values of $\mathbf{x}_t$ and a parameter vector β. Then, it makes sense to choose β so that the vector $(\varepsilon_1, \ldots, \varepsilon_T)$ becomes as small as possible.

This problem already looks very much like an optimization problem under the classical paradigm. However, in order to be tackled by some optimization method, the objective function has to be defined explicitly as a mapping from the parameter space to the reel line. At least two distinct mappings are used in econometrics for this purpose. One is to minimize $f(\beta) = \sum_{t=1}^{T} |\varepsilon_t|$, thus resulting in the least absolute deviation estimator, and the other is to minimize $f(\beta) = \sum_{t=1}^{T} \varepsilon_t^2$, which gives the least squares estimator. One solution seems as good as the other, and in fact, both versions are used in practice according to their specific advantages. The least absolute deviation estimator is robust to outliers in the data, but is hard to compute. In fact, it cannot be tackled within the classical optimization paradigm, in particular, if some censoring on the values y_t is present.[3] This case is studied in more detail in Chapter 14. The least squares estimator is the maximum likelihood estimator under the assumption that the residuals are independently, identically normal distributed (Dennis and Schnabel, 1989, p. 60). Its main advantage, however, is that it is easy to compute, in particular, if the model is linear in the parameters.

In fact, if $g(\mathbf{x}, \beta)$ is linear in β, least squares estimation corresponds to minimizing a quadratic objective function. Consequently, the first–order conditions are all linear, and optimization is equivalent to the solution of a system of linear equations, which can be achieved in $\mathcal{O}(Tk^2)$ algebraic operations, where k denotes the number of parameters in β (Dennis and Schnabel, 1989, p. 60). However, even if the model is nonlinear, least squares estimation can be executed efficiently by using gradient methods, as described in the next section, provided that the first and second derivatives of g with regard to β are available and the residuals are not too large (Judd, 1998, p. 118).

5.1.2 Gradient methods

The case of linear least squares estimation is a particular case of an optimization problem, since it reduces to the solution of a system of linear equations. However, a more general method relies on a similar principle. If the univariate objective function $f(x)$ is twice continuously differentiable,[4] it can be locally, i.e close to some point x_0, approximated by a quadratic polynomial as

$$p(x) = f(x_0) + f'(x_0)(x - x_0) + \frac{f''(x_0)}{2}(x - x_0)^2. \qquad (5.2)$$

---

[3] Censoring means that y_t is observed only when it lies within a given range. Otherwise, only the information that y_t is larger (smaller) than a certain threshold is available. Censored data are typical in income data from social security systems or for duration data models.

[4] This condition is sufficient but not necessary for the following Taylor approximation argument.

Now, an approximate solution for the minimization problem for f can be found by using the point x_1, which minimizes $p(x)$. This point is easy to calculate. As with the linear least squares estimator, only a (system of) linear equation(s) has to be solved. If p is a good approximation to f, then x_1 should be closer to the minimum of f than the starting point x_0. If the approximation restarts with the new point x_1, a close approximation of the minimum of f might finally result. In fact, this is the approach followed in Newton's method.[5]

If Newton's method converges, which is the case if x_0 is close enough to a real optimum, it converges at a quadratic rate to some x^{opt} with $f'(x^{opt}) = 0$ (Judd, 1998, pp. 96ff.). In order to make sure that x^{opt} represents a minimum, the second–order condition, $f''(x^{opt}) > 0$, has to be checked. To complete the algorithm, a stopping rule is required. The iterations stop if the change of x_i to x_{i+1} is small compared to x_i and $f'(x)$ is close to zero. These two criteria are provided in terms of the convergence parameters ε and δ, respectively.

Given that the assumptions about f are fulfilled and Newton's method converges, the error of approximation is smaller than ε unless f is nearly flat and/or has a rapidly changing curvature in a large neighbourhood of the solution (Judd, 1998, p. 98). However, the method approximates only some local minimum of f. If f exhibits several local minima, there is no guarantee that a global minimum will be found. Only if f is convex or x_0 is chosen sufficiently close to a global minimum, can convergence to a global minimum be expected.

Newton's method can also be applied to multivariate problems. Again, $f(\mathbf{x})$ is replaced by a local quadratic approximation. The iteration step becomes

$$\mathbf{x}_{i+1} = \mathbf{x}_i - H(\mathbf{x}_i)^{-1}(\nabla f(\mathbf{x}_i))', \tag{5.3}$$

where $\nabla f(\mathbf{x})$ denotes the gradient of f, i.e. the vector of partial derivatives, evaluated at $\mathbf{x}$, and $H(\mathbf{x})$ is the Hessian, i.e. the matrix of second partial derivatives. If f is convex, then $H(\mathbf{x}_i)$ is positive definite and the same arguments as in the univariate case apply. However, Newton's method is also used when $H(\mathbf{x}_i)$ is not positive definite. In the Gauss–Newton algorithm, the Hessian is approximated using the Jacobian and thus avoiding the calculation of second–order derivatives. Conditioning problems of this approximation step are tackled by the Levenberg–Marquardt algorithm (Judd, 1998, p. 119). The problem of a not positive definite Hessian is also addressed by some more refined gradient methods, such as the Broyden–Fletcher–Goldfarb–Shanno (BFGS) algorithm (Dennis and Schnabel, 1989, pp. 29ff.). An introduction to such refined methods is provided in Judd (1998, pp. 109ff.). [6]

The problems related to Newton's method and other gradient methods also apply to the multivariate case and the refinements. Convergence is not

---

[5] See also Polak (1997) and Judd (1998, pp. 96ff.).
[6] See also Polak (1997). The convergence of methods based on Taylor series expansions is analysed in Dennis and Schnabel (1989, pp. 14ff.).

guaranteed, and if the algorithms converge, the result is not necessarily an approximation to a global minimum, but maybe to a local minimum of poor quality. A further problem of the multivariate methods is the computing time devoted to the calculations of gradients and Hessians. In fact, if these values are approximated by using finite difference methods, huge computational burdens or inaccuracies may result (McCullough and Vinod, 1999). On the other hand, the analytical computation of these expressions require tedious, error–laden calculations. However, the development of mathematical software, which allows the calculation of analytical derivatives for some function classes, certainly reduces the costs associated with this task in terms of labour input. Nevertheless, sometimes such a task is completely infeasible, e.g. if the objective function includes integrals which have to be evaluated numerically.

5.1.3 Enumeration

The classical optimization methods briefly outlined in the above two sections apply only to smooth objective functions on reel spaces. Some of these can be generalized to cover also less smooth functions. A straightforward generalization to problems defined on discrete sets, however, is not possible. For these problems, which also arise quite often in real applications as well as in statistic and econometric modelling, two distinct approaches seem feasible.

First, if both the discrete set can be embedded easily within a reel space and the definition of the objective function can be naturally extended over this whole reel space, continuous optimization problems can be applied to the problem. In general, the resulting optimum will not belong to the original discrete set. Either it is replaced by the next neighbour in the discrete set, or optimization is performed under side constraints. If the first option is chosen, the resulting point may even be far off from a local optimum. The second option, in contrast, leads to a highly complex optimization problem, which often is not accessible by standard tools.

Secondly, discrete optimization procedures can be used. This is the only feasible choice, if there exists no natural embedding of the discrete set $\mathcal{X}$ in a reel space, which also works for the objective function. Then, the most primitive algorithm consists in evaluating the objective function f for any element $\mathbf{x} \in \mathcal{X}$ and choosing the minimum. Of course, this algorithm works for any discrete set and any objective function, as long as f can be evaluated for all $\mathbf{x} \in \mathcal{X}$. In contrast to the standard methods for continuous optimization, it delivers the global optimum of the problem with certainty. If the global optimum is not unique, it even provides all elements of the set of global optima. However, at least for most of the discrete optimization problems referred to in Chapters 1 to 4 and for the applications considered in Part III, enumeration is seldom a useful approach as soon as the problem instances become larger (Cook et al., 1998, p. 2). For example, the travelling salesman problem considered in Chapter 1 allows for $(n - 1)!/2$ different possible solutions, if n is the number of cities to visit. Thus, for $n = 10$, 181 440 different solutions have to be evaluated — a possible task. However, for $n = 30$, the number has increased dramatically. Even with a computer, which allows us to evaluate 10^{12} solutions per second, it would take more than 4.2×10^{12} centuries to solve the problem by enumeration! Consequently, enumeration is not a feasible algorithm for real–world problems and complex discrete optimization problems in econometrics and statistics.

5.1.4   Grid search

Both for the analysis of continuous functions, which might exhibit several local minima, and for the optimization over a discrete set, when enumeration is not feasible due to its size, grid search is an often used instrument. In fact, grid search might even be considered as the most primitive procedure used to minimize a function (Judd, 1998, p. 100).

   The search space $\mathcal{X}$ is overlaid with a — in most applications — rectangular grid $\mathcal{G} \subset \mathcal{X}$. Then, the objective function f is evaluated at all elements of $\mathcal{G}$ and the minimum is picked and used as an approximation to the unknown global minimum. If we allow the grid to become finer and finer by increasing the elements in $\mathcal{G}$, it will eventually give a good approximation to the true optimum.[7] However, grid search is usually applied to situations of a high–dimensional search space or a huge discrete search space $\mathcal{X}$. Then, even a very fine grid, i.e. using a huge number of points, still covers only a small fraction of the search space and, therefore, will not necessarily result in a good approximation. In fact, the number of grid points required to obtain a given approximation quality grows exponentially in the dimension of the search space. Nevertheless, grid search might provide general information about the 'landscape' of the optimization problem, i.e. regions with high and low values of $f(\mathbf{x})$, which is useful for further analysis.

   Furthermore, grid search can be improved by combining it with other methods. For example, for maximum–likelihood estimation problems, when multiple local optima might exist, an often followed procedure consists in restarting some gradient algorithm from different starting values, which are delivered by a grid.[8] Instead of combining grid search with standard optimization algorithms, it can also be used in connection with classical or heuristic local search algorithms. Furthermore, instead of taking a rectangular grid, quasi–Monte Carlo methods can be used to generate more uniformly scattered grid points and in due course to improve the expected

---

[7] Hansen (2000, p. 578) proposes the use of grid search for threshold estimation, when sample size becomes too large for enumeration.

[8] Judd (1998, p. 99) proposes the use of as many restarts as feasible. However, the results may still depend on the specific grid chosen for this restarting procedure.

approximation.[9] Finally, an iterative refinement of grid search — possibly using quasi–Monte Carlo point sets — as described in Fang et al. (1996) can be performed.

5.1.5 Local search

Classical local search or neighbourhood search are iterative methods based on the concept of exploring the vicinity of the current solution in each step of the algorithm (see also Ausiello et al. (1999, pp. 61ff.)). One possibility of describing the behaviour of a function in the neighbourhood of a given point consists in providing information about derivatives, if the function is smooth enough and defined on a continuous set. Therefore, the class of local search procedures also includes the iterative gradient schemes, like Newton's method, as special cases.[10] In general, a local search algorithm has to be initialized with any element of the search space as the current solution or starting value. Then, in each iteration step, a generation mechanism proposes some element in the neighbourhood of the current solution. A selection criteria defines whether this new element becomes the current solution. Otherwise, the former current solution is kept and a new generation step is then performed.

As some of the notions used in this general description of classical local search are reused for some of the optimization heuristics described in the next section, a formal description of the framework seems adequate. First, as in the previous chapter, an optimization problem can be given by a set $\mathcal{X}$ and a mapping $f : \mathcal{X} \to \mathbb{R}$.[11] Now, let f_{opt} denote the minimum of f on $\mathcal{X}$ if it exists, i.e.

$$f_{\mathrm{opt}} = \min_{\mathbf{x} \in \mathcal{X}} f(\mathbf{x}) . \qquad (5.4)$$

As pointed out in Section 5.1.3, such a minimum always exists for discrete optimization problems over some finite set $\mathcal{X}$. Then, the goal of any optimization routine is to find elements of the minimizing set

$$\mathcal{X}_{\mathrm{min}} = \{\mathbf{x} \in \mathcal{X} \mid f(\mathbf{x}) = f_{\mathrm{opt}}\} \qquad (5.5)$$

or, even better, all elements of this set.

In order to describe the local search behaviour, it is necessary to add some local structure on the search space $\mathcal{X}$. This is achieved by defining a mapping $\mathcal{N} : \mathcal{X} \to 2^{\mathcal{X}}$ from the search space $\mathcal{X}$ to the set of subsets of $\mathcal{X}$. This mapping defines for each $\mathbf{x} \in \mathcal{X}$ a set $\mathcal{N}(\mathbf{x})$ of elements, which are considered as being close to $\mathbf{x}$. $\mathcal{N}(\mathbf{x})$ is called the *neighbourhood* of $\mathbf{x}$ and

---

[9] See Section 11.6.3 later for a discussion of quasi–Monte Carlo methods.
[10] See Section 5.1.2.
[11] For some problems, a distinction between the set $\mathcal{X}$ and a subset satisfying some imposed constraints, the set of feasible solutions, is helpful. Osman and Kelly (1996, p. 4) present such a representation.

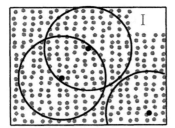 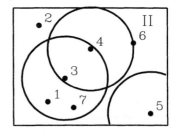

Figure 5.1   Neighbourhood concepts for local search.

all elements of $\mathcal{N}(\mathbf{x})$ are called *neighbours* of $\mathbf{x}$. If $\mathcal{X}$ is a subset of some reel vector space $\mathbb{R}^n$, neighbourhoods can be defined, e.g. by choosing some $\varepsilon > 0$: $\mathcal{N}(\mathbf{x}) = \{\mathbf{x}^n \in \mathcal{X} |\ \|\ \mathbf{x}^n - \mathbf{x}\ \| < \varepsilon\}$. For discrete sets, it is less evident, how to define $\mathcal{N}$. Even if the discrete sets can be embedded in some reel space, a generalization of the ε–spheres just introduced is often not suitable. Figure 5.1 provides a rough idea of the problem.

In the left–hand panel (I), a situation is represented where the discrete set (the grey points) can be embedded in the reel plane, thus resulting in a quite dense covering, although not in the strict mathematical sense. Then, the projections of ε–spheres depicted by the circles around the black points comprise a large number of elements of $\mathcal{X}$ and lead to a functional neighbourhood concept. In high–dimensional problems, however, the situation shown in panel II seems to be the more typical, namely that the elements of $\mathcal{X}$ are distributed very coarsely. Then, even large ε–spheres, such as those shown in the figure, may comprise only a few or even only a single element. Obviously, in such a situation, local search will not work properly. For example, starting the local search at a point like the one with the number 5 and using the depicted ε–spheres as neighbourhoods track the algorithm down to this starting point. A more detailed analysis of this problem of defining neighbourhoods on discrete sets and some solution approaches are provided later in Sections 6.1 and 9.3, and for the applications in Part III.[12]

Given any neighbourhood structure on $\mathcal{X}$ it becomes possible to define *local minima* with respect to $\mathcal{N}$. An element $\tilde{\mathbf{x}} \in \mathcal{X}$ is a local minimum with respect to $\mathcal{N}$ if

$$f(\tilde{\mathbf{x}}) = \min_{\mathbf{x} \in \mathcal{N}(\tilde{\mathbf{x}})} f(\mathbf{x})\,. \tag{5.6}$$

Obviously, the definition of local minima depends on the neighbourhood structure.

Once a local structure is imposed by a neighbourhood mapping $\mathcal{N}$, a

---

[12] For some methods for constructing neighbourhood structures, see also Aarts and Lenstra (1997, pp. 4ff.).

generation mechanism is required, which proposes some element $\mathcal{N}(\mathbf{x})$ to be considered for the next iteration. This mechanism can be a random one, i.e. choosing any element in $\mathcal{N}(\mathbf{x})$ with the same probability or a probability depending on its distance to the current solution $\mathbf{x}$. The generation procedure can also consist of a purely deterministic procedure if there exists some natural order of the elements within each neighbourhood. Then, for a given $\mathbf{x}$, the elements of $\mathcal{N}(\mathbf{x})$ are proposed according to this natural ordering, until the algorithm turns to some other $\mathbf{x}$. One natural order of the elements of $\mathcal{N}(\mathbf{x})$ is given by the function values $f(\tilde{\mathbf{x}})$ for each $\tilde{\mathbf{x}} \in \mathcal{N}(\mathbf{x})$.

The next ingredient of a local search algorithm is the selection criterion, which determines whether the proposed new element $\mathbf{x}^n$ in a neighbourhood of the current solution will become the current solution $\mathbf{x}^c$ or not. Usually, this selection criterion is based solely on a direct comparison of $f(\mathbf{x}^c)$ and $f(\mathbf{x}^n)$. If only selection rules based on a direct comparison are considered, the first neighbour $\mathbf{x}^n$ satisfying the criterion is accepted and becomes the new current solution $\mathbf{x}^c$. The acceptance criterion of standard local search is simply whether the move from $\mathbf{x}^c$ to $\mathbf{x}^n$ improves the value of the objective function, i.e. $f(\mathbf{x}^n) \leq f(\mathbf{x}^c)$.[13] Alternatively to this first–accept strategy, as already mentioned a best–accept strategy can also be implemented (Osman and Kelly, 1996, p. 5). In this case, several neighbours are generated at a time and in the subset of neighbours satisfying the criterion — if it is non–empty — the best one is selected as the new current solution.

To complete an implementation of standard local search, a stopping rule has to be implemented. Since it is not possible, in general, to check whether the current solution $\mathbf{x}^c$ is an element of the set of global minima, some other criteria have to be used. A simple rule is to stop the algorithm after a predetermined number of iterations. Then, the computing time is known a priori. However, the algorithm operates independently from any problem-specific information in this case. Alternative stopping criteria require, e.g. that no improvement can be obtained in the neighbourhood of the current solution. This stopping rule guarantees that the algorithm ends in a local minimum with regard to the given neighbourhood structure. However, the number of iterations until such a local minimum is reached can be very large or even infinite, when optimization works on a continuous set.

Figure 5.2 gives the pseudo code for the classical local search with an a priori fixed finite number of iterations $I_{\max}$. The algorithm starts with some arbitrary element $\mathbf{x}^c \in \mathcal{X}$. The counter of iterations i is set to zero. Then, in each step, a new element $\mathbf{x}^n \in \mathcal{N}(\mathbf{x}^c)$ is chosen in the neighbourhood of the current element. The value of the objective function is compared for both elements. If an improvement can be achieved by moving from $\mathbf{x}^c$ to $\mathbf{x}^n$ ($\Delta f \leq 0$), the move is performed and $\mathbf{x}^n$ becomes $\mathbf{x}^c$. This iteration is

---

[13] Therefore, this class of algorithms is also called 'greedy'.

Initialization	Choose initial $\mathbf{x}^c$ and set $i = 0$
Step 1	Set $i = i + 1$ and choose some $\mathbf{x}^n \in \mathcal{N}(\mathbf{x}^c)$
Step 2	Calculate $\Delta f = f(\mathbf{x}^n) - f(\mathbf{x}^c)$
Step 3	If $\Delta f < 0$, set $\mathbf{x}^c = \mathbf{x}^n$
Step 4	If $i < I_{\max}$, go to step 1. Otherwise, $\mathbf{x}^c$ is the output of the algorithm

Figure 5.2   Pseudo code for a classical local search algorithm.

repeated until the maximum number of iterations is exhausted. The final $\mathbf{x}^c$ is the output of the local search algorithm. If $I_{\max}$ is large, the probability that this final $\mathbf{x}^c$ is a local minimum with regard to the given neighbourhood structure is also large. However, even if the local search procedure ends up with a local minimum, there is no information about the quality of this minimum (Aarts and Lenstra, 1997, p. 10). In fact, it can be arbitrarily far from global optimality!

For this reason, local search is often combined with other optimization approaches. For example, starting the local search algorithm on a grid of points in $\mathcal{X}$ gives a restart version of the heuristic. If the number of grid points can increase without bounds, this version has eventually a chance of reaching the global optimum. Other search heuristics described in the next section are also refined versions of this standard local search algorithm. In fact, this approach is one of the dominant principles in global optimization, in particular on discrete sets. Aarts and Lenstra (1997) mention three developments in combinatorial optimization which rationalize this high impact. First, local search algorithms have *appeal* due to the analogies with processes in nature: genetic algorithms, neural networks or the Great–Deluge algorithm highlight this close relationship by their actual names. Secondly, *theoretical insights* in the behaviour of local search algorithms give an asymptotic justification for good performance in many applications.[14] Thirdly, the large increase in computational resources has offered the possibility to use local search and its variants as a standard optimization tool in *practice*. The ease of implementation and the competitive performance (see Chapter 7) strongly encourage their use.

---

[14] See Section 7.2 for related results for simulated annealing and Section 7.3 for threshold accepting.

5.2 Optimization Heuristics

Some of the classical methods described in the previous section are based on a solid mathematical base. For example, Newton's method will eventually converge to the global optimum if the objective function is convex and smooth enough and the search space is convex. However, a theoretical proof of convergence to the global optimum can be given only for some circumstances, while the algorithms may fail to converge to the global optimum in other cases. Therefore, as well as coming under the heading of classical optimization methods, they also belong to the class of optimization heuristics if applied — as is often done in practice — to problem instances where the necessary conditions for good performance are not fulfilled. In Sections 2.3.2 and 4.3 above, optimization heuristics have been introduced in a more restricted understanding. An optimization method, which is based on some intuitive idea or on the solution methods for similar problems, is called an optimization heuristic if it fulfils certain requirements. First, *it should give nearly optimum or close to optimum solutions* for many instances by using only a reasonable amount of computer resources. Secondly, an optimization heuristic should be *robust to changes in problem characteristics*. Finally, it should be *easy to implement*. Given this understanding of optimization heuristics, the standard local search procedure described above, is also an optimization heuristic. However, applied to problems with multiple local optima, there is not only no guarantee that it will eventually end up with the global optimum, but its performance will be poor for many complex problem instances.

Nevertheless, the idea of local search is used in other optimization heuristics as an optimization principle which is combined with some leading principle. The leading principles are derived from biological evolution, physical sciences, the analysis of the functioning of the nervous system or statistical mechanics. Therefore, these optimization heuristics are also described as 'meta–heuristics'.[15] Of course, instead of local search, these leading principles can also be combined with other classical optimization approaches such as gradient methods or grid search. Since the applications studied in this present volume are essentially based on discrete search spaces, local search seems to be the most adequate optimization principle.

Such optimization heuristics are applied to problems where classical methods fail to be effective and efficient. This is the case, e.g. for combinatorial optimization problems classified as intractable according to the criteria described in Section 3.3. Meanwhile, there exist many different optimization heuristics which have been successfully applied in quite different areas of

---

[15] Osman and Kelly (1996, p. 3) define meta–heuristics as follows. 'A meta–heuristic is an iterative generation process which guides a subordinate heuristic by combining intelligent concepts for exploring and exploiting the search spaces using learning strategies to structure information in order to find efficiently near–optimal solutions'.

science ranging from physics to operational research. It is clearly beyond the scope of this section to provide an extensive overview.[16] Instead, only a few widely used concepts will be outlined for two reasons. First, they represent a background against which the concept and distinctive advantages and disadvantages of threshold accepting might be judged. Secondly, the comparative analysis in Section 7.4 below assumes that the reader is familiar at least with the basic notions of these algorithms.

Classical local search and most of the refined local search techniques outlined in the following sections share the feature that a new candidate solution is generated which is solely based on the actual one. No information about former configurations is used. However, genetic algorithms and tabu search deviate from this picture, since they are set–based algorithms, i.e. in each iteration step a whole set of candidate solutions is considered and used to produce a new set of candidate solutions in the case of genetic algorithms, or a single new candidate solution for the tabu search heuristic.

5.2.1 Simulated annealing

A straightforward refinement of classical local search is given by the simulated annealing (SA) heuristic. This combines the local search optimization idea with a leading principle from statistical mechanics, where it is also known as the Metropolis algorithm. Metropolis, Rosenbluth, Rosenbluth, Teller and Teller (1953) present an algorithm for the estimation of equations of state for substances consisting of interacting individual molecules. This algorithm already contains the acceptance rule used in simulated annealing. Furthermore, the tendency of this acceptance rule to approximate a stable distribution is essential for the interpretation of the results in the context of chemical physics but also for the proof of the convergence property of simulated annealing.[17] The implementation of this idea in a refined local search algorithm for (combinatorial) optimization problems can be dated back to the contributions by Kirkpatrick et al. (1983) and Cerny (1985) (Osman and Kelly, 1996, p. 8). However, until very recently its applications were restricted to problems in operational research, such as the travelling salesman problem, and in physics.[18]

The simulated annealing heuristic is based on an analogy between combinatorial optimization problems and the annealing process of solids

---

[16] For a bibliography of heuristic search research through to 1992, see Stewart, Liaw and White (1994). A more recent bibliography on meta–heuristics is given by Osman and Laporte (1996).

[17] See Section 7.2.

[18] For a biography of applications, see Osman and Laporte (1996, pp. 532–541). A description from a practician's point of view is given by Pirlot (1992, pp. 13ff.) and Pham and Karaboga (2000, pp. 11ff.), who also provide applications to problems from engineering.

(Aarts et al., 1997, pp. 96ff.). Cerny (1985, p. 51) stresses this analogy to thermodynamics as motivation for the good performance of simulated annealing: 'It might be surprising that our simple algorithm worked so well in the examples [. . .]. We believe that this is caused by the fact that our algorithm simulates what Nature does in looking for the equilibrium of complex systems. And Nature often does its job quite efficiently'. Annealing denotes a process in which a solid is first melted and then gradually cooled down until it solidifies again. A rapid cooling scheme will result in solids with an inhomogeneous crystal structure, including many irregularities and defects. If the cooling is carried out slowly, then at each temperature the melted solid can form an equilibrium structure, finally ending up in an optimum or nearly optimum crystal structure.

In order to mimic this process, the SA heuristic differs from the classical local search basically in its acceptance criterion. For a given value T, which is interpreted as a temperature in the so-called cooling scheme, a proposed neighbour $\mathbf{x}^n$ of the current solution $\mathbf{x}^c$ is accepted

$$\begin{array}{ll} \text{always,} & \text{if } f(\mathbf{x}^n) < f(\mathbf{x}^c) \\ \text{with probability } e^{(-\Delta f/T)}, & \text{if } \Delta f = f(\mathbf{x}^n) - f(\mathbf{x}^c) > 0. \end{array} \quad (5.7)$$

Hence, an improvement from $\mathbf{x}^c$ to $\mathbf{x}^n$ is always accepted as in the classical local search heuristic. However, a worsening, i.e. a move uphill on the objective function 'landscape', is also accepted, although with a probability decreasing as the extent of the worsening. Due to this property the simulated annealing algorithm may escape suboptimal local minima, in contrast to the classical local search. At each temperature T, the standard iteration of the local search procedure is repeated many times. Then, the probability of $\mathbf{x}^c$ being some specific element of the search space $\mathcal{X}$ should resemble the theoretical distribution. This feature of the algorithm is described as 'simulation', whereas the aspect of 'annealing' arises from the fact that the cooling scheme gives a sequence of temperatures T tending to zero. Consequently, at the very end of an optimization run by simulated annealing, $T = 0$, and the algorithm will accept a move from $\mathbf{x}^c$ to $\mathbf{x}^n$, if and only if, $f(\mathbf{x}^n) < f(\mathbf{x}^c)$, i.e. if it represents a move downhill as in the standard local search procedure.

Section 7.2 below will summarize some findings on the asymptotic behaviour of simulated annealing which indicate that under certain circumstances, including particular forms for the cooling sequence, this feature allows us to conclude that simulated annealing will approximate the global optimum with high probability as the number of iterations tends to infinity. Recall that this is not the case for the classical local search heuristic!

The performance of simulated annealing in real applications depends crucially on the definition of the neighbourhood structure on the search space $\mathcal{X}$ and on the cooling schedule. This aspect is considered in some detail for the threshold accepting algorithm in Chapter 8. Since threshold accepting is quite similar to simulated annealing, some of the arguments put forward in this chapter also apply to simulated annealing. Besides the definition of neighbourhoods and the cooling schedule,[19] the performance of simulated annealing might be increased by more specific refinements and a parallelized implementation (Fox, 1995).

5.2.2   Genetic algorithms

The leading principle of genetic algorithms is imitated from the process of evolution.[20] In contrast to standard local search and simulated annealing, they do not operate on a single current solution, but on a set of current solutions, which is called the current *population*. Given such a set of current solutions, the new candidates are generated through mutation and cross–over from the current population. The decision whether a resulting new candidate solution becomes part of the new current population depends on the objective function, i.e. mutations and cross–overs are more likely to be accepted if they improve the objective function. While it is not possible to describe the generation of new candidates as a selection of an element in a neighbourhood of some current element, it is possible to formulate genetic algorithms as local search procedures on the power set $2^{\mathcal{X}}$, i.e. on the set of all possible populations on the search space $\mathcal{X}$.

In order to apply the principles of mutation and cross–over on the current population it is efficient to code all elements as binary strings. Now, cross–over is performed according to the fitness — measured by the value of the objective function — of the elements (chromosomes). The offsprings of the cross–overs replace some weaker members of the last generation or the last generation as a whole. In addition to cross–over, some chromosomes could be subject to random mutations of individual bits.

Today, many efficient implementations of genetic algorithms exist. However, it proved to be necessary to combine the basic principle of genetic algorithms with other elements such as local search or one of the algorithms presented in the following sections in order to obtain satisfactory results. An implementation to the travelling salesman problem is given by Potvin (1996), while Dorsey and Mayer (1995) present one of the few applications to problems in econometrics and statistics, namely the maximum–likelihood estimation of canonical disequilibrium models for the USA loans market, which exhibits several local maxima and, therefore cannot be solved efficiently by standard

---

[19] Bölte and Thonemann (1996) propose the use of genetic programming for optimizing the cooling schedule. This approach requires large amounts of computing resources. For the quadratic assignment problems analysed in their paper, non–monotonic cooling schedules seem to outperform standard geometrically decreasing cooling schedules.

[20] Mitchell (1996) and Pham and Karaboga (2000) present an introduction and some applications. A bibliography on theory and applications is provided by Osman and Laporte (1996, pp. 518–525).

gradient methods.[21] Genetic algorithms are also used in economic theory as a means of modelling learning behaviour, which eventually leads to a rational expectations equilibrium (Arifovic, 1994).

5.2.3 Tabu search

Tabu search can also be described as a set–based local search procedure. However, in each iteration only a single current solution is considered. The basic implementation of tabu search proceeds as follows. In each iteration step, it selects the best element of the neighbourhood of the current solution, which then becomes the new current solution. The difference to a standard local search algorithm with such a best–accept strategy consists of the fact that the neighbourhood of some $\mathbf{x}^c$ change during the proceeding of the algorithm. Elements of the search space, which have been visited by the algorithm recently, can be eliminated from the neighbourhoods by giving them tabu–active status for a certain number of iterations. The tabu list has to be managed in such a way as to guide the system away from parts of the solution space which were already explored. At the same time, it has to guarantee that a good solution can be achieved despite of tabu entries.

Practical implementations of tabu search differ in the methods used for updating the resulting tabu lists and in design parameters such as tabu–tenure, i.e. the number of iterations that a tabu–active status remains on some element $\mathbf{x}$. Furthermore, a sort of inverse tabu can be introduced by adding tabu solutions to the neighbourhoods, e.g. elements with very good fitness in terms of the objective function f. Pham and Karaboga (2000, pp. 8ff.) provide an introduction to tabu search, while a reference to applications can be found in Osman and Laporte (1996, pp. 514–545).

5.2.4 Great–Deluge algorithm

The Great–Deluge algorithm introduced by Dueck (1993) is based on some real–world analogy such as simulated annealing. Since this analogy is more easily explained for a maximization problem, let us assume that an element of the set $\{\mathbf{x} \in \mathcal{X} | f(\mathbf{x}) = \max_{\mathbf{x} \in \mathcal{X}} f(\mathbf{x})\}$ is required. This corresponds to searching a highest elevation in a landscape, which is given by the multivariate surface generated by the objective function on the data set for a given neighbourhood definition. The Great–Deluge algorithm proceeds following its biblical model: it is assumed that an endless rain is falling on the landscape. Consequently, the waters increase gradually, while the moves of the algorithm from some $\mathbf{x}^c$ to any $\mathbf{x}^n \in \mathcal{N}(\mathbf{x}^c)$ are constrained by the additional requirement that the algorithm is not allowed to 'get its feet wet'.

---

[21] See Section 15.3 below for more details.

This constraint can also be interpreted as some special tabu scheme. In due course, the moves become more and more restricted, until the algorithm eventually ends up on a high elevation of the landscape. The moves are performed in predefined neighbourhoods of the current solution, while the increase of the water level depends on improvements of the objective function. An implementation of this algorithm in a comparison study including the simulated annealing algorithm and other heuristics, can be found in Sinclair (1993).

5.2.5 Noising method

The noising method was proposed by Charon and Hudry (1993). This is also a variant of classical local search. This heuristic takes the robustness as a leading principle. In order to obtain some robustness of the standard local search algorithm, in each iteration some noise is added to the data of the problem before evaluating the objective function at a given element. For example, for a travelling salesman problem, the noise can be introduced by randomly moving the coordinates of the cities. The amount of noise added to the data decreases to zero at the last iteration. Therefore, while in the beginning the values of the objective function might differ substantially from their true values, i.e. without noise added, they converge to their true values at the very last iteration. Using the data with noise, a classical local search is performed in each iteration step, i.e. the algorithm moves down hill as long as possible. The real implementation used by Charon and Hudry (1993) for comparison with simulated annealing is taken from graph theory. In their application to the clique partitioning problem for artificially generated graphs, the algorithm performs several descent moves at each level of noising, interchanging moves for the data with noise and similarly for the data without noise.

5.2.6 Ruin and recreate

The ruin and recreate principle proposed by Schrimpf et al. (1998) can be combined with most of the local search algorithms working on a single candidate solution at a time. The difference consists in the changes applied to a current solution. Instead of experimenting with local, i.e. small changes, considerable parts of the existing solution, e.g. a tour of the travelling salesman problem, are completely removed and rebuilt afterwards. If rebuilding takes place in some random fashion, ruin and recreate would correspond to the local search technique using larger neighbourhoods. However, the recreate step consists in adding the deleted elements sequentially in a locally optimal way. For the travelling salesman problem, for example, the removed cities are added one by one at the place within a tour which gives the shortest overall tour length. Schrimpf et al. (1998) report very encouraging results by using this approach in combination with threshold accepting, in particular for highly

complex problem instances with many side constraints or discontinuities of the
objective function for any reasonable local neighbourhood definition.

In this present volume, an application of ruin and recreate is presented for
a problem instance when standard threshold accepting, although improving
considerably over existing data, fails to provide completely satisfying results.
This problem is the uniform and orthogonal design problem. The ruin and
recreate implementation is analysed below in Section 11.5. Obviously, some
other problems treated or just outlined in this book could also be tackled
by the ruin and recreate approach. The implementation, however, is more
demanding than for the standard threshold accepting heuristic, because the
methods for recreating a locally optimum solution differ fundamentally from
problem to problem. Thus, for some of the problems discussed in Part III it
is not evident at all how a ruin and recreate step could be implemented.

The ruin and recreate step can be combined with any one of the refined
local search algorithms mentioned before. Then, in each iteration, instead of
some local change a complete ruin and recreate step is performed. Afterwards,
based on the difference of the objective function for the solution before and
after ruin and recreate and given the acceptance criterion of the local search
heuristic, a decision is made whether to proceed with the old candidate or to
accept the new one.

Ruin and recreate belongs to a class of algorithms which try to combine
ideas from local search heuristics with constructive approaches.[22] Osman and
Kelly (1996, pp. 13f.) denote such combined approaches as 'problem–space
methods'.

5.2.7   Hybrid methods

Ruin and recreate and the more general class of problem–space methods are
based on a combination of different optimization ideas and leading principles.
Algorithms resulting from such combinations are often called hybrid methods.
Although the optimization ideas and leading principles for meta–heuristics
presented in this section are not exhausting by far , their combination —
possibly including some constructive features like in the ruin and recreate
heuristic — allows for an almost illimitable number of hybrids. Several hybrid
versions of some of the presented algorithms have already been successfully
used in practice.

The advantage of hybrid methods is that they allow for more degrees
of freedom when implementing the algorithm to a specific problem class.
Therefore, the chance of finding a suitable implementation is increased. At
the same time, however, this possible advantage of hybrid methods becomes
a disadvantage due to the increase in complexity. In general, a larger set

---

[22] See Ausiello et al. (1999, pp. 321ff.) for some examples of constructive approaches.

of parameters has to be set a priori and the algorithm becomes much more tailored to a specific problem instance, which might explain the high performance of some hybrid method implementations. Consequently, they lack the ease of implementation of a pure refined local search algorithm such as simulated annealing or threshold accepting and require much more expert knowledge both on algorithms and the specific problem instance.

As the aim of this book is to motivate the use of global optimization heuristics as a standard tool for complex problems in statistics and econometrics, the focus on the most simple concepts might be justified. However, hybrid methods should be kept in mind as a tool for problem instances when elementary global optimization heuristics fail to provide satisfactory results even after some tuning of their parameters.

5.2.8 Neural networks

The main application of neural networks is not optimization. They were developed for simulating the brain behaviour in the context of artificial intelligence research. However, some particular forms of neural networks lend themselves easily to specific optimization problems.[23] The example of discrete Hopfield nets will highlight this aspect (Pirlot, 1992, pp. 54ff.).

A discrete Hopfield net can be described by a graph with N vertices, called *neurons*, and an associated state value of minus one or one. The connections between the neurons — the edges of the graph — are weighted. Let w_{ij} denote the weight of the connection between neurons i and j ($w_{ij} = 0$ if there is no edge between i and j in the graph). The state of the network at time t is given by a vector of states $\mathbf{x}(t)$. Now, the updating rule of the neural network can be written as

$$\mathbf{x}_i(t+1) = \operatorname{sgn}\left(\sum_{j=1}^{N} w_{ij} x_j(t) - t_i\right), \tag{5.8}$$

where t_i is a threshold value associated with neuron i. Equation (5.8) is interpreted in the neural network context as follows: a neuron is in its upper level (1) if the sum of all inputs ($w_{ij} x_j(t)$) passes a given threshold t_i.

Each iteration step corresponds to a complete update of the state vector. This can be done in asynchronous mode, i.e. by using the latest available state information for each neuron, or in synchronous mode, i.e. by always using state information from time t for evaluating equation+(5.8) in $t + 1$. If the matrix $W = (w_{ij})$ is symmetric and its diagonal elements are zero, it can

---

[23] See Pham and Karaboga (2000, pp. 15ff.) and Osman and Kelly (1996, pp. 7f.) along with the cited literature.

be shown that the energy of the system

$$E[\mathbf{x}(t)] = -\frac{1}{2} \sum_{i=1}^{N} \sum_{j=1}^{N} w_{ij} x_i x_j + \sum_{i=1}^{N} t_i x_i \qquad (5.9)$$

is non–increasing with t. The update rule (equation (5.9)) eventually drives the system into a stable state $(\mathbf{x}(t+1) = \mathbf{x})$ which corresponds to a local minimum of equation (5.9). Now, in order to exploit this feature, a combinatorial optimization problem has to be presented in a form suitable for neural networks. A few examples are described in Pirlot (1992, pp. 55ff.). However, the use of neural networks for optimization problems is still limited by the possibility of transferring the problem into this form.

6

The Global Optimization Heuristic Threshold Accepting

Dueck and Scheuer (1990) introduced threshold accepting as a deterministic analogue to simulated annealing in an application to the travelling salesman problem.[1] However, they also consider non–deterministic versions of the algorithm, which in their application produced slightly better results. Therefore, the version of threshold accepting presented and discussed in this chapter is the non–deterministic version, which shares many aspects with the simulated annealing algorithm introduced in the previous chapter. Nevertheless, it is a marked feature of threshold accepting that it is possible to present a completely deterministic implementation with similar performance.

After the description of the algorithm and its parameters, which are relevant for its performance in real applications, in Section 6.1, Section 6.2 delivers a graphical presentation of the process of threshold accepting for a simple numerical example, while Section 6.3 provides references to some applications of threshold accepting.

6.1   The Algorithm

Let $f : \mathcal{X} \to \mathbb{R}$ be an instance of an optimization problem. For this present chapter, it is assumed that $\mathcal{X}$ is a discrete set. The application of threshold accepting to continuous optimization problems is discussed later in Chapter 15. The main features are similar to those of the discrete case. The algorithm should find at least one of potentially many different optimal solutions, i.e. an element of $\mathcal{X}_{\min}$, where

$$\mathcal{X}_{\min} = \{\mathbf{x} \in \mathcal{X} \mid f(\mathbf{x}) = f_{\mathrm{opt}}\} \tag{6.1}$$

and

$$f_{\mathrm{opt}} = \min_{\mathbf{x} \in \mathcal{X}} f(\mathbf{x}) \,. \tag{6.2}$$

---

[1] For a short presentation, see also Aarts et al. (1997, pp. 92ff.).

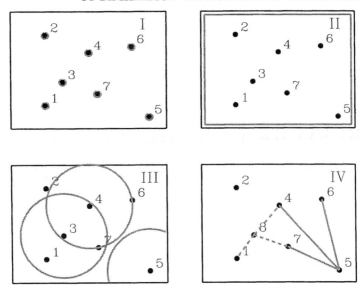

Figure 6.1 Neighbourhood concepts for threshold accepting.

Threshold accepting is a refined version of the standard local search procedure described above in Section 5.1.5. In particular, it shares two elements with the standard local search procedure. First, the search space $\mathcal{X}$ has to be given some neighbourhood structure by a mapping $\mathcal{N} : \mathcal{X} \longrightarrow 2^{\mathcal{X}}$, where $2^{\mathcal{X}}$ denotes the set of all subsets of $\mathcal{X}$, which defines for each $\mathbf{x} \in \mathcal{X}$ a neighbourhood $\mathcal{N}(\mathbf{x}) \subset \mathcal{X}$. Secondly, the algorithm starts with some arbitrary element $\mathbf{x}^c \in \mathcal{X}$. The algorithm differs from a standard local search by its acceptance rule, which completes the description of the threshold accepting algorithm.

It is difficult to describe the selection of a neighbourhood structure at this most general level. Therefore, Figure 6.1 provides only a schematic representation. A deeper discussion of the aspects linked to the neighbourhood structure is postponed to the chapters given later in Part III of this book which present applications. Furthermore, Section 9.3 contains a general guideline for the valuation of neighbourhood concepts.

In Figure 6.1, the numbered dots mark the elements of the discrete set $\mathcal{X}$. The grey–shaded areas provide a graphical representation of the neighbourhoods. In the upper panels, two trivial neighbourhood mappings $\mathcal{N}$ are represented. In I, the only neighbour of any element is the element itself, i.e. $\mathcal{N}(\mathbf{x}) = \{\mathbf{x}\}$. Any local search algorithm based on such an atomistic neighbourhood structure gets stuck in the initial element $\mathbf{x}^c$. The other extreme case is given in II, where all elements are neighbours of all elements, i.e. $\mathcal{N}(\mathbf{x}) = \mathcal{X}$ for all $\mathbf{x} \in \mathcal{X}$. Consequently, a local search becomes a random search for this neighbourhood concept.

The lower panels of the figure show two concepts which are more similar to those used in typical examples of local search or refined local search algorithms. In III, the discrete space $\mathcal{X}$ is embedded in the Euclidean plane $\mathbb{R}^2$. Then, a standard neighbourhood concept consists in considering ε-spheres around the points. The grey circles indicate such ε-spheres. The elements of the neighbourhood comprise the projection of these spheres on $\mathcal{X}$, e.g. the neighbourhood of the point with label 4 comprises the points with labels 3, 6 and 7. The plot shows a typical problem of this concept, namely that the spheres have to be quite large in order to make sure that the resulting neighbourhoods are non-trivial. In fact, the sphere around point 5 contains only this point, although the spheres are already quite large. We come back to this problem of projected ε-spheres for the applications given in Chapter 11. Another approach is shown in panel IV of Figure 6.1. Here, each point is assigned three neighbours, e.g. $\mathcal{N}(5) = \{4, 6, 7\}$ or $\mathcal{N}(3) = \{1, 4, 7\}$. These neighbours are chosen to be the next neighbours according to some distance measure. In the simple example, this distance is just the Euclidean distance. However, in real applications, often quite different measures of distance are used. For example, in the travelling salesman problem, two different paths are considered as being close to each other, if one can be obtained from the other by exchanging only two cities. The neighbourhoods for all applications of threshold accepting in Part III are constructed according to this principle.

To summarize this short description of the meaning of neighbourhood structures for the performance of a threshold accepting implementation, it is stressed that the applications of threshold accepting and — to a limited extent also the example used for tuning in Chapter 8 — indicate that the right 'size' of neighbourhoods is crucial for a good performance of threshold accepting.

With regard to the initialization of the algorithm with some $\mathbf{x}^c$, some general remarks seem necessary. Threshold accepting shares the requirement of choosing some starting element $\mathbf{x}^c$ with most other search heuristics. In practice, different approaches can be observed with regard to this initialization. The most commonly used one is to select $\mathbf{x}^c$ randomly from $\mathcal{X}$, i.e. to choose any $\mathbf{x} \in \mathcal{X}$ with equal probability $1/ \mid \mathcal{X} \mid$, if $\mathcal{X}$ is a finite set. Another choice is to use some 'natural' starting element, which does not necessarily has to be a good one. Examples of such natural starting values are results obtained by some ad hoc method.[2] A third choice, often advocated by people not familiar with the working of optimization heuristics from the class of refined local search algorithms, is to choose a 'good' starting element or even the best known one, i.e. $\mathbf{x} \in \mathcal{X}$ with $f(\mathbf{x})$ small.

At least from a theoretical point of view, it is difficult to judge the merits of

---

[2] For example, in the application to the optimal aggregation of time series in Chapter 13 below, the official grouping provided by the statistical agency represents such a 'natural' element.

these different methods of initialization of the threshold accepting algorithm. The theoretical results on the convergence properties of threshold accepting — and similarly for simulated annealing — gathered in Chapter 7 will show that asymptotically it makes no difference which $\mathbf{x}^c$ was chosen, as long as some fundamental requirements on $\mathcal{N}$ and other parameters of the algorithms are fulfilled.

Nevertheless, the choice of a random starting element is strongly recommended for the following heuristic reasons. First, if one starts with a good solution, which might have been found by some other optimization heuristic or ad hoc optimization approach, this element might correspond to a local minimum with regard to the given neighbourhood structure. Then, the ability of the refined local search heuristics such as threshold accepting is challenged to a larger extent from the beginning of the optimization run. In fact, parameters have to be chosen in a way to make sure that the algorithm has a chance to escape such a local minimum. This goal is realized by implementing a very generous acceptance rule at the start of the algorithm. Consequently, almost any move is accepted and the algorithm tends to some almost random part of the objective function landscape. The potential efficiency gain by using a good starting solution is lost, and even worse, the algorithm needs some iterations to diverge to some almost randomly selected region of $\mathcal{X}$. This aspect is substantiated by the experience gained from many implementations of simulated annealing and threshold accepting, which indicates that it is hardly ever a good choice to select a specific starting element instead of choosing it at random.

Given some initial element $\mathbf{x}^c$, the threshold accepting algorithm proceeds in a similar way to the classical local search, simulated annealing and most of the refined local search algorithms discussed in the previous chapter. It performs a large number of iterations or exchange steps. In each iteration, the algorithm first generates a new element $\mathbf{x}^n$ in the neighbourhood $\mathcal{N}(\mathbf{x}^c)$ of the current solution. Then, based on an acceptance criterion, it is decided as to whether the new element becomes the current one for the next iteration, or whether the current element is kept and a new exchange step is tried. Finally, a stopping criterion has to be checked after each iteration.

The aspects related to the generation of new elements in the neighbourhood $\mathcal{N}(\mathbf{x}^c)$ are the same as for standard local search or simulated annealing. Although the neighbourhoods are fixed in advance once and for all,[3] different strategies can be followed for the generation of $\mathbf{x}^n \in \mathcal{N}(\mathbf{x}^c)$. First, the new element might be chosen randomly out of the neighbourhood which gives an equal selection probability $1/ \mid \mathcal{N}(\mathbf{x}^c) \mid$ to each element. Secondly, some elements might be favoured by attributing them with a higher probability of getting selected. Such a weighted selection scheme could be based, e.g. on the

---

[3] This contrasts threshold accepting in its standard version from a tabu search implementation.

values of the objective function $f(\mathbf{x}^n)$. However, such an approach requires additional evaluations of the objective function, which might be quite time consuming. Finally, some deterministic scheme can be set up to select the elements of a neighbourhood in a specific order. Such a scheme has been proposed by Dueck and Scheuer (1990) for the purely deterministic version of threshold accepting. There, some a priori ordering is introduced in each neighbourhood and the new elements are selected following this ordering one after the other. A further possibility for a deterministic choice is to evaluate the objective function f on all elements of the neighbourhood and to select the element with the smallest function value as a test case.

Once having defined a test case $\mathbf{x}^n$ in the neighbourhood of the current solution $\mathbf{x}^c$, the differences between the different local search algorithms come into play. While the classical local search algorithm accepts the test case $\mathbf{x}^n$ only if it represents an improvement compared to the current solution $\mathbf{x}^c$, i.e. if $f(\mathbf{x}^n) \leq f(\mathbf{x}^c)$, the acceptance criteria of refined local search algorithms are more complex. As already outlined in Section 5.2.1, simulated annealing uses a probabilistic acceptance rule for $\mathbf{x}^n$. A test case is always accepted if it corresponds to an improvement in the objective function, and is accepted with positive probability in all other cases where the probability is a decreasing function of the degree of worsening, i.e. of $f(\mathbf{x}^n) - f(\mathbf{x}^c)$. Threshold accepting is similar to simulated annealing, as test cases with higher values of the objective function might be accepted. On the other side, it is similar to a classical local search, as the acceptance criterion is purely deterministic. Threshold accepting will accept $\mathbf{x}^n$ as a new current solution if and only if $f(\mathbf{x}^n) - f(\mathbf{x}^c) \leq T$ for some predefined non–negative threshold value T.

Given a threshold height $T > 0$ in one iteration of threshold accepting, the algorithm will either move from the current solution $\mathbf{x}^c$ to a new one $\mathbf{x}^n$, or stick to the current one. It accepts $\mathbf{x}^n$ if it is better than the current solution in terms of the objective function, or at least, if it is not much worse, where the meaning of 'much' is given by the threshold height T.

While a run of the classical local search algorithm is completely described by the generation method for test cases, the acceptance criterion and a stopping rule, e.g. a fixed number of iterations, the more sophisticated heuristics, i.e. simulated annealing and threshold accepting, require some additional parameter settings. In particular, the sequences of the acceptance probabilities for simulated annealing and of the threshold heights for threshold accepting have to be defined. For this description of threshold accepting, it is assumed that the sequence of threshold heights has one entry per iteration of the algorithm. Practical implementations tend to use shorter lists of threshold values and perform several exchange steps for one threshold value. At least for the simulated annealing algorithm, there is some rationale in doing so, based on the analogy to the annealing process. One value for the acceptance probability corresponds to a certain temperature of the heat bath in the annealing experiment. Now, given such a temperature, the system needs some

Initialisation	Choose a threshold sequence T_i, $i = 0, \ldots, I_{\max}$, set $i = 0$ and generate an initial $\mathbf{x}^c$
Step 1	Choose some $\mathbf{x}^n \in \mathcal{N}(\mathbf{x}^c)$
Step 2	Calculate $\Delta f = f(\mathbf{x}^n) - f(\mathbf{x}^c)$
Step 3	If $\Delta f \leq T_i$, set $\mathbf{x}^c = \mathbf{x}^n$
Step 4	If $i < I_{\max}$, set $i = i + 1$ and go to step 1. Otherwise, $\mathbf{x}^c$ is the output of the algorithm

Figure 6.2   Pseudo code for the threshold accepting algorithm.

time (iterations) to adopt its optimal structure for that given temperature.

The choice of threshold sequences will be discussed in some detail in Sections 8.4 and 9.4, as well as for some specific applications in Part III. Here, it might suffice to assume that the threshold heights decrease to zero as the algorithm proceeds. Eventually, i.e. for i larger than some i_0, all T_i equal zero. Consequently, for $i \geq i_0$, test cases $\mathbf{x}^n$ will be accepted if and only if they constitute a real improvement over the current solution $\mathbf{x}^c$. If a large number of iterations is performed for this zero threshold height, the probability of ending up with a local minimum for the given neighbourhood structure is large and will tend to one as the number of iterations at this final threshold height tends to infinity.

Now, after introducing all central aspects of the threshold accepting algorithm, Figure 6.2 summarizes the description by providing the pseudo code for the threshold accepting heuristic. For this version of the algorithm, it has been assumed that the stopping criterion is given by a predefined maximum number of iterations $I_{\max}$. Furthermore, it has been assumed that the number of threshold values is equal to the number of iterations $I_{\max}$. Of course, some of the threshold values T_i can be equal, thus corresponding to the case of several iterations per threshold value discussed above.

A comparison of the pseudo code for threshold accepting presented in Figure 6.2 with the pseudo code of the classical local search algorithm given above in Figure 5.2 indicates the close relationship between both approaches. However, the difference in the acceptance rule is crucial. Threshold accepting differs from the simulated annealing algorithm presented in Section 5.2.1 also solely by the acceptance rule. Figure 6.2 becomes a pseudo code for the simulated annealing algorithm if the acceptance rule $\Delta f \leq T_i$ is replaced by the rule presented in equation (5.7). In fact, threshold accepting can be interpreted as a limiting case of the simulated annealing algorithm with a

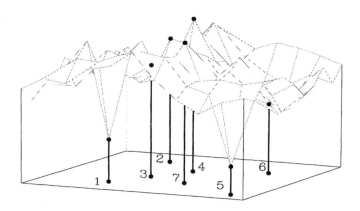

Figure 6.3   A minimization problem.

deterministic acceptance rule.[4]

6.2   Proceeding of Threshold Accepting: An Example

In order to provide a more intuitive understanding of the proceeding of threshold accepting, this section provides a simple example. Thereby, the search space $\mathcal{X}$ consists of a set of seven points as shown above in Figure 6.1. The neighbourhoods are assumed to comprise exactly three elements for each $\mathbf{x} \in \mathcal{X}$ according to the next neighbour criterion as depicted in panel IV of Figure 6.1. Furthermore, to each $\mathbf{x} \in \mathcal{X}$ there corresponds a function value $f(\mathbf{x})$, which is represented by the length of the vertical lines in Figure 6.3. Thus, only the bold–faced points on the landscape are relevant. The grey–shaded landscape is provided solely for reference purposes, i.e. in order to show the similarity of the simple example with the problems analysed above in Section 1.2.

The problem consists in finding the $\mathbf{x} \in \mathcal{X}$ with smallest $f(\mathbf{x})$. A short look at the figure indicates that the minimum is given for the point with label 5. However, the latter does not present an algorithm, which will work well in higher–dimensional problems with much larger sets $\mathcal{X}$, as studied in Part III. Although the grey–shaded landscape seems to indicate some local minima, this impression might be misleading, since the plotted landscape does not correspond one–to–one to the local structure provided by the neighbourhoods

---

[4] See Althöfer and Koschnik (1991) and the discussion below in Section 7.3.

based on next neighbours. Therefore, this landscape is made almost invisible in Figure 6.4, which shows a run of threshold accepting for this specific simple problem.

In this figure, minimization begins in panel I with the selection of an initial $\mathbf{x}^c$, which is given by the point with label 3. The neighbours of this point are the points with labels 1, 4 and 7, respectively. The mapping $\mathcal{N}(\mathbf{x}^c)$ is indicated by the dashed lines on the plane from the point with label 3 to its neighbours. Next, from the three neighbours of the point with label 3, one is randomly selected. In this case, the selected point is the one with label 7. This is indicated by the arrow from the point with label 3 to the one with label 7. Now, the value of f for the point with label 7 has to be compared with the one for the original $\mathbf{x}^c$. This is achieved in the figure by putting a grey–shaded copy of the vertical line at point 3 close to the vertical line at point 7. As can be seen from this comparison, $f(\mathbf{x}_3) < f(\mathbf{x}_7)$, where the indices of $\mathbf{x}$ denote the labels of the points in the figure. Finally, the current threshold value is added to $f(\mathbf{x}_3)$ before comparison with $f(\mathbf{x}_7)$. The value of the threshold is indicated by the dashed grey line starting at the top of the solid grey line. An arrow at the end of this dashed line indicates that the real value is even larger than the one shown on the figure, but censored in order to fit on to the plot. A comparison of both lines shows that the condition

$$f(\mathbf{x}_7) \leq f(\mathbf{x}_3) + threshold$$

is clearly fulfilled. Therefore, at the end of the first try, the point with label 7 becomes the current solution $\mathbf{x}^c$.

Now, the procedure is repeated starting with $\mathbf{x}_7$, as shown in panel II. The next neighbours of $\mathbf{x}_7$ are the points with labels 1, 3 and 4. From these, the point with label 1 is randomly selected as the new candidate solution, as indicated by the arrow in the plane. Again, the comparison of $f(\mathbf{x}_1)$ with $f(\mathbf{x}_7)$ and the threshold is made easier by putting a copy of the vertical line at point $\mathbf{x}_7$ close to the point with label 1. Since $f(\mathbf{x}_7) > f(\mathbf{x}_1)$, the necessary condition for acceptance of $\mathbf{x}_1$ as the new current solution is fulfilled independently from the threshold value. Thus, the algorithm moves to $\mathbf{x}_1$.

From panel III, it can be detected that $\mathbf{x}_1$ is a local minimum with regard to the neighbourhood structure imposed by the three next–neighbours rule. In fact, all $f(\mathbf{x}_3)$, $f(\mathbf{x}_4)$ and $f(\mathbf{x}_7)$ are larger than $f(\mathbf{x}_1)$. Nevertheless, $f(\mathbf{x}_1)$ is not the global minimum that the algorithm is looking for, as $f(\mathbf{x}_5) < f(\mathbf{x}_1)$. However, the algorithm is not aware of this situation, but routinely continues its exchange attempts. From the three neighbours of the point with label 1, the point with label 4 is selected as the new candidate. As in panel I, the threshold value becomes important for the decision whether to accept this move or not. The grey line at the point with label 4 indicates $f(\mathbf{x}_1)$, while the dashed line on top gives the current threshold value. Obviously,

$$f(\mathbf{x}_4) = f(\mathbf{x}_1) + threshold,$$

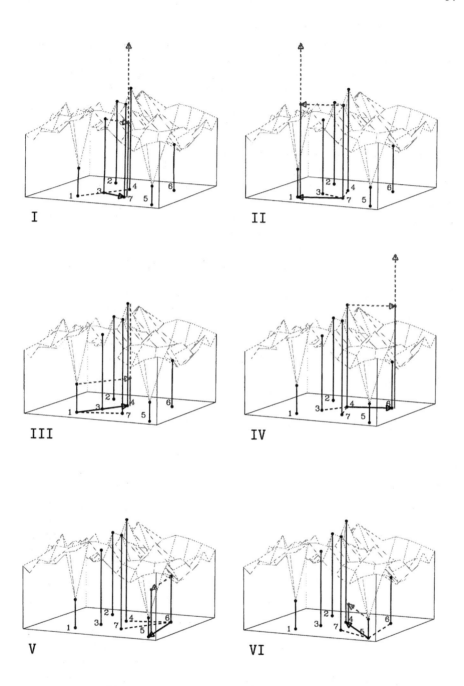

Figure 6.4 Minimization by threshold accepting.

and x_4 is accepted as the new current solution x^c.

The move from x_4 to x_6, which is one of the three neighbours of x_4, imposes no problem as indicated in panel IV, since $f(x_6) < f(x_4)$. It should be noted, however, that the last moves from x_1 to x_4, and now to x_6, represent a typical example of the hill–climbing behaviour of threshold accepting. In fact, without the property of accepting a worsening of the objective function up to a given threshold, it would have been impossible to move back from the point with label 1, which represents a local minimum.

In panel V, another decreasing move is chosen, namely from x_6 to x_5, which belongs to the neighbourhood $\mathcal{N}(x_6)$. While for the first two iterations shown in the upper panels of the figure a large threshold value was chosen, it was already reduced for the next exchange steps shown in panels III and IV, and finally decreased to zero for the last two tries. Therefore, at the very end the algorithm accepts only improvements of the objective function. Consequently, the tentative move from x_5 to x_4, which is the last exchange try of this short run of threshold accepting, is rejected and the algorithm ends in the global optimum x_5.

Although threshold accepting typically behaves as in this simple example, the much higher complexity of real applications makes it unlikely that a global optimum is achieved within a few iterations. It may also happen that a global minimum is visited while the threshold value is still large, and therefore left again afterwards. Therefore, usually not only the last current solution x^c is kept, but also the best solution x_{opt} encountered during the optimization procedure. If $f(x_{opt}) < f(x^c)$, the result is improved by using this method. However, at the same time such an outcome indicates that the threshold accepting algorithm is not well tuned yet.

6.3 Some Applications

Although the use of simulated annealing is more widespread in operational research than the use of threshold accepting, a few implementations are already documented in the literature, which cover both applications in traditional operational research areas as well as new applications in statistics and econometrics. The applications of threshold accepting include the travelling salesman problem (Dueck and Scheuer, 1990), multi–constraint 0–1 knapsack problems (Dueck and Wirsching, 1991), Portfolio Optimization (Dueck and Winker, 1992), Optimal Aggregation (Chipman and Winker, 1995a; 1995b), the quadratic assignment problem (Nissen and Paul, 1995), scheduling (Siedentopf, 1995), model selection (Winker, 1995; 2000b) problems in experimental design (Winker and Fang, 1997; 1998; Fang, Lin, Winker, and Zhang, 2000) and censored quantile regressions (Fitzenberger and Winker, 1998). A few more applications are cited in the bibliography provided by Osman and Laporte (1996, p. 547).

The first application of threshold accepting presented by Dueck and

Scheuer (1990) was to a travelling salesman problem with 442 points. In fact, this problem is of the travelling salesman type, but is derived from a chip wiring problem. Dueck and Scheuer (1990) found that the threshold accepting implementation improved results for this problem compared to implementations of standard local search and simulated annealing documented in the literature. Since this improved performance could be achieved at lower computational costs, the authors concluded that threshold accepting may not only be competitive with other optimization heuristics, but even superior at least for some problem instances. The 442–point–problem studied by Dueck and Scheuer (1990) is also used as a bench–mark case for the tuning of threshold accepting described later in Chapter 8.

The second implementation of threshold accepting, described by Dueck and Wirsching (1991), covers multi–constraint 0–1 knapsack problems. A simple representative of this problem class can be described as follows. Given a finite set of objects of different size and a knapsack of given volume, the aim is to exhaust the volume of the knapsack as far as possible by filling it with some of the given objects. It turns out that this problem, like the travelling salesman problem, is of high inherent complexity.[5] Since knapsack problems — like the travelling salesman problem — are standard bench–mark cases in the operational research literature, they often serve for comparative studies. Hanafi, Freville and Abdellaoui (1996) performed such a comparative study, including threshold accepting for knapsack problems. The main findings are reported later in Section 7.4.

Dueck and Winker (1992) have applied threshold accepting to the portfolio optimization problem. While most standard portfolio optimization approaches are restricted to the well–known return–variance–analysis introduced by Markowitz (1952), these authors were interested in calculating optimal portfolios of securities for different objective functions. In particular, measures such as shortfall risk or worst–case return were of central interest. As these objective functions do not lead to well behaved optimization problems, they could not be solved by the usual optimization tools. Threshold accepting offered an opportunity to overcome this obstacle and to calculate at least good approximations to optimal portfolios under alternative objective functions. Therefore, this application can be considered as the first application of threshold accepting to a problem from economics which does not fall into the range of standard operational research problems. Furthermore, the standard optimization paradigm, which is based on the mean–variance approach, is challenged when other objective functions are considered. Recent developments in the finance literature indicate that the solution to problems as described by Dueck and Winker (1992) becomes more important. Recently, Gilli and Këllezi (2000) have used threshold accepting for similar portfolio

---

[5] In fact, these problems are NP–complete; see Section 13.9 for an introduction to the concept of NP–completeness.

optimization problems.

Another application of threshold accepting away from the area of operational research is presented by Chipman and Winker (1995a; 1995b). They analyse the problem of aggregation of time series arising in econometrics. It turns out that this problem corresponds to a highly complex large–scale integer optimization problem. The details of the problem and the threshold accepting implementation are given below in Chapter 13.

While portfolio optimization and optimal aggregation are not typical operational research problems, the quadratic assignment problem studied by Nissen and Paul (1995) is one. It is related to real–world implementations of threshold accepting to problems in job–shop scheduling (Siedentopf, 1995) and chip– or machine–layout, which have been solved by using threshold accepting in technical consulting.

Further problems of statistics and econometrics are discussed later in Part III. These applications of threshold accepting include experimental design (Chapter 11), model selection (Chapter 12), and censored quantile regressions (Chapter 14).

Of course, implementations of threshold accepting are not restricted to the areas just cited or to the other applications introduced in this book. Therefore, the presentations in the following chapters, in particular Chapter 9, advocate the use of threshold accepting to complex optimization problems, which are ubiquitous in statistics and econometrics. In particular, if the problems cannot be tackled successfully by standard methods, a threshold accepting implementation may provide a bench–mark. Threshold accepting is a general purpose algorithm and, consequently, well suited for an exploratory analysis of optimization problems (Aarts, Lenstra, van Laarhoven and Ulder, 1994, p. 118).

7

Relative Performance of Threshold Accepting

7.1 Introduction

The examples given in Part I and the description of the algorithm in the previous chapter demonstrate that threshold accepting is well suited to provide solutions or at least high–quality approximations for some complex optimization problems. This property is necessary in order to consider threshold accepting to be a reasonable choice for these kinds of optimization problems. However, it is not sufficient to make threshold accepting an imperative choice.

Given the heuristic feature of threshold accepting on the one side and the complexity of some optimization problems in econometrics and statistics on the other, it is not easy to answer the question 'Which algorithm is the best?' for all optimization problems. In fact, a general answer is infeasible as any optimization problem exhibits its own specific features and even different problem instances might differ significantly with regard to the appropriate algorithm. Of course, one would always prefer an algorithm which provides the exact solution to any problem instance with certainty by using almost no computing resources. However, such an expectation is unrealistic given the fact that many, if not most, optimization problems are highly complex. Consequently, none of the heuristics introduced in Chapters 5 and 6 can fulfil this unrealistic need. Therefore, other measures of performance have to be considered to motivate our choice of a specific optimization routine.

In this present chapter, two different concepts are used to compare the performance of algorithms. First, the hypothetical case is considered, when the number of iterations of the algorithms is allowed to tend to infinity. In this asymptotic consideration, one is interested whether the algorithm will eventually end up in the global optimum and, if so, at which rate this goal will be achieved. Of course, a simple enumeration algorithm will always asymptotically solve any optimization problem over a finite set. However, the number of iterations needed to obtain at least a useful approximation may grow at a tremendously high rate. Therefore, this algorithm, in general, is not well suited. The asymptotic behaviour of threshold accepting is less simple. Nevertheless, Section 7.4 presents asymptotic results for threshold accepting, which, essentially, are based on results for the simulated annealing heuristic described in Sections 7.2 and 7.3.

The second approach for ranking different algorithms consists in comparing their performance for a specific problem and a given amount of computational resources. Results of such comparisons from the literature are summarized in Section 7.4. The tuning of threshold accepting described in Chapter 8 is also based on this second approach of performance measurement. Furthermore, it contains a comparison with classical local search and simulated annealing for a specific instance of the travelling salesman problem. The drawback of this approach of relative performance measurement is that all results obtained by such measures, in principle, allow only conclusions for the specific problem instances included in the study. A careful experimental design might provide some trust that the results carry over to other problem instances of the same optimization problem. Furthermore, repeated findings of good performance by different authors, including authors more related to other algorithms, allow for the hypothesis that the results are robust.

7.2 Asymptotic Behaviour

Before providing some results on the asymptotic performance of simulated annealing and threshold accepting, the question might be raised as to why the asymptotic behaviour of an algorithm is of any interest at all for practical purposes. In fact, the good asymptotic performance of complete enumeration in the sense that it will always provide the exact global optimum for discrete optimization problems, does not tell us that it is a good choice for real applications. However, on the other side we might consider a heuristic to be at least a little bit doubtful if it does not reach the global optimum even with an infinite number of iterations. Nevertheless, this is the case for most of the classical optimization methods mentioned above in Section 5.1, if the problems exhibit several local minima.

For classical local search, it is not possible to give any non–trivial convergence results. As the number of iterations goes to infinity, classical local search will eventually reach a local minimum for the given neighbourhood structure. As long as this neighbourhood structure is not trivial, this local minimum will depend on the starting point. It might be far away from some global optimum in both of the senses described in Section 5.1.5, i.e. both with regard to the value of the objective function as well as the distance to some optimum element. Of course, if classical local search is combined with grid search in the form of a multi–start version, it can be shown that it will converge to a global optimum with a probability of one as the number of restarts tends to infinity.[1]

The convergence results reported for simulated annealing and threshold accepting in the next two sections are stronger. It can be shown that simulated annealing reaches the global optimum with probability as close to one as one wishes with the number of iterations going to infinity. For threshold accepting, the result is that the global optimum can be approximated as close as required with a probability as close to one as we would wish with the number of iterations going to infinity. Of course, in order for these results to hold, the parameters of the algorithm, in particular the cooling schedule for simulated annealing and the threshold sequence for threshold accepting, have to be chosen in the right way. Unfortunately, the proofs of the convergence results are not constructive. Hence, it is not possible to derive direct hints for the set–up of the algorithms for a specific problem instance. Therefore, this practical aspect of tuning is considered in Chapter 8 on a more heuristic base.

A suitable mathematical model for the functioning of threshold accepting on a finite search space $\mathcal{X}$ are nonstationary Markov chains.[2] Introductions to the theory of Markov chains are given, e.g. by Kemeny and Snell (1976), Isaacson and Madsen (1976), Seneta (1981), Freedman (1983), Revuz (1984), and Norris (1997) to mention just a few.

Let the search space be given by $\mathcal{X} = \{\mathbf{x}_1, \ldots, \mathbf{x}_n\}$, and let f denote the objective function to be minimized over $\mathcal{X}$. Then, a local search algorithm can be described in terms of Markov chains by:

1. The initial probability distribution $\pi^{(0)}$, where

$$\pi_i^{(0)} \;=\; P(\mathbf{x}^{(0)} = \mathbf{x}_i) \geq 0, \quad \text{for all } i = 1, \ldots, n \qquad (7.1)$$

$$\sum_{i=1}^{n} \pi_i^{(0)} \;=\; 1, \qquad (7.2)$$

which describes the choice of an initial $\mathbf{x}^c = \mathbf{x}^{(0)}$.

2. The one–step transition probabilities

$$p_{ij}^{(k)} \equiv \tilde{p}_{ij}^{(k-1,k)} = P(\mathbf{x}^{(k)} = \mathbf{x}_j \mid \mathbf{x}^{(k-1)} = \mathbf{x}_i), \quad \text{for all } i,j = 1, \ldots, n, \quad (7.3)$$

which provide the probabilities that a certain exchange step is performed at the kth iteration of the algorithm.

---

[1] See Aarts et al. (1997, p. 93). These authors use the same example to demonstrate that threshold accepting may fail to find a good solution with high probability, while simulated annealing does not. This is not too surprising given the slightly superior asymptotic properties discussed in the following sections. However, the results for a real life problem given below in Table 8.1 indicate that these asymptotic results have to be interpreted with some care.

[2] In fact, most of the theory can also be applied to countable infinite sets. However, as the applications in this present book are restricted to finite sets, only this case is considered in this section. An introduction to Markov chains on continuous state spaces is provided in Meyn and Tweedie (1993) and Norris (1997, pp. 60ff.).

Following the description of threshold accepting in the previous chapter, the selection of $\mathbf{x}^{(0)}$ is performed at random uniformly on $\mathcal{X}$. Hence $\pi_i^{(0)} = 1/|\mathcal{X}| = 1/n$ for all $i = 1, \ldots, n$. Before giving the defining equations for the transition matrices the proposal matrix $Q = (q_{ij})$ is introduced. This summarizes the information from the neighbourhood definition on $\mathcal{X}$ and is defined by

$$q_{ij} = \begin{cases} \frac{1}{|\mathcal{N}(\mathbf{x}_i)|}, & \text{if } \mathbf{x}_j \in \mathcal{N}(\mathbf{x}_i) \\ 0, & \text{if } \mathbf{x}_j \notin \mathcal{N}(\mathbf{x}_i). \end{cases} \tag{7.4}$$

Hence, $q_{ij} = 0$, if $\mathbf{x}_j$ is not in the neighbourhood $\mathcal{N}(\mathbf{x}_i)$ of $\mathbf{x}_i$. Otherwise, under the assumption that each element of the neighbourhood of $\mathbf{x}_i$ is selected as a candidate solution with equal probability, $q_{ij} = 1/|\mathcal{N}(\mathbf{x}_i)|$. Using this proposal matrix, the transition probabilities $(p_{ij}^{(k)})$, which are summarized in the transition matrices $P^{(k)}$ can be defined for a given threshold value T_{k-1}:

$$p_{ij}^{(k)} = \begin{cases} q_{ij} = 0, & \text{if } j \notin \mathcal{N}(\mathbf{x}_i) \\ q_{ij}, & \text{if } j \in \mathcal{N}(\mathbf{x}_i), j \neq i, \text{ and } f(\mathbf{x}_j) - f(\mathbf{x}_i) \leq T_{k-1} \\ 0, & \text{if } j \in \mathcal{N}(\mathbf{x}_i) \text{ and } f(\mathbf{x}_j) - f(\mathbf{x}_i) > T_{k-1} \\ 1 - \sum_{j' \neq i} p_{ij'}^{(k)}, & \text{if } j = i. \end{cases}$$
$$\tag{7.5}$$

Given the initial probability distribution $\pi^{(0)}$ and the one–step transition matrices $p_{ij}^{(k)}$, it is — in principle — possible to calculate the probability distribution after k iterations of the algorithm:

$$\pi^{(k)} = \pi^{(0)} P^{(1)} P^{(2)} \cdots P^{(k)}. \tag{7.6}$$

Of course, for practical applications, it would be very interesting to know the $\pi^{(k)}$. Even more interesting are derived measures such as the probability to have passed by a member of the optimizing set or by a member of the set of satisfying results, i.e. sufficiently near to some global optimum $\mathbf{x}^{\mathbf{opt}}$ during the first k steps. Unfortunately, such calculations become very complex as soon as $\mathcal{X}$ contains more than a few elements and also for arbitrary objective functions f.[3] Therefore, a general solution by using analytical tools is hardly feasible. For this reason, the asymptotic analysis of Markov chains in an optimization context is mainly concentrated on $\sum_{\mathbf{x}_i \in \mathcal{X}_{\min}} \pi_i^{(k)}$ as k tends to infinity. Nevertheless, the convergence of the probability $\sum_{\mathbf{x}_i \in \mathcal{X}_\delta} \pi_i^{(k)}$ will also be analysed, where $\mathcal{X}_\delta$ denotes the set of $\mathbf{x} \in \mathcal{X}$ such that $f(\mathbf{x}) - f_{\mathrm{opt}} \leq \delta$ for some $\delta > 0$. Of course, one is interested in having these probabilities as close as possible to, or equal to one.[4]

---

[3] Althöfer and Koschnik (1991, pp. 194f.) present results obtained by simulation for a very simple problem instance with $|\mathcal{X}| = 5$.

[4] For practical applications, it is less relevant to obtain a good final solution, but to have passed by a good solution during optimization. Althöfer and Koschnik (1991, p. 193)

Given the random initial distribution $\pi_i^{(0)} = 1/|\mathcal{X}|$, a necessary condition for obtaining good results in the sense just defined is that there exists a *path* from any $\mathbf{x}_i$ to any $\mathbf{x}_j$ in $\mathcal{X}$, i.e.

$$\forall \mathbf{x}_i, \mathbf{x}_j \in \mathcal{X} \ \exists l \in \mathbb{N}_{>0}, \ k_0, \ldots, k_l \in \{1, \ldots, n\} \quad \text{such that}$$
$$\mathbf{x}_i = \mathbf{x}_{k_0}, \ldots, \mathbf{x}_{k_l} = \mathbf{x}_j \text{ and } \forall l' \geq 1 \ \mathbf{x}_{k_{l'}} \in \mathcal{N}(\mathbf{x}_{k_{l'-1}}) \,. \tag{7.7}$$

If this condition was not fulfilled, starting with a 'bad' $\mathbf{x}^{(0)}$ can preclude the convergence to an optimal solution by any local search procedure. If the neighbourhoods $\mathcal{N}(\mathbf{x})$ are constructed such that the condition given by equation (7.7) is met, $\mathcal{X}$ together with the sets $\mathcal{N}(\mathbf{x})$ is called *connected* or *irreducible*.

Let d_{ij} be the length of the shortest path joining $\mathbf{x}_i$ and $\mathbf{x}_j$ and $D = \prod_{i,j \in 1, \ldots, n} d_{ij}$; then straightforward calculations show that for all $i, j \in \{1, \ldots, n\}$

$$(Q^D)_{ij} > 0 \,, \tag{7.8}$$

i.e. the proposal probability for a path leading from $\mathbf{x}_i$ to $\mathbf{x}_j$ of length less or equal to D is positive. Obviously, this construction does not take into account real transition probabilities yet, but assumes that any proposed transition is accepted by the algorithm.

For a connected set $\mathcal{X}$,[5] it is possible to calculate the 'threshold to cross' going from $\mathbf{x}_i$ to $\mathbf{x}_j$ on a path $\mathbf{x}_{k_0}, \ldots, \mathbf{x}_{k_l}$ as

$$T_{\mathbf{x}_{k_0}, \ldots, \mathbf{x}_{k_l}} = \sum_{l'=1}^{l} \max\{0, f(\mathbf{x}_{l'}) - f(\mathbf{x}_{l'-1})\} \,. \tag{7.9}$$

In general, there will exist more than one path from $\mathbf{x}_i$ to $\mathbf{x}_j$ in $\mathcal{X}$. Let $\Lambda_{i \to j}$ be the set of all such paths. Then, it seems reasonable to define the 'minimum threshold to cross' on any path from $\mathbf{x}_i$ to $\mathbf{x}_j$ to be given by

$$T_{i \to j} = \min_{\Lambda_{i \to j}} T_{\mathbf{x}_{k_0}, \ldots, \mathbf{x}_{k_l}} \,. \tag{7.10}$$

The intuitive meaning of the minimum threshold to cross is the amount of uphill moves in the landscape given by f and $\mathcal{N}$, which are unavoidable when moving from $\mathbf{x}_i$ to $\mathbf{x}_j$, i.e. even if $f(\mathbf{x}_j) < f(\mathbf{x}_i)$ the landscape might force some uphill moves. In fact, Figure 6.3, 95, in the preceding chapter provides an instance, where such uphill moves are required. Finally, the quantity

$$\tilde{T} = \max_{i,j} T_{i \to j} \tag{7.11}$$

---

argue that the convergence result for threshold accepting presented also apply to this situation.

[5] The term 'connected set' is always used in the meaning 'for a neighbourhood mapping $\mathcal{N}$'. Nevertheless, if there is no risk of mixing up different sets of neighbourhoods, the short expression will be used throughout this chapter.

denotes the 'maximum threshold to cross' going from any point $\mathbf{x}_i$ in $\mathcal{X}$ to any $\mathbf{x}_j$ also in $\mathcal{X}$.

The values $T_{i \to j}$ and $\tilde{T}$ can be used to illustrate the problems in a convergence analysis of optimization heuristics such as simulated annealing or threshold accepting. In a connected set $\mathcal{X}$ with a random uniform initial distribution $\pi^{(0)}$, the probability of being in some state $\mathbf{x}_i$ after a finite number of iterations, in general, will be strictly positive. Now, if the current value of the threshold T becomes smaller than $T_{i \to j}$ it is not possible to jump directly from $\mathbf{x}_i$ to $\mathbf{x}_j$ even if $\mathbf{x}_j \in \mathcal{N}(\mathbf{x}_i)$. The length l of a path from $\mathbf{x}_i$ to $\mathbf{x}_j$ is finite if we disregard any cyclical parts, i.e. segments of a path starting and ending in the same element of $\mathcal{X}$. Hence, if T decreases monotonically to zero, one eventually risks ending up in a situation with strict positive probability for some $\mathbf{x} \notin \mathcal{X}_{\min}$.

From a theoretical point of view, the situation is somewhat better for simulated annealing, since due to the probabilistic acceptance rule there is always a positive probability to go from $\mathbf{x}_i$ to $\mathbf{x}_j$, as long as there actually exists such a path. Therefore, in the next section the convergence of simulated annealing will be analysed first, before deducing results for threshold accepting in Section 7.4.

7.3 Convergence of Simulated Annealing

Recall that by using the proposal matrix Q from equation (7.4) the transition matrices for simulated annealing are given by

$$
p_{ij}^{(k)} = \begin{cases} q_{ij} = 0\,, & \text{if } j \notin \mathcal{N}(\mathbf{x}_i) \\ q_{ij}\,, & \text{if } j \in \mathcal{N}(\mathbf{x}_i), j \neq i, \text{ and } f(\mathbf{x}_j) - f(\mathbf{x}_i) < 0 \\ q_{ij} \exp\left(\frac{f(\mathbf{x}_i) - f(\mathbf{x}_j)}{C_{k-1}} \right), & \text{if } j \in \mathcal{N}(\mathbf{x}_i), j \neq i, \text{ and } f(\mathbf{x}_j) - f(\mathbf{x}_i) \geq 0 \\ 1 - \sum_{j' \neq i} p_{ij'}^{(k)}, & \text{if } j = i\,, \end{cases}
$$

$$(7.12)$$

where C_{k-1} is the value of the cooling sequence at time $k - 1$. Let us assume for the moment that we hold $C_k = C$ constant. Then, all transition matrices $P^{(k)}$ are equal. Let $\mathrm{P} = P^{(k)}$ denote this transition matrix. For connected $\mathcal{X}$, equation (7.8) holds. Furthermore,

$$
p_{ij}^{(k)} > 0 \iff q_{ij} > 0\,. \tag{7.13}
$$

Consequently, $\mathrm{P}^D > 0$ with D from equation (7.8) and the inequality holds componentwise. Following Kemeny and Snell (1976, p. 69), this condition is equivalent to P being a regular transition matrix. Then, from the Markov ergodic convergence theorem for regular transition matrices (Kemeny and Snell, 1976, p. 71), it follows that

$$
\pi^{(k)} = \pi^{(0)} P^{(1)} \dots P^{(k)} = \pi^{(0)} P^k
$$

converges to some equilibrium probability distribution π independent of $\pi^{(0)}$ as k tends to infinity. Furthermore, $\mathbf{P}^k$ converges to a probability matrix with all rows being equal to π and we find

$$\lim_{k \to \infty} \pi^{(k)}(\mathbf{x} \in \mathcal{X}_{\min}) = \sum_{\mathbf{x} \in \mathcal{X}_{\min}} \pi(\mathbf{x}), \qquad (7.14)$$

where $\pi^{(k)}$ denotes the probability distribution after k applications of the transition matrix P on the initial distribution $\pi^{(0)}$. By choosing the constant C to be small, the probability in equation (7.14) can be made arbitrarily close to one (Hajek, 1988, p. 312).

This result reflects the basic property of the simulated annealing algorithm to settle down in high–quality solutions, while still maintaining a positive probability for visiting any point in the search space. This property guarantees asymptotical convergence to the global optima as the number of iterations tends to infinity. The discussion in Aarts et al. (1997, pp. 102ff.) shows that this result can be generalized to a broader set of transition matrices and acceptance probabilities, respectively. However, the fastest convergence may be achieved by using the standard acceptance probabilities presented in equation (7.12). Although providing a first convergence result on simulated annealing, using a constant cooling parameter C means that an infinitely long Markov chain has to be run in order to obtain a high probability of achieving a global optimum. In order to make this probability converge to one, the parameter C should decrease, although still allowing for an infinite number of transitions for each value C_k(Aarts et al., 1997, pp. 104ff.). Obviously, such a construction is not possible.

Therefore, the exposition of the convergence of a homogeneous Markov chain for fixed C has served for illustration purposes only. Nevertheless, it provides a general idea on how to obtain the following results for the case which consists of a finite sequence of segments of Markov chains of finite length. Such a sequence of finite–length segments of Markov chains with different transition matrices due to different values of the cooling parameter C_k is called an inhomogeneous Markov chain. Let the transition probabilities $p_{ij}^{(k)}$ be defined as in equation (7.12), but assume that the sequence of cooling values C_k, $k = 0, 1, 2, \ldots$ satisfies the following conditions:

$$\lim_{k \to \infty} C_k = 0 \qquad (7.15)$$

$$C_k \geq C_{k+1}, \qquad k = 0, 1, \ldots. \qquad (7.16)$$

In order to obtain the required convergence result, it has to be shown that the inhomogeneous Markov chain defined by these transition matrices is strongly ergodic, which means that the probability distribution $\pi^{(k)}$ converges in distribution. Theorem 7 in Aarts et al. (1997, p. 107) states that this condition is fulfilled, given that the cooling sequence C_k satisfies

$$C_k \geq \frac{\Gamma}{\log(k + k_0)} \ , \quad k = 0, 1, \ldots$$

for some $\Gamma > 0$ and $k_0 > 2$. In this case

$$\lim_{k \to \infty} P(\mathbf{x}_k \in \mathcal{X}_{\min}) = 1 \,,$$

which means that asymptotically, i.e. with the number of iterations going to infinity, a suitable implementation of simulated annealing provides a global optimum with a probability converging to one.

In the literature, several estimates of Γ are provided which depend on specific assumptions on the underlying optimization problem (Gidas (1985); Hajek (1988); Gelfand and Mitter (1991)). In the work by Hajek (1988), Γ is linked to the depth of the deepest local, nonglobal minimum. Thereby, depth defines the maximum threshold to cross when moving from the local minimum to some element of the search space with a smaller function value.[6]

Nevertheless, even if a good approximation of Γ and k_0 can be obtained, it is not evident how to use the asymptotic convergence result for a simulated annealing implementation with a finite number of iterations. A few remarks on this aspect can be found in Aarts et al. (1997, pp. 109ff.). In particular, it is noted that most cooling schedules used in practice are heuristic schedules.

7.4   Convergence of Threshold Accepting

Based on the convergence results for the simulated annealing algorithm, Althöfer and Koschnik (1991) provide some results for threshold accepting. In fact, they approximate sample paths of a simulated annealing implementation by sample paths of a threshold accepting implementation and show that in this sense simulated annealing can be described as belonging to the convex hull of threshold accepting. By using this approximation scheme, the following convergence result for threshold accepting can be obtained from the convergence results for simulated annealing (Althöfer and Koschnik, 1991, p. 188). For every tuple $(\varepsilon, a, n, \delta)$ with $\varepsilon > 0$, $a > 0$, $n \in \mathbb{N}$, $n \geq 3$, and $\delta > 0$, there exists a threshold sequence $(T_k)_{k=0}^{K-1}$ in $\{i\delta/[(n-1)(n-1)]| i \in \mathbb{N}\}$ of some length $K = K(\varepsilon, a, n, \delta)$ such that an application of threshold accepting to any problem instance, which can be described by a finite, undirected, and connected graph with less than n vertices $\mathcal{X}$ and a reel–valued objective function $f : \mathcal{X} \to [0, a]$, yields

$$P(f(\mathbf{x}_K) - f_{\mathrm{opt}} \leq \delta) \geq 1 - \varepsilon \,.$$

---

[6] See also Ryan (1995).

To put this in other words, given an instance size n, a range of the objective function $[0, a]$ and 'quality' parameters δ and ε, there exists a finite threshold sequence such that running a threshold accepting with this threshold sequence results in a solution which is not worse than the true optimum than δ with probability $1 - \varepsilon$, i.e. a solution arbitrarily close to a global optimum is obtained with a probability arbitrarily close to one provided that the number of iterations is large enough and the right threshold sequence is chosen.

Unfortunately, since this result for threshold accepting is obtained by using the results for simulated annealing it is not constructive either. In fact, the number of iterations for the threshold accepting algorithm required to obtain a good enough approximation of the simulated annealing sample paths is, in general, quite large. Consequently, this result neither provides an argument in favour of using threshold accepting compared to simulated annealing, nor does it deliver a guideline on how to select a threshold sequence for a given problem instance.[7]

Despite these shortcomings of the convergence results for simulated annealing and threshold accepting, the proof of asymptotic convergence makes them different from classical local search, which does not lead to a high–quality approximation of the global optimum with strictly positive probability as long as several local minima are present.

7.5 Results of Comparative Implementations

The first sections within this present chapter provided results on the asymptotic convergence of simulated annealing and threshold accepting. This theoretical analysis shows that both heuristics eventually converge to a high–quality result provided that cooling or threshold sequence, respectively, are adequately chosen. However, since the proofs of these results are by no means constructive, the usefulness of these algorithms cannot be based solely on these asymptotic findings.

Two alternative approaches might be considered. First, optimal and worst–case performance of the algorithms can be analysed for very small scale problems, when enumeration of all alternatives seems feasible. Besides the already mentioned example in Althöfer and Koschnik (1991, p. 194), the contribution by Aarts et al. (1997, pp. 94ff.) has to be considered within this context. They provide a worst–case example of an optimization problem, where threshold accepting may fail to find the global optimum even with the

---

[7] For a very small problem, Althöfer and Koschnik (1991, p. 194) provide the optimal threshold sequences obtained by enumeration of all sequences of length from 1 to 24. It turns out that for this example, the use of monotonically decreasing threshold sequences does not result in the highest probability of ending in the global minimum. Bölte and Thonemann (1996, p. 410) obtain similar results by using heuristic optimization by genetic programming for the cooling schedule of simulated annealing.

number of iterations going to infinity. However, the result depends on the assumption of a non–increasing threshold sequence, while the result provided by Althöfer and Koschnik (1991) shows that a non–increasing sequence is not always optimal.

The second approach consists in comparing the performance of real applications of different algorithms in computational experiments. This is the approach followed for the rest of this section, since more evidence including the threshold accepting heuristic is available for this kind of analysis. Unfortunately, comparative studies based on experimental results for different algorithms display some problems. Most of them can be traced back to the fact that the experience with and knowledge about the algorithms can be quite different. Consequently, some algorithms may be better tailored to a specific problem instance. Then, a superior performance may indicate either higher efficiency of the algorithm itself or simply a better implementation due to better knowledge and more experience. This aspect should be kept in mind when looking at the sparse empirical findings summarized in this section. The experience with threshold accepting is still limited (Aarts et al., 1994, p. 118). Consequently, the number of comparative studies is small. Nevertheless, there exist a few such studies, and some of these include both simulated annealing and threshold accepting. A further potential drawback for interpreting the results for the applications in this book is that they are mainly chosen from standard operational research problems. However, even with a large number of comparative studies, including different kinds of problems, a general statement on the relative performance does not become possible. Thus, the results presented in this section should be taken as empirical evidence in favour of some algorithms, rather than providing a general ranking.

Probably the first comparative study including threshold accepting, besides the original contribution by Dueck and Scheuer (1990) is the application of Ulder, Aarts, Bandelt, van Laarhoven and Pesch (1990) to travelling salesman problems.[8] Although their emphasis is on genetic local search algorithms, they include simulated annealing and threshold accepting as bench–marks. The results indicate that when using the same neighbourhood structure, threshold accepting slightly outperforms simulated annealing and, for larger problem instances, also the genetic local search algorithm. The genetic local search algorithm can be improved by using a different topology. However, for this different topology no results are presented for threshold accepting or simulated annealing, thus rendering the comparison difficult. In fact, Barr et al. (1995, p. 13) state that it is 'unfair' to use different neighbourhoods for the comparison of local search heuristics.

Aarts et al. (1994) have studied the performance of different algorithms for some job shop scheduling problems. The algorithms include classical local

---

[8] For a discussion of travelling salesman problems and the implementation of threshold accepting, see Chapter 8.

search, simulated annealing, threshold accepting and a genetic algorithm. For threshold accepting, the threshold sequence is generated rather ad hoc, based on the sequence used in Dueck and Scheuer (1990) (Aarts et al., 1994, p. 120). While threshold accepting and simulated annealing give similar results if the solution of simulated annealing is close to the true optimum, simulated annealing performs better for the other cases, as does the genetic algorithm. It is not clear whether this result may at least partially depend on the choice of the threshold sequence. All three algorithms clearly outperform the classical local search and are dominated by a problem–specific implementation of tabu search.

Siedentopf (1995) compares several algorithms for a standard scheduling problem. The alternatives include branch and bound approaches, as well as genetic algorithms. She finds threshold accepting to be the fastest algorithm providing the exact optimum for the selected problem instances.

Hanafi et al. (1996) compare the performance of different local search approaches for the 0–1 multidimensional knapsack problem. The search procedures include a greedy algorithm, simulated annealing, a noising method (Charon and Hudry, 1993), and threshold accepting.[9] Among these multiple purpose optimization heuristics, threshold accepting gave the best results in less computing time than simulated annealing and the noising method, but with a significantly larger time consumption than the greedy algorithm. These authors also study some problem–specific optimization algorithms which seem to outperform threshold accepting.

Lin, Haley and Sparks (1995) compare the performance of three variants of simulated annealing and threshold accepting for three scheduling problems in a single–server system. In addition to the standard versions of the algorithm, the variants use adaptive schemes for the neighbourhoods or cooling schedules. The adaptive schemes for neighbourhoods are problem–specific and assign different a priori probabilities for parts of the search space. The heuristic procedure for adapting the threshold or cooling schedule can be applied to other problems as well.[10] For all variants, threshold accepting achieves a better or comparable performance to simulated annealing.

In Glass and Potts (1996), the comparison of algorithms is performed for flow shop scheduling problems. The algorithms studied include a multi–start descent method, threshold accepting and simulated annealing, tabu search and two versions of genetic algorithms. For each problem instance, all algorithms were allowed to use the same amount of computing time. From the four local search algorithms, simulated annealing performed the best. However, the differences to threshold accepting were rather small. The

---

[9] The first application of threshold accepting to the 0–1 multidimensional knapsack problem is described in Dueck and Wirsching (1991).

[10] A similar approach has been put forward by Hu, Kahng and Tsao (1995) — see below in Section 8.4.

standard genetic algorithm was worse than the local search procedures, while the best results were obtained for a genetic algorithm with some local search ingredients. The authors mention that it is a commonly found result that genetic algorithms outperform multiple purpose local search heuristics such as simulated annealing and threshold accepting only if some problem–specific structure is imposed. A similar result was obtained in an earlier study by Glass, Potts and Shade (1994) for the unrelated parallel machine scheduling problem. Unfortunately, in this study only classical local search, simulated annealing, tabu search and genetic algorithms are used. Again, the simple implementation of a genetic algorithm is worse than simulated annealing, tabu search and a problem–specific implementation of the genetic algorithm.

A different area of applications has been studied by Tajuddin and Abdullah (1994). These authors use sixteen different optimization heuristics for the optimization of a neural network structure. Threshold accepting exhibits the highest overall success rate. However, the number of updates per neuron is larger than for simulated annealing. Consequently, it is not evident whether it might be concluded from this application that threshold accepting is more efficient as indicated by the authors: 'The results underline the superiority of threshold accepting in terms of the success rate, although this algorithm requires a huge number of updates per neuron for convergence'.

To sum up the limited evidence from comparative studies on the performance of local search heuristics including threshold accepting, three aspects deserve attention. First, there is not one 'winner' of the contest, i.e. an algorithm outperforming the others for all problem instances. However, as long as the number of iterations is small compared to the size of the search space, the refined local search heuristics, i.e. simulated annealing and threshold accepting, outperform genetic algorithms.[11] Furthermore, they are always better than standard local search. Secondly, the performance of genetic algorithms seems to depend to a larger degree on a problem–specific adaption. Thirdly, the ranking of the other local search heuristics depends both on the problem instances and on the specific tuning. In particular for simulated annealing and threshold accepting, the convergence results of Sections 7.3 and 7.4 suggest that no unequivocal superiority of one of them over the other could be expected. Hence, it is less the choice of a specific heuristic optimization algorithm, but its careful implementation combined with knowledge and experience of the user which makes it a valuable tool for improving results or for tackling problems regarded as intractable by conventional approaches.

---

[11] Brind, Muller and Prosser (1995) support this statement with results of an extensive comparison study for resource management problems, including genetic algorithms and simulated annealing. They conclude that when computer time is restricted, simulated annealing is always to be preferred. Pham and Karaboga (2000) also state that simulated annealing outperforms genetic algorithms as long as the number of iterations is small.

8

Tuning of Threshold Accepting

The results of the previous chapter may be summarized in two statements. First, for a good choice of parameters, threshold accepting will eventually provide a very close approximation to the global optimum with a probability close to one for many optimization problems. Secondly, its performance seems to be better than simulated annealing for most of the few problems for which comparative studies have been performed.

However, the number of iterations in real applications is limited by available computer resources and sometimes is quite small when compared to the size of the problem instances. Therefore, asymptotic convergence results may provide a sound theoretical base for the choice of a specific algorithm, but cannot provide any guarantee for good performance with a limited number of iterations. Comparative studies, on the other hand, allow for clear statements about relative and — if the true global optimum is known — absolute performance. Unfortunately, these performance results apply solely to the specific problem instances considered in the studies. For any other instance, even of the same problem, the results may differ. This situation is quite familiar to applied econometricians, since a single estimation result can never be used to 'verify' some theoretical statement. Nevertheless, a large and growing number of estimation results, which do not contradict the theory, make the theory an empirically robust one. The same holds true for the performance analysis of optimization heuristics.

Therefore, this chapter contributes empirical evidence on the performance of threshold accepting, in particular on the dependency of performance on the central parameters of the algorithm. Unless the findings for the tuning of threshold accepting derived from this performance analysis are rejected by future experiences, they provide a base for the tuning techniques applied for the threshold accepting implementations in Part III.

As for any real implementation of optimization heuristics, there are several degrees of freedom in performing such an experiment, including the choice of problem instances, the coding of the algorithm, the choice of parameters and the reporting of results (Barr et al., 1995, p. 11). Therefore, before diving into the computational results, a general description of the experiment and its main goals seems in order.

First, as it is difficult and not very intuitive to discuss the tuning of the algorithm in a purely theoretical way, a classical example from the integer programming literature is chosen to demonstrate the effects, namely the travelling salesman problem, which is introduced in Section 8.1. As already pointed out, the use of one specific problem instance inhibits, at least in principle, a general interpretation of the results. Nevertheless, some tentative conclusions can be drawn and tested on the applications of interest without repeating the complete and expensive experiment.

Secondly, a detailed description of the implementation of threshold accepting is given, in particular with regard to the choice of neighbourhoods, in Section 8.2. The analysis of the impact of different choices of neighbourhoods is not a goal of the experiment. Some results for different topologies are presented only in order to motivate that the choice of neighbourhoods is crucial for the performance of threshold accepting and that the choice of trivial neighbourhoods, in general, produces poor results.

Finally, the experiment is directed towards assessing the impact of the threshold sequence and the number of iterations on the result quality. All findings should be interpreted under this premise.

8.1   The Travelling Salesman Problem

An outline of the travelling salesman problem has already been given in the introduction in Section 1.2, where it serves as an example of a complex discrete optimization problem. In fact, travelling salesman problems are often used as examples in this context. At least three reasons can be put forward for this choice. First, the problem is very intuitive and can be exposed without any formal effort. Secondly, the solution of travelling salesman problems is of great practical importance, as it represents the simplest form of any routing problem (forwarding agencies, chip layout, etc.). Thirdly, from a theoretical point of view, travelling salesman problems are known to be highly complex optimization problems. In fact, they belong to the class of so–called NP–complete problems,[1] which means that a huge set of other complex optimization problems could be solved, once an efficient algorithm for the solution of the travelling salesman problem is known.

The travelling salesman problem can be formulated as follows. A salesman has to visit a given, finite set of cities. His point of departure is one of those cities, and he has to end his tour by returning to this city. In doing his round trip he has to choose the shortest way. If the cities are characterized by their geographic location, any tour results in a specific tour length. Alternatively, a measure of costs related to the means of transport between any two cities, or

---

[1] See Section 13.9.2 for a definition of NP–completeness.

a measure of time required to travel from one city to any other, could be used. Then, instead of looking for the shortest way, the salesman has to look for the cheapest or fastest way to visit all cities. However, the formal description of the travelling salesman problem as a discrete optimization problem remains the same:

$$f : \mathcal{X} \longrightarrow \mathbb{R}_{>0} \, ,$$

where $\mathcal{X}$ is the (finite) set of all tours visiting all cities and f maps each tour to its length, costs or time requirements, respectively. Given a number of n cities to visit, $\mathcal{X}$ is equivalent to the set of all permutations of the numbers $1, \ldots, n$. Of course, neither the choice of the point of departure nor the direction of the round trip is important for the length of the tour. Consequently, the size of the search space $\mathcal{X}$ can be slightly reduced by forming equivalence classes.

The travelling salesman problem belongs to a class of highly complex problems, the so–called NP–complete problems. Consequently, it is unlikely that there exists a deterministic algorithm solving any instance of the problem to optimality with consumption of computing resources growing 'only' polynomially in the problem size.[2] The simple enumeration algorithm, which calculates f for any element of $\mathcal{X}$ and chooses the best one, requires computing resources which are proportional to the size of the set $\mathcal{X}$, which itself grows exponentially in the number of cities. Therefore, the use of optimization heuristics seems quite reasonable for these kinds of problems. Many different heuristics have been applied to the travelling salesman problem as it represents a standard touchstone.[3] This is a further reason to discuss the tuning of threshold accepting by using this problem as an example. Another reason is the fact that computations are very fast for modestly sized instances of the travelling salesman problem. Hence, it is possible to perform many replications for different parameter settings in order to obtain a sort of response surface for the performance of the threshold accepting implementation.

It has to be noted, however, that this present chapter does not aim at providing an extensive analysis of heuristic approaches for approximative solutions of travelling salesman problems. In fact, much more is known about this specific problem than can be treated within this tuning context.

8.2 An Example with 442 Points

The travelling salesman problem used for the tuning experiment in the next sections comprises 442 cities or points. In fact, the points do not correspond to the geographical situation of some cities, but are the coordinates of some

---

[2] See Section 13.9.2.

[3] See also the results presented in Section 8.3, Johnson and McGeoch (1997), and Ausiello et al. (1999, pp. 329ff.).

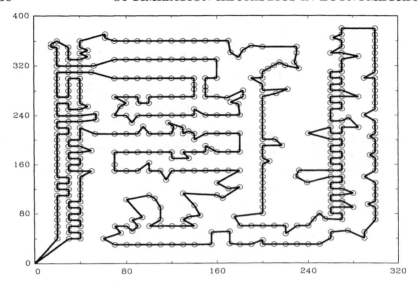

Figure 8.1   Coordinates of the 442–points problem.

industrial drilling problem for chip layout. The coordinates of the problem
and an optimal solution for this problem are provided by Holland (1987).[4]
Holland used a combination of a branch and bound approach with methods
of polyhedric combinatorics to reduce the set of possible solutions in order
to obtain the optimal solution. In order to achieve a good performance, the
method of Holland depends on a starting tour quite close to the optimal one.
This starting tour can be obtained by optimization heuristics.

Figure 8.1 shows the 442 points and the shortest tour given by Holland
of length 5080.51. The layout of the problem does not look too complicated.
However, given only the points to be connected in Figure 8.1 it is far from
being obvious how to tackle this task by using the threshold accepting
heuristic. The different steps for performing this task are described in the next
sections, namely describing the discrete optimization problem as optimization
over a landscape of function values and providing the points on this landscape,
which correspond to different tours in the original layout, with some local
structure in form of neighbourhoods. Section 8.3 describes this step, before the
tuning of other parameters of the threshold accepting algorithm is described
in Section 8.4.

---

[4] This is also the first problem analysed in Dueck and Scheuer (1990), introducing the
threshold accepting heuristic.

8.3 Choice of Neighbourhoods

From the discussion of theoretical convergence properties in the previous chapter, it can be concluded that the local structure imposed on a discrete optimization problem by the choice of a neighbourhood mapping $\mathcal{N}$ is important, if not crucial for the performance of a refined local search heuristic. Unfortunately, this theoretical analysis does not provide a practical guideline for the selection of neighbourhoods, except for a few general requirements. In particular, the *neighbourhoods have to be large enough* in order to reach any element of $\mathcal{X}$ from any other at least as long as the threshold value is large. In this section, some neighbourhood concepts for the travelling salesman problem are considered and heuristic arguments are provided for the specific choice made, which is the base for the tuning of other parameters of threshold accepting in the next section.

Unfortunately, it is difficult to imagine and impossible to draw the topology of the set of tours connecting 442 points introduced by a specific definition of neighbourhoods. Therefore, some approximative presentations have to be used. First, two rather trivial definitions of neighbourhoods are considered, which have already been outlined in Section 6.1 (see Figure 6.1).

The discrete topology is given by neighbourhoods $\mathcal{N}^d(\mathbf{x}) = \{\mathbf{x}\}$ for all $\mathbf{x} \in \mathcal{X}$, i.e. the only neighbour of a tour is the tour itself. Obviously, the threshold accepting algorithm will be tied to the initial tour $\mathbf{x}^{(0)}$ and in terms of the Markov chain representation, the probability distribution in step k will be equal to the initial distribution $\pi^{(0)}$ for all k. Of course, it makes no sense to use any iterations in this case. Under these circumstances, threshold accepting will stick to the initial tour. If a restart version is employed, it coincides with a trivial random search algorithm. The fact that no convergence to any local or global optimum is given in this case is not in contradiction with the global convergence property mentioned in Section 7.3, as one of the necessary conditions for the global convergence property to hold is the connectivity of $\mathcal{X}$ for the neighbourhood mapping $\mathcal{N}$. However, with regard to the discrete topology, $\mathcal{X}$ is not connected.

Another trivial case is given by $\mathcal{N}^g(\mathbf{x}) = \mathcal{X}$ for all $\mathbf{x} \in \mathcal{X}$, i.e. all tours are adjacent to all others.[5] Of course, for this definition of neighbourhoods, the connectivity of $\mathcal{X}$ holds. Consequently, convergence to the global optimum, in principle, is possible if the other prerequisites are fulfilled. Nevertheless, a detailed analysis of the convergence results indicates that the rate of convergence may be faster if smaller neighbourhoods are used. Furthermore, the heuristic argument illustrated by Figure 8.2 may give an idea why in practical applications, when the number of iterations is finite and rather small compared to $|\mathcal{X}|$, this crude topology on $\mathcal{X}$ is far from being a good choice.

---

[5] Fox (1995, p. 489) states that it is 'impractical (and [...] nonsensical except for some contrived problems) to let every state be an explicit neighbour of every other state'.

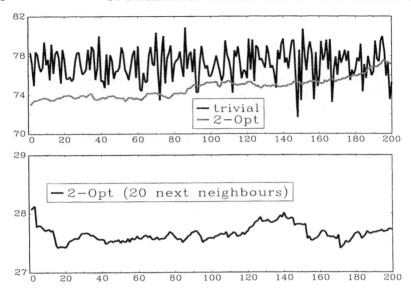

Figure 8.2   Local structure for different topologies.

Figure 8.2 depicts the values of the objective function f, i.e. tour lengths, for three 'random walks' in $\mathcal{X}$.[6] In the upper panel, the black line corresponds to the second trivial case just mentioned, i.e. when all tours are assumed to be neighbours to all other tours. It provides tour lengths (divided by 1000) for 200 tours, which have been obtained by the following set–up. First, a starting tour $x_1 \in \mathcal{X}$ is randomly generated. Then, for $i = 2, \ldots, 200$, a neighbour $x_i \in \mathcal{N}(x_{i-1})$ is randomly selected. For the resulting sequence of tours $x_1, \ldots, x_{200}$, the tour lengths $f(x_1), \ldots, f(x_{200})$ are calculated and plotted.

Although some neighbourhood structure has been used for obtaining the sequence $x_1, \ldots, x_{200}$, it is evident that it corresponds to a purely random selection of elements in $\mathcal{X}$, as for all $x \in \mathcal{X}$, $\mathcal{N}(x) = \mathcal{X}$. Hence, it does not come as a big surprise that the tour lengths jump without any structure, at least without any structure which might be identified from the plot, when moving from x_i to x_{i+1}. In addition to this information, the values of $f(x)$ provided by the black line in Figure 8.2 give an idea about the order of magnitude one has to expect for randomly chosen tours, i.e. for the trivial random search algorithm. The tour lengths are about 15 times longer than the length of the shortest tour shown in Figure 8.1.

The two other lines shown in the upper and lower panel of Figure 8.2

---

[6] Greenwood and Hu (1998) point out that the use of such simple random walks is not optimal for assessing the local structure of the search space if this local structure exhibits marked changes, e.g. due to penalties for infeasible solutions. They propose that random walks should be confined to small areas within the search space in such cases.

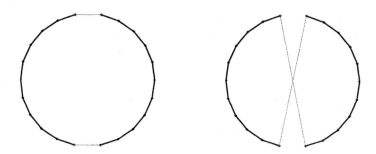

Figure 8.3   A $2 - Opt$ move (see Johnson and McGeoch (1997, p. 230)).

are obtained by a similar approach as the first one. Only the neighbourhood concept used for generating $\mathbf{x}_i \in \mathcal{N}(\mathbf{x}_{i-1})$ changes. For the grey line in the upper panel, a very commonly used neighbourhood definition for travelling salesman problems, namely neighbourhoods defined by $2 - Opt$ moves, is applied.[7] These neighbourhoods are described by a construction method, i.e. $\mathcal{N}_{2-Opt}(\mathbf{x})$ consists of all elements of $\mathcal{X}$, which can be constructed from $\mathbf{x}$ by using $2 - Opt$ moves. Thereby, a $2 - Opt$ move consists of breaking the tour $\mathbf{x}$ at any two edges in two separate paths and reconnecting those paths in the other possible way. Figure 8.3 gives a schematic idea of an $2 - Opt$ move.[8] The grey line in the upper panel of Figure 8.2 related to neighbourhoods generated from $2 - Opt$ moves indicates local behaviour of the tour lengths. Moving from one tour to the next by performing a $2 - Opt$ move, in general results only in small changes of the tour length, as could have been expected. The implementations of threshold accepting and simulated annealing based on this local structure on $\mathcal{X}$ will be called the standard versions. In contrast to the trivial topology, the definition of two tours being close to each other if they can be generated by a $2 - Opt$ move from one another clearly introduces some local structure, as the smooth development of the grey line exhibits.

However, a $2 - Opt$ move as shown in Figure 8.3 would seldom be a good choice. Hence, a restricted version of $2 - Opt$ moves considers only breaking points close to each other under the relevant metric on the space where the points of the tour are located. The corresponding neighbourhoods are

---

[7] See Johnson and McGeoch (1997, p. 230), Cook et al. (1998, pp. 246f.), and Ausiello et al. (1999, pp. 64ff.). In Chapter 7 of Cook et al. (1998), some other heuristics are described, including $3 - Opt$ moves, the Lin–Kernighan tour improvement method and Christofides' heuristic. For the approximation properties of the latter, see also Ausiello et al. (1999, pp. 94ff.).

[8] A $3 - Opt$ move allows the breaking up of the tour at three edges and its recombination in one of the various feasible ways.

denoted $\mathcal{N}_{2-Opt-NN}(\mathbf{x})$. Thereby, an element of $\mathcal{N}_{2-Opt-NN}(\mathbf{x})$ is given by all tours which can be generated from $\mathbf{x}$ by a $2 - Opt$ move applied to a pair of close neighbours. Obviously, the neighbourhoods $\mathcal{N}_{2-Opt-NN}(\mathbf{x})$ are smaller than the neighbourhoods induced by applying any $2 - Opt$ moves. For the simulation of local behaviour, these constraint neighbourhoods add a minor complication when starting with an arbitrary random tour. This tour might include connections between points which are not close neighbours. Such connections can be broken by a $2 - Opt$ move, but added only with a small probability. Consequently, even without any local search implementation, the application of such constraint $2 - Opt$ moves reduces the length of a tour considerably in the beginning. Therefore, the simulation in the lower panel of Figure 8.2 shows a segment of a random walk using this neighbourhood definition with 20 next neighbours after this effect has settled down. The versions of simulated annealing and threshold accepting working with this neighbourhood definition will be called *distance versions.*

As it is not the main subject of this chapter to study good or optimal neighbourhood structures for the travelling salesman problem, these heuristic arguments may be sufficient to highlight the importance of the choice of a problem–adequate neighbourhood structure for the performance of any local search heuristic. Furthermore, although it is probably not the best choice for the travelling salesman problem, the neighbourhood definition based on $2-Opt$ moves seems not unreasonable at first sight. Hence, the tuning of other parameters, i.e. number of iterations and threshold sequence, is analysed given this neighbourhood structure.

Finally, the choice of the distance version of threshold accepting can be advocated by the overview of results achieved with different optimization heuristics given in Table 8.1, which provides results for different implementations of simulated annealing and threshold accepting. The results in the first row for the classical local search method introduced above in Section 5.1.5 serve for comparison purposes only. The number of tries in the second column indicates how many times the different algorithms were run using different seeds for the initialization of the random number generator used for generating new candidate solutions. The last column gives the shortest tour length obtained in one of these tries.

It is evident that the classical local search algorithm, based on the $2 - Opt$ moves as the other implementations, is far from being a good choice for larger instances of the travelling salesman problem. It performs worse than any of the other implementations although an unlimited number of iterations was allowed until no further improvement was possible, i.e. a local minimum was reached. The best out of 100 tries was still more than 6% worse than the real optimum. Most of the tries of the classical local search heuristic resulted even in tour lengths more than 10% longer than the shortest tour. In contrast, the best published solutions with threshold accepting deviate only about 0.3% from the global optimum. By using the methods described in the next section

Table 8.1 Results for the 442–points problem

Algorithm	Trials	Iterations per try	shortest tour length
Classical Local Search[a]	100	until no further improvement[b]	5419
SA standard[c]	—	2 000 000	5330
SA distance[c]	—	2 000 000	5176
TA standard[d]	> 25	2 000 000	5194
TA distance[d]	> 25	2 000 000	5097
TA deterministic[d]	100	884 000	5134
TA tuned[a]	40	2 000 000	5093
	40	4 000 000	5088

[a] Implementation by the author. The classical local search algorithm is described in Section 5.1.5. The overall best result found during the simulation study described in Section 8.4 is 5082.63.

[b] Classical local search needed 30 000 to 100 000 iterations until a local minimum was reached. Typically, 40 000 to 50 000 iterations were sufficient.

[c] Rossier, Troyon and Liebling (1986).

[d] Dueck and Scheuer (1990).

for the tuning of threshold accepting, these deviations can be reduced to less than 0.25% for the same number of iterations and to less than 0.04% if the number of iterations is further increased.

Potvin (1996, pp. 362 and 366) summarizes some more results for the 442–city problem obtained by different optimization heuristics, in particular genetic algorithms. However, only one source reports a best result with a distance to the real optimum of less than 0.1% for 1 out of 50 runs. As it is not clear whether the problem instance is exactly the same as the one used in this chapter,[9] the results are not included in Table 8.1.

Despite the fact that the specific choice for the travelling salesman problem might seem natural and appeared to produce high–quality results in several applications, it should be kept in mind that the choice of $\mathcal{N}$, in general, is crucial for a good performance of any refined local search heuristic. However, sometimes it is not the most obvious definition which produces the best results! The choice of neighbourhoods for the other applications of threshold accepting presented in Part III provides further evidence in this aspect.

---

[9] The optimal tour length reported in Potvin (1996) differs from the one calculated for the solution given in Holland (1987).

8.4   The Choice of Parameters for Threshold Accepting

Keeping in mind that it may not be an optimal choice, the neighbourhood structure imposed on the travelling salesman problem by $2 - Opt$ moves and the restriction of these moves to points close to one another is taken as granted in this section.[10] Now, the interest concentrates on the remaining parameters which possibly determine the performance of threshold accepting. These parameters consist of the threshold sequence and the number of iterations performed for each threshold. In contrast to most of the applications of threshold accepting cited in Section 6.2, the choice of these parameters is based on simulation results.

Two slightly different versions of the threshold accepting algorithm are used. In the first version, the thresholds are defined as absolute values, as in the original version described in Section 6.1. This version will be called the *difference version* of threshold accepting. The difference of the *quotient version* consists of the fact that the thresholds are defined as factors on the current solution.

To begin with, a linear threshold sequence was chosen for the first simulation runs. The starting values of the threshold sequence were increased from zero, corresponding to the trivial local search algorithm, to 20 in 40 steps. At each step, 40 runs of threshold accepting with different randomly chosen starting configurations were performed. A randomly chosen starting configuration, thereby, is given by a seed used for the initialization of the random number generator used in the algorithm. Hence, both the starting tour and the random selection of neighbours throughout the optimization will differ for runs with a different seed. For each threshold sequence, the mean, standard deviation and minimum of the resulting 40 tours were calculated. This experiment was repeated for different numbers of iterations ranging from 200 000 to 4 000 000.

Figure 8.4 provides a first impression of the behaviour of the resulting tours with regard to the choice of parameters. It plots the mean values of the 40 tours obtained for each parameter choice against the parameter pair of starting threshold and number of iterations. In order to keep some details visible, the results for the four smallest values of the threshold sequence starting values are not included in the figure.

It is obvious that the threshold accepting algorithm outperforms any trivial local search algorithm, which would correspond to a threshold sequence starting value equal to zero, over the whole range of numbers of iterations used. Choosing a linear threshold sequence with a starting value slightly higher than zero decreases the mean value of the length of resulting tours from about 5800 to less than 5250 for a starting value in the range between 11 and 14.5 and

---

[10] For the simulations, the neighbourhoods were restricted to $2 - Opt$ moves between a pair of points for which the second belongs to the 20 nearest neighbours of the first with regard to the Euclidian distance.

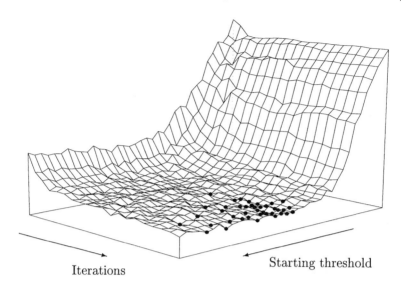

Iterations   Starting threshold

Figure 8.4   Mean tour length for different parameter choices I.

for at least 200 000 iterations. This descent is rather steep until the minimum
is reached. Increasing the starting threshold further leads to a slight increase
in the resulting mean tour length to about 5260 for a starting value of 20.

As a first result, it can be stated that the loss when choosing too high a
value for the starting threshold is much lower than in the case with too small a
value. Of course, if one was to choose a very high starting value, the threshold
accepting algorithm would start as a random walk through the search space
and only later on, as the thresholds become smaller, a local search behaviour
would result. This feature is responsible for the loss in efficiency when choosing
too high a starting value.

The second observation from Figure 8.4 concerns the influence of the
number of iterations. The decrease of the mean value of the tour length is
monotone in the number of iterations for a 'reasonable' choice of the starting
value for the threshold sequence, whereas for a local search algorithm or too
small a value of the threshold an increase in the number of iterations leads to
no visible improvement of the results. This last finding is not too surprising
since once a simple local search algorithm gets stuck in a local minimum, any
further increase of the number of iterations cannot have any influence on the
result.

For positive starting values for the threshold sequence, the improvement of
the results for an increasing number of iterations is not linear. As there exists
a fixed and finite optimum, it can only descend at a diminishing rate.

Finally, the solid dots in the plot mark parameter pairs resulting in a mean
tour length of less than 5150. These points are concentrated near the kink

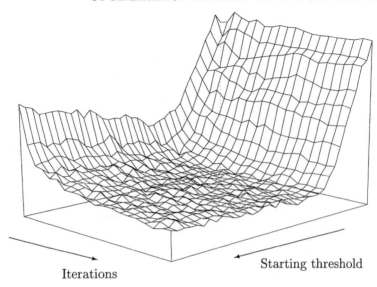

Iterations   Starting threshold

Figure 8.5   Mean tour length for different parameter choices II.

of the curves along the thresholds for high numbers of iterations. Thus, in order to achieve very good results it seems to be helpful to identify this kink. It is evident that the identification is very expensive in terms of computer resources if it has to be done for a high number of iterations. Fortunately, the plot indicates that there might be some hope in using information about the kink for a very small number of iterations as an indicator for the value of the parameters at the kink for a high number of iterations. Then, it is possible to identify the kink by using only a limited amount of computing resources. Before analysing this feature in some more detail, a short glance is given to the results of an alternative implementation of the threshold sequence.

First, a linear threshold sequence can be used as in the first case. However, the modified decision rule of the quotient version of threshold accepting is applied, i.e. a test solution is accepted as a new current solution if the value of the tour length does not increase by more than a multiple of the current tour length given by the threshold value. Of course, it is no longer reasonable to study starting values of the order of magnitude of 20 as before. Instead, Figure 8.5 is based on 40 starting values between zero and 0.4%.

A first inspection of this figure exhibits no clear–cut differences to Figure 8.4. In fact, both the rapid decline of mean tour lengths and the slow increase after the kink appear in the same way. To anticipate some findings of the applications studied in the next chapters it should be noted that this coincidence of the results for the difference and quotient version of threshold accepting cannot always be expected. It turns out that, in general, the quotient version performs better if the changes of the objective function

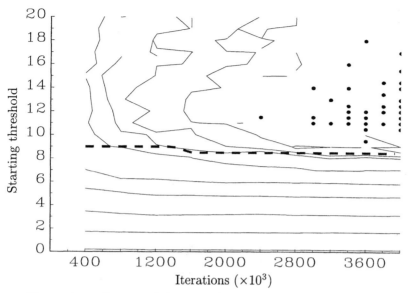

Figure 8.6   Contour of mean tour length for different parameter choices I.

inside a neighbourhood cover a large range, whereas the difference version
is superior if the order of magnitudes of such local changes within a given
neighbourhood are similar.

The next step of the analysis consists in identifying the optimal value for
the starting threshold for a given number of iterations. A regression analysis
(see Table 8.2 below) indicates that the dependence of the mean tour length
on the threshold value can be well approximated by a broken linear trend. In
addition, the regression equations include a constant and a dummy variable for
standard local search (starting threshold equal to zero, i.e. $ST = 0$). The value
of the starting threshold, where the estimated broken linear trend reaches its
minimum, is identified as the 'optimal starting threshold'. Figures 8.6 and 8.7
are contour plots of the data already depicted in Figures 8.4 and 8.5, i.e.
in the parameter space spanned by the number of iterations and the value
of the starting threshold. The lines mark parameter combinations resulting
in identical mean tour lengths, which increase from the upper right to the
lower left parts of the parameter space. Again, the dots mark parameter
constellations for which a mean tour length of less than 5150 has been found.
In addition, the dashed line in Figure 8.6 indicates the 'optimal starting
thresholds' identified by the regression analysis just described.

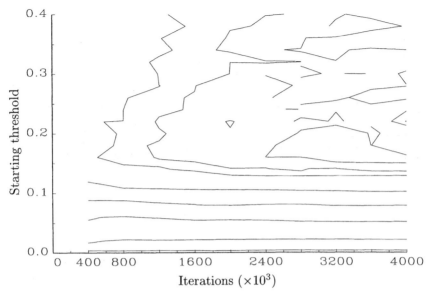

Figure 8.7 Contour of mean tour length for different parameter choices II.

From this representation it becomes evident that the optimal choice of a starting threshold depends only weakly, if at all, on the number of iterations performed. If the number of iterations is increased substantially, the optimal starting threshold is reduced only slightly. Therefore, these results imply that it is possible to identify the kink for a relatively small number of iterations and to apply threshold accepting afterwards with a larger number of iterations by using this optimal parameter value.

Despite the recommendation by Barr et al. (1995, p. 22), regression analysis is still a rather atypical tool in the evaluation of the results of computational experiments. Nevertheless, the extensive set of data obtained by the simulation study lend themselves easily to a regression analysis. As a dependent variable, the difference between the mean tour length and the optimal tour length is chosen. The explanatory variables comprise a Constant (Const.), a dummy for trivial local search ($ST = 0$), the starting threshold ST and a suitably normalized starting threshold trend for $ST > 8$, which allows us to model a broken linear trend in ST. Besides these variables for modelling the impact of the starting threshold, the number of iterations ($iter$) is included. Two different models are used for this last variable. First, if the number of iterations has to grow exponentially for a linear improvement of the approximation quality, it is adequate to use $1/\log(iter)$. This is the case which can be guaranteed based on the convergence results from Section 7.4. Secondly, if a quadratic increase is sufficient, $1/\sqrt{iter}$ is more appropriate.

Both models are estimated by using different break points for the modelling of the impact of the starting threshold ST. The optimal break point was

Table 8.2 Regression analysis of tuning experiment[a]

	Dependent variable: (mean tour length – optimal tour length)	
Const.	456.664 (94.79)	574.652 (285.81)
$I[ST = 0]$ [b]	144.645 (36.32)	144.645 (36.89)
ST	−62.394 (-224.16)	−62.394 (-227.67)
$(ST - 8) \times I[ST > 8]$ [b]	63.050 (152.62)	63.050 (155.01)
$1/\log(iter)$	1156.409 (35.06)	—
$1/\sqrt{iter}$	—	1491.551 (35.97)
R^2	0.992	0.992
SEE	16.106	15.858
log(likelihood)	−3439.981	−3427.234

[a] t–values in parentheses.
[b] $I[\cdot]$ denotes the indicator function.

determined as the one minimizing the sum of squared residuals under the constraint that the total impact of the starting threshold exhibits a non-monotone behaviour. The optimal points obtained this way were equal to 8.5 for both specifications.[11] The estimation results are summarized in Table 8.2.

Both models result in a good description of the data. The adjusted R^2 is larger than 0.99 in both cases. Although the standard error of estimation (SEE) is slightly smaller for the quadratic model, it is not significantly so. Hence, it is not possible to differentiate between the two models based on the available data.

The second option for generating threshold sequences is based on the local characteristics of the optimization problem. The thresholds are selected according to the differences of tour lengths between two neighboured tours. In order to obtain an overall impression of the height of these steps, an empirical distribution is generated along the following lines. A random tour is generated a large number of times. For each of these tours, one element of its neighbourhood is randomly chosen. Then, the absolute value of the

---

[11] Without the additional constraint, an optimum value of 8 would result.

difference between the lengths of both tours defines a step height. The empirical distribution of these step heights represents an approximation of the true — discrete — distribution. Now, a threshold sequence is obtained by using these step heights in descending order. If for each threshold the same number of iterations is performed, this procedure gives more weight — in terms of total iterations — to step heights which are more likely to occur. In general, for a good choice of local structure, these will be step heights close to zero.

Of course, there exist at least three alternative procedures for obtaining an empirical step distribution. The first consists in choosing some elements of the search space at random and calculating the absolute differences in objective function value to all neighbours. For the second alternative, only the smallest of these steps is kept for the empirical distribution, i.e. only the depth of the local minimum is considered. Finally, the third alternative consists in performing a random walk through the search space and calculating the differences in each step. In the limit, the empirical distributions generated by the first and third method will converge to the true step distribution, while the second method approximates the depth distribution. Hence, it appears to be difficult to make an a priori choice based on theoretical arguments. In fact, some preliminary tests indicated that there is no distinctive advantage of using one of the methods compared to the others.

Depending on the local structure of the problem, the empirical step distributions obtained this way may show high mass close to zero and at the same time a long tail, i.e. several large and very large step heights. As the results for the linear sequences indicated that choosing too high a starting threshold may reduce the efficiency, the simulation was repeated now by using only some lower percentile of the empirical step distribution as the threshold sequence.[12] Of course, the number of iterations per threshold is adopted to make sure that the total number of iterations remains unchanged. Figure 8.8 shows the results for the mean tour length.

The overall impression does not differ much from the figures for the linear threshold sequence. If only a small fraction of the empirical step distribution is used, however, the surface becomes more bumpy. This phenomenon has to be attributed to the fact that the number of different threshold values becomes small. Furthermore, there are less parameter constellations resulting in a mean value of the objective function smaller than 5150. Nevertheless, the best solution found by all tries was obtained by using this implementation of threshold accepting.

Figure 8.9 finally presents the results for a threshold sequence obtained from a relative step distribution. This is generated analogously to the previous one. The sole difference consists in the calculation of relative step heights, i.e. the

---

[12] The chosen percentile can be interpreted as the initial probability of acceptance of cost increasing moves. See Nurmela (1993, p. 21f.) for a related argument.

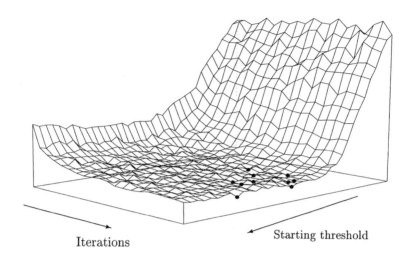

Iterations   Starting threshold

Figure 8.8 Mean tour length for different parameter choices III.

step heights are divided by the length of the first tour. Then, a quotient version
of threshold accepting is applied by using these threshold sequences. Again,
the differences to the quotient version with a linear threshold sequence are
negligible.

To summarize the findings on the threshold sequences, it is not possible
to provide a strong argument for one of the four methods described in this
section. However, the last two methods are data–driven, i.e. they can be
applied routinely without any prior knowledge of the orders of magnitude
to be expected for the objective function or the differences of values of the
objective function in a neighbourhood. Therefore, these data–driven methods
are used in the applications in Part III.

8.5   Restart Threshold Accepting

In the previous section, the mean performance of threshold accepting for
different parameter settings was analysed. In real applications, however, often
the best result obtained by a sequence of tries is more important than the
mean performance of the algorithm. In fact, Table 8.1 is a typical example, as
only the best results are presented. Sometimes it is not even mentioned how
many attempts have been necessary to obtain these best results.

Of course, it is possible to generate similar plots to the one presented in the
previous section for the minimum tour length out of the 40 tries performed
for each parameter constellation. These plots are less smooth, but still show
the crucial influence of a good choice of the threshold parameter. However,

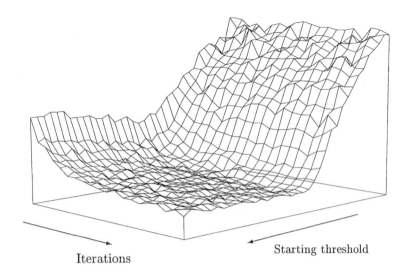

Iterations   Starting threshold

Figure 8.9   Mean tour length for different parameter choices IV.

the empirical minimum is not a robust estimator of the true minimum
obtained by an infinite number of tries, which in these applications always,
i.e. for all parameter constellations, coincides with the global optimum. Hence,
reasonable results have to be based on robust estimates. The lower 5 and 10
percentiles are chosen for this purpose.

If the above analysis of the simulation results is repeated with regard to such
lower percentiles, results may differ from the outcomes of the mean analysis
for two reasons. First, the distribution of the percentiles relative to the mean
may change for different threshold sequences. Secondly, the distribution can
be affected by differing number of iterations. The impact of this second aspect
on the distribution is analysed in the following under the heading of 'restart
threshold accepting'.[13]

This analysis can also be motivated by another look at the shortcomings of
the classical local search procedure. In the standard version of local search,
an increase in the number of iterations will improve the results only up to the
point when the first local minimum is reached. Further increases of the number
of iterations are useless. If instead the additional iterations are used for new
tries departing from another random start configuration, an improvement of
the best overall solution may result. In the limit, i.e. for an infinite number of
restarts, even a standard local search will find the global optimum.

---

[13] For some theoretical arguments on restart implementations, see Fox (1994). A heuristic
    argument in the context of genetic algorithms is provided by Farley and Jones (1994,
    p. 170).

Table 8.3 Restart threshold accepting I ($T_0 = 12$)

	Iterations per try		
	100 000	1 000 000	10 000 000
Runs	10 000	1000	100
Mean	5317.07	5170.52	5138.22
SD	52.83	28.69	21.81
10%[a)]	5251.11	5135.45	5112.22
5%	5234.50	5124.83	5107.41
1%	5204.20	5109.90	5098.23

[a)] 10, 5 and 1% denote the lower 10, 5 and 1 percentiles, respectively.

This observations results in the following experimental setting for a restart version of threshold accepting. One threshold sequence is fixed a priori.[14] For each threshold sequence, the threshold accepting algorithm was run with 100 000, 1 000 000 and 10 000 000 iterations. One run with 10 000 000 iterations corresponds to 10 runs with 1 000 000 or 100 runs with 100 000 iterations. In order to obtain good estimates of the lower percentiles, 100 runs were performed with the largest number of iterations. Consequently, the number of runs with different random starting configurations was 1000 and 10 000 for 1 000 000 and 100 000 iterations, respectively. Table 8.3 summarizes the results obtained for a starting threshold T_0 of 12.

A first tentative interpretation of the findings is that the trade–off between more tries with different initial configurations and a higher number of iterations per try is in favour of the latter. In fact, both the mean and the lower percentiles become smaller as the number of iterations per try increases while holding the total use of computer resources (tries times number of iterations) constant. It should be noted, however, that users are mainly interested in the overall best results, i.e. the very small percentiles of the distribution. In Table 8.3, the 1% percentile for the tries with 10 000 000 iterations is estimated based on solely 100 observations. Consequently, the lower right entry of the table has to be interpreted with some care.[15]

A slight increase of the starting threshold T_0 to 14 does hardly change the results, as could have been expected from the analysis in the last section. Table 8.4 summarizes the findings.

---

[14] For this experiment, only linear sequences with different starting values are considered.
[15] Methods from extreme value theory could probably be used to assess the behaviour in the tails in more detail.

Table 8.4   Restart threshold accepting II ($T_0 = 14$)

	Iterations per try			
	100 000	1 000 000	10 000 000	40 000 000
Runs	10 000	1000	100	25
Mean	5314.14	5170.92	5138.67	5127.57
SD	52.13	26.37	20.75	21.31
10%[a]	5248.63	5137.77	5111.75	5091.15
5%	5230.84	5129.70	5107.40	5087.85
1%	5205.27	5115.68	5091.38	—

[a] 10, 5 and 1% denote the lower 10, 5 and 1 percentiles (8 and 4 percentiles for the last column), respectively.

In addition to the first simulation, 25 runs with 40 000 000 iterations each were performed. Again, a decline in the mean tour length results. 10 and 5 percentiles cannot be estimated based on 25 observations. Instead, the 8 and 4 percentiles are given in the respective lines of the table. Finally, the experiment was repeated for a starting threshold T_0 of 8. Table 8.5 summarizes the results.

As a starting threshold of 8 is a little bit smaller than the optimum value found by the simulation experiment in the previous section, it does not come as a surprise that the overall performance is worst in this case. However, the overall best result in the lower right cell of the table is far better than in the previous table. As mentioned before, the estimate of the lower 1% percentile in this case is based on one observation and for the lower 5% percentile on five observations. Consequently, this outcome might be regarded as 'pure luck'.

The above analysis gives valuable information on the dependence of mean and percentiles on the number of iterations and restarts. However, it does not answer yet the practitioner's question whether it is preferable to perform a single run with a very high number of iterations (Bölte and Thonemann, 1996, p. 412), a few restarts with a high number of iterations, or many tries with a low number of iterations given a limit of available computing resources. The outcomes of the three experiments can be used also to address this question. Therefore, the tries are grouped to 100 subexperiments each consisting of 100 tries with 100 000 iterations, 10 tries with 1 000 000 iterations and one try with 10 000 000 iterations. Then, only the best results are compared. As mentioned above, this is not a valid approach from a theoretical or statistical perspective. Nevertheless, it represents the usual approach in comparative studies of optimization heuristics[16] as it gives the answer to the practitioners'

---

[16] See Section 7.4 and Table 8.1.

Table 8.5   Restart threshold accepting III ($T_0 = 8$)

	Iterations per try		
	100 000	1 000 000	10 000 000
Runs	10 000	1000	100
Mean	5383.15	5209.77	5158.24
SD	61.23	40.31	30.65
10%[a)]	5304.86	5160.88	5118.73
5%	5286.31	5147.49	5106.49
1%	5248.98	5131.78	5086.73

[a)] 10, 5 and 1% denote the lower 10, 5 and 1 percentiles, respectively.

question. The upper part of Table 8.6 shows the number of times, the respective combinations of iterations per try and repetitions gave the best result. In the lower part the mean departure of the best results for the three categories from the best results of these three categories are depicted.

If one has only computer resources for a total of 10 000 000 iterations, these results clearly indicate that one should neither perform many tries with a very low number of iterations, nor only one huge run. Instead, the best expected performance is given by a choice which falls between the two extremes, i.e. restarting the threshold accepting a few times with a large number of iterations.

To sum up the findings of this section on restart threshold accepting, two aspects seems noteworthy. First, the usual practice in computational experiments to consider mainly the best outcome of several tries can be misleading as it presents a nonrobust statistic. The analysis of percentiles seems more adequate. However, as the central interest of practitioners is best–case performance, a comparison of restart versions of threshold accepting should also be based on this criterion. Secondly, the results indicate that mean value, standard deviation and low percentiles decrease, all things being equal, with an increasing number of iterations. Nevertheless, in order to obtain high–quality results, such as those reported in Table 8.1, the combination of some restarts with a medium–sized number of iterations outperforms the single start/huge number of iterations choice.

While the finding of the previous section on the non–monotone impact of the starting threshold can be based on theoretical arguments, it seems more difficult for the restart criterion. In fact, one would have to know something about the distribution of optimization results, themselves dependent on the number of iterations, in order to draw conclusions. However, the only hard

Table 8.6 Restart threshold accepting IV (comparison)

	Iterations per try		
	100 000	1 000 000	10 000 000
Starting	Runs		
threshold	100	10	1
Times best in 100			
8	0	56	44
12	0	65	35
14	0	58	42
Mean departure from best			
8	99.71	11.46	16.54
12	73.56	5.16	14.48
14	73.62	6.93	13.68

information at hand is that this distribution will eventually, i.e. with the number of iterations tending to infinity, converge in probability to the one–point global optimum distribution. Consequently, the results in this section have to be taken with care and tested on other problem instances, before drawing strong conclusions.

8.6 Conclusions

In this chapter, the performance of the threshold accepting algorithm was analysed in some detail for a travelling salesman problem with 442 points. The simulations exhibit the crucial influence of a problem with an adequate (or 'good') choice of parameters. The robust impact of parameters determining the threshold sequence allows us to draw conclusions from tries with a small number of iterations for tries with a large number of iterations. Thus, it becomes possible to tune the algorithm with a limited need for computer resources, thus enabling solutions of very high quality.

Of course, further research is required, repeating the above experiments on other instances of the travelling salesman problem or other problems tackled by threshold accepting, in order to obtain some empirical robustness. In particular, so far, only monotone threshold sequences have been analysed within this setting. Some preliminary experiments on small problem instances indicate that non–monotone sequences may be superior (Aarts et al., 1997, p. 113). In fact, Althöfer and Koschnik (1991) and Hu et al. (1995) calculate the optimal threshold sequence for very small problem instances. Their results indicate that the threshold sequence, which leads to the global optimum with highest probability for a given finite number of iterations, may be non–monotone and even contain negative threshold values. Hu et al. (1995)

also present a version of threshold accepting with a self–tuning threshold sequence which may lead to non–monotone sequences. For an implementation on travelling salesman problems with up to 200 cities, these non–monotone sequences lead to slightly better solutions during the early stages of the algorithm and for randomly generated problem instances. Hence, it may be worth experimenting with non–monotone sequences if the number of iterations, which can be performed under constraints given by available computer resources, is small.[17] Another argument, when the use of non–monotone sequences can be appropriate, is based on the local structure of the problem. Osman and Christofides (1994) study monotone and non–monotone cooling schedules for simulated annealing for a capacitated clustering problem. They propose that non–monotone sequences may be important if the problem exhibits a 'bumpy' topology.

Nevertheless, the preliminary results provided in this chapter can be and are already used for the tuning of threshold accepting implementations for the problems covered in Part III. In fact, studying the approaches which lead to improvements in this example, may be at least a good source of ideas about how best to accomplish it in other domains (Johnson and McGeoch, 1997, p. 310).

---

[17] The problem of optimal aggregation described below in Chapter 13 has been approached by using non–monotone threshold sequences. This approach improved the results slightly when only a very small number of iterations was performed.

9

A Practical Guide to the Implementation of Threshold Accepting

9.1   Three Steps for a Successful Implementation

In Part I, it was argued that many problems met in economics, statistics and econometrics are highly complex optimization problems. Since standard optimization routines often fail to provide satisfying results for these kinds of problems, the use of optimization heuristics such as threshold accepting seems warranted. The previous chapters of the second part of the book introduced the threshold accepting heuristic in some more detail. Due to its easy structure, which implies low costs in terms of manpower for its implementation, combined with strong convergence properties and high performance for some real life applications, it is a natural candidate to tackle these problems. The goal of this last chapter in Part II is to provide the reader with the relevant information and tools for their own successful implementations of threshold accepting such as the ones presented in Part III.

The way to a successful implementation of threshold accepting might be summarized in three steps, which in turn are described in some detail in the remaining sections of this chapter. The first step consists in formulating the problem. Often, a problem in econometrics and statistics comes rather in the disguise of a modelling or inference problem than as explicit optimization problems. Other applications are treated routinely by some ad hoc methodology without even noticing the underlying optimization problem. Therefore, this first step is not trivial at all. The problem has to be formulated and described as an explicit optimization problem. In particular, an objective function and the search space have to be defined. Furthermore, the formulation of the optimization problem should also comprise a complexity analysis, i.e. a check whether some standard optimization algorithm can provide satisfying results for this problem,[1] given the size of typical problem instances.

The second step is devoted to a more careful analysis of the search space or space of potential solutions. Often, side constraints require us to limit the search to some subset of a convenient space, i.e. a space which can be easily described. Then, it is important to provide a tractable characterization of the search space. Otherwise, looking for a new element in the search space, or even for the starting element, might become a difficult task. More important for the applications presented in Part III is the definition of a neighbourhood mapping or a set of possible neighbourhood mappings on the search space. Again, it is not only important to provide some 'local structure', but also to have an operational neighbourhood mapping.

Only the final step has direct reference to the threshold accepting algorithm, because in this step the parameters of the algorithm are determined. In particular, a threshold sequence has to be chosen. Furthermore, either the number of iterations is fixed for each element of the threshold sequence or some other stopping rule has to be selected. As pointed out in Chapter 8, the choice of the parameters of the threshold accepting algorithm can be made data–dependent. First, an empirical jump distribution is generated, which provides estimates of typical step sizes. Then, a small tuning experiment (as discussed in Section 8.4) is used to determine which part of the corresponding sequence of thresholds is actually used. Finally, the analysis of the results obtained by the implementation is summarized within this last step of the analysis. Thereby, results can be compared with known solutions from standard algorithms, optimal solutions obtained for small scale problem instances or ad hoc solutions usually applied for this problem.

The example of multivariate lag structure identification introduced in Section 3.2.2 and covered in detail in Chapter 12 is used throughout this chapter to present a specific choice for each step of the implementation of threshold accepting discussed in this practical guide.

9.2   Make it an Optimization Problem

In Section 4.1, we assumed that typical optimization problems arising in econometrics and statistics can be expressed by the simple formula

$$\max_{\mathbf{x} \in \mathcal{X}} f(\mathbf{x}) \,, \tag{9.1}$$

where the set $\mathcal{X}$ is a — possibly discrete — subset of some n–dimensional reel vector space, i.e. $\mathcal{X} \subset \mathbb{R}^n$. However, while some problems, e.g. maximum–likelihood estimation, are naturally given in this notation, it is less evident for others, e.g. the travelling salesman problem. In fact, for the latter it is more convenient to use the set of possible tours through all n cities as

---

[1] For test procedures for this purpose, see also Section 15.1.

search set $\mathcal{X}$ instead of some imbedding in a reel vector space. Nevertheless, both maximum–likelihood estimation and the travelling salesman problem are explicitly stated as optimization problems: 'find the parameters maximizing the likelihood' or 'find the shortest tour connecting all n cities', respectively.

Many other tasks in econometrics and statistics do not appear as explicit optimization problems. Nevertheless, these problems are often the more complex ones, which cannot be tackled by standard procedures, but might be candidates for the application of threshold accepting. Hence, it is important to detect and describe these problems as optimization problems. A few examples may provide an idea on how to find the hidden optimization problems:

- estimate parameters β_0, β_1 and σ for the model $y_t = \beta_0 + \beta_1 x_t + \varepsilon_t$, $\varepsilon \sim \mathcal{N}(0, \sigma^2)$;

- select the lag length in a vector autoregressive model;

- choose parameters for a simulation study for some estimator with nonstandard distribution;

- solve a rational expectations model for given parameter values.

No explicit reference to optimization is made in any of the above examples. Thus, it is necessary to analyse any single step of research in statistics and econometrics with regard to inherent optimization problems.[2] In fact, all given examples are optimization problems. One solution to the first example would be the ordinary **least** squares estimator. The lag length selection can be performed by **minimizing** an information criterion (see also Chapter 12). The third example describes a problem of **optimal** experimental design (see also Chapter 11). Finally, the last example seems a bit odd at first sight, as it refers to the solution of a nonlinear system of equations. However, this problem is also tackled as an optimization problem. In fact, the algorithms used for this purpose look for solutions which bring the errors of all equations of the system **as close to zero as possible.**

Of course, these examples are far from being comprehensive or exhaustive. Nevertheless, in statistics and econometrics, the keywords 'estimate', 'choose', 'select', and 'solve', as well as their synonyms, almost always hint to an underlying optimization problem. Once the fact is acknowledged that we have to deal with an optimization problem, the next step consists of identifying or selecting an objective function. As far as problems of modelling, estimation and inference are concerned, statistical theory often provides concepts which can easily be translated into terms of objective functions, i.e. least squares, maximum–likelihood, minimum absolute deviations, information criteria, size or power of tests, etc., to name just a few. Furthermore, problems arising in

---

[2] See also the examples provided above in Chapter 3.

experimental design and the set–up of simulation studies result in objective functions, which are described in the experimental design literature (see Chapter 11). Finally, for problems related to the solution of systems of equations, which may also arise from game theoretic problems, objective functions can be obtained by considering standard algorithms for their solution, such as Newton's method and its refinements.[3] These algorithms also have to define an objective function, which measures the deviation from a real solution, in order to provide a stopping rule.

For other problems, which do not seem to fit into one of these categories, an objective function might still be derived as an answer to the following questions:

1. Which properties of a solution are most important?
2. How can these properties be evaluated and measured by some reel valued variables?
3. Which values of the reel valued variables correspond to a good solution (high values, small values or values close to zero)?

Finally, if the objective comprises several dimensions, a weighting function has to be used to obtain a univariate measure of the quality of the obtained solutions. In the example of multivariate lag structure identification, two objectives have to be considered: obtain a good fit to the data and use as few lags as possible. The information criteria described in Section 12.3 provide a weighting of these two factors based on a theoretical asymptotic argument.

Besides the objective function, the set of potential solutions $\mathcal{X}$ — the search space — also has to be defined. For applications in statistics and econometrics, the answer to this question is often straightforward. Nevertheless, sometimes additional constraints have to be taken into account. This can be done either by reducing the search space or by adding a penalty term to the objective function, which punishes derivations from the imposed constraints. The second approach is to be preferred if otherwise the search space — taking the constraints into account — becomes very sparse and not connected under some natural topology. This aspect is important both for the construction of the neighbourhoods discussed in the next section and for the construction of candidate solutions within these neighbourhoods. If tests on whether a certain element satisfies all side constraints exactly, i.e. belongs to the constrained set $\mathcal{X}$, become very complicated, the performance of a threshold accepting implementation decreases. Thus, it is important to obtain a simple description of the elements of $\mathcal{X}$. Otherwise, no efficient implementation of any local search heuristic is possible. If this condition was not crucial, one could simply start local search on the set of optimal solutions of the problem!

Reconsider the example of multivariate lag order selection in a vector

---

[3] See Section 5.1.2 for further references.

autoregressive model given by

$$X_t = \begin{pmatrix} x_{1t} \\ x_{2t} \\ \vdots \\ x_{dt} \end{pmatrix} = \sum_{k=1}^{K} \Gamma_k X_{t-k} + \varepsilon_t, \tag{9.2}$$

where $\varepsilon_t = (\varepsilon_{1t}, \ldots, \varepsilon_{dt})$ are identically and independently distributed as $\mathcal{N}_d(0, \Sigma)$, Σ denotes the $d \times d$ covariance matrix of the error distribution, and the matrices

$$\Gamma_k = \{\gamma_{i,j}^k\}_{1 \leq i,j \leq d}$$

provide the autoregression coefficients. Two choices of the search space can be observed in practical applications. Either only lag structures including all lags up to a maximum lag order K are considered. Then, the search space is given by $\{1, \ldots, K\}$ and optimization can be done through enumeration of all potential solutions. Otherwise, all possible lag structures with a lag length up to K are considered. Then, a total of 2^{Kd^2} different lag structures are potential solutions. This search space can be represented by $\{0, 1\}^{Kd^2}$, i.e. by the set of $K \times d^2$ $\{0, 1\}$ matrices. Thereby, each potential solution is represented by such a matrix, where every entry equal to one means inclusion of the corresponding lag in the model.

A last argument is connected both to the choice of neighbourhoods and the search space. As the threshold accepting, in general, performs a large number of iterations and in each iteration compares the value of the objective function for the current and the new candidate solution, its performance depends on these objective function evaluations. Consequently, objective functions, which are easily evaluated, are to be preferred to more complicated ones (including for example the evaluation of multivariate integrals or the inversion of large matrices). Furthermore, the efficiency of a threshold accepting implementation increases tremendously if the value of a new candidate solution in a neighbourhood of the current one can be calculated by some updating rule instead of a complete reevaluation of the objective function. For example, the $2 - Opt$ step described for the travelling salesman problem in the previous chapter allows for such a fast update. Unfortunately, for many applications in statistics and econometrics — including the ones presented in Part III — the development of such updating rules might require a substantial amount of manpower in the first instance. Hence, it is recommended only in cases when a threshold accepting implementation seems to provide good results, is planned to be used often in the future, and is not efficient enough for larger instances, when the objective function has to be reevaluated for any new candidate solution.

9.3 Give it Local Structure

Threshold accepting is a refined local search algorithm. Therefore, its performance depends on local properties of the landscape generated by the objective function and the neighbourhood mapping $\mathcal{N}$, which assigns the neighbouring points to each element $\mathbf{x}$ of the search space $\mathcal{X}$. The difficulties of defining a suitable neighbourhood mapping $\mathcal{N}$ have already been discussed in the context of classical local search in Chapter 5 and for the threshold accepting algorithm in Chapter 6.[4] To sum up the arguments, which are important for the performance of a real application:

1. Neighbourhoods have to be chosen in a way to introduce local structure with regard to the objective function. Neighbourhoods should neither be too large nor too small.
2. Neighbourhoods should be easy to construct or, put differently, it should be easy to generate $\mathbf{x}^n \in \mathcal{N}(\mathbf{x}^c)$ for any $\mathbf{x}^c \in \mathcal{X}$.

The choice of a neighbourhood mapping satisfying all these conditions is certainly one of the most demanding tasks when setting up a threshold accepting implementation. As it is difficult to provide a general method for this purpose, three approaches, which have been applied successfully, are outlined. Furthermore, a heuristic method is proposed which can be used to assess the features of the chosen neighbourhood structure.

The first and most straightforward neighbourhood mapping can be applied when $\mathcal{X}$ can be embedded naturally in some reel valued vector space, i.e. $\mathbb{R}^d$. Then, for each $\mathbf{x} \in \mathcal{X}$, any $\varepsilon \geq 0$ defines a neighbourhood in $\mathbb{R}^d$, the so–called ε–spheres, given by

$$\mathcal{N}_\varepsilon(\mathbf{x}) = \{\tilde{\mathbf{x}} \in \mathbb{R}^d \mid \|\mathbf{x} - \tilde{\mathbf{x}}\| < \varepsilon\},$$

where $\|\mathbf{x}\|$ denotes the usual Euclidian metric on $\mathbb{R}^d$. Projecting these ε–spheres on $\mathcal{X}$, a neighbourhood mapping for $\mathcal{X}$ results:

$$\mathcal{N}(\mathbf{x}) = \{\tilde{\mathbf{x}} \in \mathcal{X} \mid \|\mathbf{x} - \tilde{\mathbf{x}}\| < \varepsilon\}.$$

This method works well, if the dimension d is not too large and $\mathcal{X}$ is scattered 'densely' in $\mathbb{R}^d$, i.e. for small ε the neighbourhoods comprise several elements of $\mathcal{X}$. This is the case, e.g. for continuous optimization problems. In Figure 6.1, p. 90, the other case is depicted, when even quite large ε cannot guarantee that the neighbourhoods contain at least two elements. If $\mathcal{X}$ is a finite set as for most of the examples discussed in Part III, this neighbourhood definition is not very well suited for two reasons. First, $\mathcal{X}$ is not scattered densely in this case. Secondly, while it is easy to generate randomly an element of $\mathcal{N}_\varepsilon$ in $\mathbb{R}^d$, it is not clear a priori, how to do this in the projection on $\mathcal{X}$. Generating elements

---

[4] See also Figures 5.1 and 6.1 above.

of $\mathcal{N}_\varepsilon$ and verifying whether they happen to belong to $\mathcal{X}$ will take infinitely long, if $\mathcal{X}$ is a discrete subspace. Therefore, the use of projected ε–spheres seems to be well suited mainly for optimization problems on a continuous set in some $\mathbb{R}^d$. Chapter 15 provides a discussion of the potential of applications of threshold accepting to such problems.

The idea for the second neighbourhood mapping is adopted from the ε–spheres. However, a different metric is employed, which can be applied directly to a discrete set $\mathcal{X}$ without an embedding in some reel vector space. Instead, it is required that $\mathcal{X}$ can be described as some subset of a d–dimensional binary space, i.e. $\mathcal{X} \subset \{0,1\}^d$. Typical examples are matrices with entries zero and one, which mirror, for example, whether a specific variable is included in a model or not. To such sets, the Hamming distance can be applied (Hamming, 1950). The Hamming distance d_H between two elements $\mathbf{x}_1$ and $\mathbf{x}_2$ of $\mathcal{X}$ is given by the number of differing entries

$$d_H(\mathbf{x}_1, \mathbf{x}_2) = \sum_{j=1}^{d} \mid \mathbf{x}_{1j} - \mathbf{x}_{2j} \mid, \qquad (9.3)$$

where $\mathbf{x}_i = (\mathbf{x}_{i1}, \dots, \mathbf{x}_{id})$ for $i = 1, 2$, respectively. Now, neighbourhoods can be defined as for the ε–spheres simply by setting a maximum Hamming distance d_H^{max}. Then, the neighbourhood mapping is defined by $\mathcal{N}(\mathbf{x}) = \{\tilde{\mathbf{x}} \in \mathcal{X} \mid d_H(\mathbf{x}, \tilde{\mathbf{x}}) \leq d_H^{max}\}$. If $\mathcal{X}$ is rather dense in $\{0,1\}^d$, this approach satisfies all requirements. First, for a suitable choice of d_H^{max} (typical choices in applications are $2 - 8$), local behaviour is introduced and the neighbourhoods are not too small. Secondly, the Hamming distance also provides a construction method for candidate solutions in $\mathcal{N}(\mathbf{x})$. In fact, one randomly selects some $\mathbf{x}_j$, $j \in \{1, \dots, d\}$ and replaces these by $(1 - \mathbf{x}_j)$. Keeping the number of such changes smaller than d_H^{max} guarantees that the Hamming distance between old and new elements are in the admissible range. Furthermore, as long as $\mathcal{X}$ is dense in $\{0,1\}^d$, it is likely that a new element in $\mathcal{X}$ can be found this way. Otherwise, one repeats the random exchange try until an element of $\mathcal{X}$ results. Due to these good properties of neighbourhoods defined by the Hamming distance, Ausiello and Protasi (1995, p. 76) consider it the natural choice for defining neighbourhoods on $\{0,1\}$–spaces. The applications presented in Chapters 12 and 13 are based on this neighbourhood concept.

Obviously, the Hamming distance concept is also a suitable choice for the multivariate lag order selection problem, which can be described as minimization of an information criterion over the search space $\{0,1\}^{Kd^2}$. In this case, using the Hamming distance has even an intuitive interpretation. For example, two multivariate lag structures, which have a Hamming distance of two, differ in exactly two entries. Consequently, a neighbouring element to a given lag structure can be generated by removing one lag present in this lag structure and adding another one not yet present. Given this intuitive meaning of Hamming distance, a further refinement of the neighbourhood concept can be proposed. Instead of allowing the exchange of any two lags, this exchange step might be constrained to lags which are close to each other on the time axis, e.g. replace lag 3 of one variable by lag 5 of the same variable. In fact, this constrained version of a neighbourhood concept based on the Hamming distance is used in the application in Chapter 12 and can also be considered in similar applications.

The third approach starts from the second required property for a good neighbourhood mapping. Thus, given some element $\mathbf{x} \in \mathcal{X}$, one tries to answer the question as to how a similar element might be obtained by simple constructive steps. One example of such a constructive approach has already been presented by the Hamming distance. However, often $\mathcal{X}$ is not easily embedded in a $\{0,1\}$ vector space. Nevertheless, the idea might still work. The $2 - Opt$ move described for the travelling salesman problem in Chapter 8 represents such a constructive solution. A different approach is presented in Chapter 11 for the uniform design problem. While the Hamming distance approach is a general method, which can be applied to any set embedded in a $\{0,1\}$–space, construction methods are always problem–dependent and no general proceeding can be advised. Compared to the Hamming distance method, direct construction approaches have the advantage that each exchange step produces a candidate solution, while the Hamming distance approach might require us to repeat the generation step if $\mathcal{X} \neq \{0,1\}^d$.

All three mentioned methods for obtaining a neighbourhood mapping are possibly a reasonable choice for a local search heuristic like threshold accepting. However, the question as to whether they satisfy all requirements has to be answered separately for each specific problem. Consequently, a few heuristic instruments are proposed which allow us to obtain some further information on the quality of a given neighbourhood definition.

First, simulations can be performed to assess the local behaviour of the objective function with regard to the given neighbourhood mapping. Since threshold accepting has to climb uphill sometimes to avoid bad local minima, the size of uphill steps is relevant for its performance. A simulation to obtain an estimate of these steps is performed as follows. One randomly selects elements $\mathbf{x}_i \in \mathcal{X}$. For each $\mathbf{x}_i$ an element $\tilde{\mathbf{x}}_i \in \mathcal{N}(\mathbf{x})$ in the neighbourhood of $\mathbf{x}_i$ is also randomly selected. The empirical distribution of the absolute differences of the objective function value for both elements $|f(\mathbf{x}_i) - f(\tilde{\mathbf{x}}_i)|$ provides information on the step size, which might be expected in a local setting. This distribution can be compared with the empirical distribution of random exchange tries, which can be obtained quite similarly with the sole difference being that $\tilde{\mathbf{x}}_i$ is selected completely at random, i.e. independently of $\mathbf{x}_i$. The chosen neighbourhood mapping implies a local behaviour of the objective function if the empirical distribution for the step size, which is obtained with regard to this neighbourhood, is much more concentrated around zero than in the random case.

An alternative simulation, which also provides some heuristic information on the local behaviour of the objective function with regard to a given neighbourhood mapping consists in simulating a random walk through $\mathcal{X}$. One starts with an arbitrary element $\mathbf{x}_0 \in \mathcal{X}$. Then, in each step a new element $\mathbf{x}_{i+1}$ is chosen randomly in $\mathcal{N}(\mathbf{x}_i)$. A plot of $f(\mathbf{x}_i)$ against i gives an idea of the local behaviour of f along such a random walk through the search space. It might exhibit several ups and downs, but should look smoother than a similar plot which is obtained by choosing any $\mathbf{x}_i$ completely at random in $\mathcal{X}$. Such a plot is depicted in Figure 8.2 for the travelling salesman problem discussed in the previous chapter.

A third indicator is given by the size of neighbourhoods. Typically, this size should depend on the size of the search space itself. Therefore, it is not possible to provide some guideline for the absolute size of neighbourhoods. It can be argued, however, that it should grow with the dimension of the problem. For finite sets $\mathcal{X}$, it is recommended that typical neighbourhood sizes should be calculated and these compared with the size of $\mathcal{X}$. A neighbourhood should not comprise a relevant part of $\mathcal{X}$, but should include a few elements which might grow at a rate of about $c^{\sqrt{d}}$, where d denotes the dimension of an embedding space.

Finally, if several neighbourhood mappings seem feasible for a given problem and their implementation does not require a huge amount of manpower, it is strongly recommended that the performance of the threshold accepting heuristic be compared for different neighbourhood concepts, at least for a few samples with a low number of iterations.

9.4   Cross the Thresholds

The choice of objective function, search space and neighbourhoods provide the essential ingredients for a local search implementation. In order to make it a threshold accepting application, the only additional input consists of the threshold sequence $\{T_i\}$, $i = 1, \ldots, I$, which provides the acceptance rule for the exchange step and also the stopping rule, if I is finite and for each threshold T_i only a limited number of exchange tries is performed. Furthermore, through the choice of I and the number of iterations to be performed for each T_i, the total running time of the algorithm is determined.

Section 8.4 describes a data–driven approach for optimal threshold sequences, exploiting the fact that the dependency of optimization results on the threshold sequence seems to be almost independent of the number of iterations performed. Consequently, extensive tuning is possible by using only a limited amount of resources. Nevertheless, the set–up and analysis of such a tuning experiment requires a non–negligible amount of manpower. Therefore, in this practical guide it is argued that such an extensive tuning experiment might be only a second step once a successful implementation has been found.

For a first implementation, the other data–driven method already described in Section 8.4 is preferred.

This is based on the simulation set–up already described for the analysis of neighbourhoods in the previous section. Again, a number of elements $\mathbf{x}_i \in \mathcal{X}$ is randomly selected. For each $\mathbf{x}_i$, an element $\tilde{\mathbf{x}}_i$ in the neighbourhood of $\mathbf{x}_i$ is chosen. Let $\Delta_i = |f(\mathbf{x}_i) - f(\tilde{\mathbf{x}}_i)|$ denote the absolute difference of the objective function between the two points. This value indicates the step height that the algorithm would have to overcome in one direction when moving from $\mathbf{x}_i$ to $\tilde{\mathbf{x}}_i$, or vice versa, respectively. The example illustrated in Figure 6.4 above highlights the importance of these step heights, as does the analysis of asymptotic convergence in Section 7.3. As the threshold sequence copies to some extent the cooling schedule of simulated annealing, a monotonic decreasing threshold sequence is a reasonable first choice, although it does not have to be the optimal one in all circumstances. Therefore, the Δ_i are arranged in decreasing order. The tuning experiment in Chapter 8 indicates that starting with too high values of the threshold might reduce performance. Hence, only a fraction α of the sequence Δ_i, comprising the lower values, is used in the algorithm. For each of these thresholds the same number of iterations is performed. Consequently, very small thresholds, which appear more often in such a sequence, if the neighbourhoods imply a really local structure, obtain more weight in terms of iterations. The rationale for this choice is that for small step heights a larger number of steps might be necessary in order to escape local minima. Finally, the lowest Δ_i being equal to zero (each $\mathbf{x}$ is a neighbour to itself), the algorithm then ends with several iterations for a threshold of zero, i.e. accepts only decreasing moves. This features increases the probability that the algorithm ends up at least with a local minimum.

Instead of considering absolute differences of objective function values, it is also possible to consider relative changes. By using the same procedure as just described, one can obtain a relative threshold sequence which can be used in a variant of threshold accepting, where the acceptance criterion is not defined in absolute, but in relative terms, i.e. $\mathbf{x}^n$ is accepted if and only if $(f(\mathbf{x}^n) - f(\mathbf{x}^c))/f(\mathbf{x}^c) < T_i$. This version seems to work better if the objective function is not very well scaled, i.e. changes by several orders of magnitude over the search space with very high spikes and deep holes.

Using this data–driven generation of a threshold sequence leaves only two tuning parameters: the share of simulated step heights, which is used as thresholds (α) and the number of iterations performed for each threshold value. While the impact of the latter should be clearly positive, the tuning experiment in the previous chapter indicates that the influence of α might look like a U–shaped function. In order to find out a reasonable value for α, as a first step the algorithm should be run with different values of α and a low number of iterations. Based on the results of these first runs, a reasonable range of α can be determined and used for the real optimization runs with a larger number of iterations.

Finally, the question about the optimal number of iterations is easy to answer: more iterations improve the expected quality of the result. Hence, only the available computer resources set a limit to this quantity. However, as analysed in some detail in the previous chapter, it is not evident whether, for a given amount of computing resources, it is better to use this all for one single run of a threshold accepting implementation with a very large number of iterations, or to split it up in a collection of runs — let us say ten to thirty — with a proportionally lower number of iterations, each starting with different initializations. The results from Section 8.5 indicate that the latter, i.e. the restart version of threshold accepting, might be preferred.

9.5 Summary

At the end of this chapter, which also concludes the second part of the book, a summary both of the practical guide to the implementation of threshold accepting and of the general discussion of the threshold accepting heuristics seems adequate.

Chapter 5 provides a short overview on classical and heuristic optimization approaches. The examples presented in Part I already indicated the limitations of classical optimization approaches and, consequently, the potential for the application of other optimization heuristics. In Chapter 6, the threshold accepting algorithm is introduced as a local search meta–heuristic. The usage of such meta–heuristics, in particular of the threshold accepting heuristic is advocated due to their simplicity, robustness and easy modification (Osman and Kelly, 1996, p. 15). Nevertheless, other optimization heuristics may also be applied successfully to some optimization problems in econometrics and statistics.

For threshold accepting, some asymptotic convergence results can be obtained based on similar results for the simulated annealing heuristic. However, the presentation of these results in Chapter 7 does not provide hints on how to implement threshold accepting efficiently for a given optimization problem. Consequently, Chapter 8 presents the results of an extensive tuning experiment, which allows for some conclusions to be made on how to deal with the parameters for a threshold accepting implementation. Given all this information, it still remains an art to implement such heuristics successfully on new problem instances, i.e. to exploit the positive features in a setting with a necessarily finite number of iterations. Some skill and experience is and will be required (Aarts et al., 1997, p. 117) as in the case — and this is an aspect often ignored — of classical heuristics omnipresent in statistical software packages.[5]

---

[5] Some fallacies when using these standard algorithms without the required skill and experience are described in McCullough and Vinod (1999).

Three steps have to be made for a successful implementation of threshold accepting as outlined in this chapter. First, the optimization problem has to be understood and described as such, thus providing a univariate objective function. At the same time, the set of potential solutions has to be defined. Secondly, the set or space of potential solutions has to be given some local structure by using a neighbourhood mapping. Besides the requirement that neighbourhoods should not be too small, given the objective function, they have to be chosen in a way to obtain rather small changes of the objective function when moving from one potential solution to a neighbour. Thirdly, threshold values have to be set which allow the algorithm to escape from bad local minima during the optimization procedure. A mainly data–driven method is proposed for this step, which is also used in several of the examples discussed in Part III.

Thus, the examples of applications of threshold accepting to problems in statistics and econometrics discussed in the following chapters combine two aspects. First, they are of interest in their own right, as solutions based on the threshold accepting algorithm are presented for problems which have either been regarded as intractable so far or for which only far from optimum results have been achieved by ad hoc algorithms or classical heuristics. Secondly, the discussion of the implementation of threshold accepting to different problem instances gives a guideline on how to exploit the power of this meta–heuristic to a maximum for further applications.

Part III

Applications in Statistics and Econometrics

10

Introduction

This third part of this book complements the previous parts by presenting some successful implementations of threshold accepting in econometrics and statistics. Each of the implementations is of its own interest as it tackles optimization problems arising from practical problems which appear to be intractable by using standard optimization tools. Furthermore, the different implementations of threshold accepting provide further insights on the functioning of the algorithm and the details which contribute to its efficient behaviour. Although the examples might seem to come from quite different areas of econometrics and statistics, they cover or touch a large part of the typical tasks encountered in these areas. Therefore, these examples also serve as a motivation for a broader use of techniques similar to threshold accepting in all parts of statistical modelling, following the arguments for a broader view of the optimization paradigm put forward in Part I.

To give only an elementary introduction and overview on the different fields of applications in statistics and econometrics studied in Part III would require at least one volume on its own. Consequently, the following chapters cannot aim at providing such a complete view on the covered research topics. Instead, a short general outline of the field and some general problems will be provided. Furthermore, some basic references are included for the reader who would like to obtain a deeper understanding of the framework. Next, the specific problem for the application of threshold accepting is introduced and some results on its complexity and other solution approaches are presented, before turning to the actual implementation of threshold accepting. Finally, an analysis of the results obtained by this implementation highlights the gains and potential limitations of this method. Each section is concluded with an outlook to future research on the specific implementation and on related areas where threshold accepting could be a useful tool. The next five chapters cover the following applications:

- Chapter 11 Experimental design;

- Chapter 12 Multivariate model identification;

- Chapter 13 Optimal aggregation;

- Chapter 14 Censored quantile–regression;

- Chapter 15 Continuous global optimization.

The first application comes from the statistical experimental design field. It is not the analysis of results, which might be obtained by using experimental design methods, which is in the centre of interest, but the design itself. Consequently, the results obtained by the threshold accepting implementation are of interest in quite distinct fields of applications. By abstracting from some a priori knowledge about the problem at hand, finding an optimal design can be interpreted as the task of finding a closest possible discrete approximation of the uniform distribution in some d–dimensional unit cube. For this purpose, the threshold accepting implementation presented in Chapter 11 is in direct competition with several other approaches, including number theoretic methods obtained from algebraic geometry. Despite this very mathematical background of the problem, there exist several applications in more applied research such as the design of experiments in the growing field of experimental economics or response surface analysis for statistics without closed–form solution which have to be simulated for each parameter constellation. Furthermore, quasi–Monte Carlo methods used for simulation are also based on ideas from experimental design. However, typically a much larger number of replications is required, thus making a threshold accepting implementation for this purpose more demanding.

The second application presented in Chapter 12 covers the modelling problem in econometrics. This uses threshold accepting to select a specific model out of a given model class, given some model selection criterion. The method is generally applicable and offers the possibility for extensive data mining. However, in Chapter 12 a more constrained application is analysed, when the model is already derived on theoretical and empirical grounds up to its dynamic specification, which often cannot be derived based solely on economic knowledge and observation. This is the case, e.g. in vector autoregressive models or vector error correction modelling, where theory might imply some hypothesis on long–run relationships but fails to propose an exact scheme for the dynamic adjustment towards this long–run relationship. Unfortunately, empirical evidence and some Monte Carlo studies indicate that the inference on the long–run relationship may depend heavily on the chosen specification for the short–run dynamics. Thus, for an objective testing methodology it seems preferable not to leave it to the discretion of the econometrician to choose the dynamic part in some ad hoc fashion, but to select it on some a priori criterion. This is the contribution of Chapter 12.

Once the model is specified up to the parameters, which have to be estimated, the question of which data to use comes up. In particular, if the model describes the behaviour of variables which may be observed at different levels of aggregation, a choice has to be made. Of course, in general the most disaggregated approach offers the most information. However, often a very disaggregated approach is infeasible, either for lack of data or for lack of degrees of freedom, if a multivariate simultaneous equation model is considered. Except for some very special cases of perfect aggregation, the use of data at a higher aggregation level introduces an additional bias to the estimates which is labelled 'aggregation bias' and, of course, is related to the theoretical problem of aggregation. Again, optimization heuristics such as threshold accepting cannot overcome this crucial problem, but may help to make the best out of the possible choices. In the case of aggregation, this means finding a mode of aggregation which minimizes the additional aggregation bias. Even this pragmatic approach is found to be intractable in its most general form. Therefore, it offers a good bench–mark for testing optimization heuristics like threshold accepting against some ad hoc methods of aggregation that are used, e.g. by official statistical agencies and, in due course, by the econometrician working with the data supplied by these agencies. This is the subject of Chapter 13, which also provides a more formal introduction to complexity issues in its appendix (Section 13.9).

A more specific estimation problem is studied in Chapter 14 which, however, is also related to shortcomings of the available database. The censored quantile regression is used for two reasons. First, censoring in the database makes it necessary to account for censored observations. This problem typically arises in individual data, e.g. income data from social security statistics. Secondly, the mean response of the dependent variable does not contain enough information as, e.g. different quantiles of the income distribution may react differently to exogenous shocks. In fact, the need to observe the reaction at several quantiles of the distribution can be seen as a special kind of dealing with the aggregation problem. However, the use of optimization heuristics comes in at a quite different point for this problem. The censored quantile regression problem can be reduced to a discrete optimization problem over a finite set of potential solutions. Imposing some natural topology on this set, e.g. the topology inherited from the continuous parameter space, i.e. ε–spheres, the objective function exhibits multiple local minima. Given the finite set of potential solutions, the estimation problem can be solved exactly by enumerating all of these candidate solutions. For practical problems, however, the set of candidate points is much too large for the enumeration algorithm to finish in reasonable time (less than a year of computing time, let us say). Therefore, several approximate algorithms have been proposed in the literature to obtain solutions.[1] These serve as a bench–mark for the threshold accepting implementation, which seems well suited for this kind of

---

[1] In fact, some of the earlier contributions to this literature assumed that the proposed algorithms will always find the true solution, or at least a very good approximation, i.e. the presence of several local optima was ignored.

optimization problems.

Finally, in Chapter 15 continuous global optimization problems are analysed as they come up, e.g. in maximum–likelihood estimation. As long as the likelihood function is well behaved, being globally convex for example, and reasonably scaled to avoid very flat regions, standard gradient methods are highly efficient in obtaining good approximations to the global optimum. Of course, due to the limited accuracy of digital computers, in general no exact solution can be given. The situation worsens dramatically if the likelihood function looses some of its good properties, exhibiting, e.g. multiple local optima, discontinuities or very flat or very spiky regions. Then, the results provided by standard algorithms may be far away from an optimal solution and even lead to erroneous conclusions about the estimated economic relationships, as some examples highlight. In such cases, optimization heuristics such as threshold accepting can help to improve results in such situations without, however, being able to guarantee convergence to the global optimum with a finite number of iterations. Unfortunately, so far no threshold accepting implementation for continuous global optimization problems is available. Hence, this chapter summarizes evidence gained for simulated annealing and genetic algorithm implementations for this purpose. Furthermore, it describes tests for detecting the presence of a nonstandard optimization problem and suggests a framework for a future threshold accepting implementation for this purpose.

To sum up the goals of the analysis of different areas of implementation, it is to show that a probably — or even just a possibly — good approximation to the exact or optimal solution of an econometric problem is, in general, preferable to a certainly bad approximation or to approaches which contain a large share of subjective judgment. Furthermore, the implementation details provided for the threshold accepting algorithm for these problem instances provides a guideline for the implementation to similar problems. Finally, the results obtained for the different applications are of interest on their own, as they improve the results obtained by other methods.

11

Experimental Design

11.1 Introduction

The use of experimental design methods is widespread in areas such as social sciences, engineering, medicine, agricultural engineering, pharmaceutical, biology, chemistry etc., but rather novel in economics and statistics. Therefore, this chapter starts with a few general remarks on what experimental design is all about, why it might be useful in economics and statistics, and, finally, why it constitutes an optimization problem suitable for a heuristic approach such as threshold accepting. Before continuing with this introduction, it should be noted that parts of the results presented in Sections 11.3 and 11.4 draw on Winker and Fang (1997; 1998), Fang, Lin, Winker, and Zhang (2000) and Fang, Ma and Winker (2000).

11.1.1 What is 'experimental design'?

Although the term 'experimental design' is often used in a broader context, including all aspects involved in the set–up of experiments,[1] in this chapter only the statistical core of the problem is considered. It applies to situations when one is interested in a mapping

$$f : \mathcal{X} \longrightarrow \mathbb{R}, \qquad (11.1)$$

where $\mathcal{X} \subset \mathbb{R}^d$ is some — potentially discrete — subset of the d–dimensional Euclidean space $\mathbb{R}^d$ and the exact functional form of the mapping f is not known. Typical examples of this situation are statistics, which depend on some nuisance parameters, or nonlinear regression functions. Sometimes, a priori knowledge on the functional class, to which f belongs, is available, e.g. it might be known that f is a polynomial of degree two, but still some parameter or

---

[1] For example, Frigon and Mathews (1997) use such a broader concept in their elementary introduction to experimental design in business and management applications. The textbook by Dean and Voss (1999) also extends the analysis to further aspects of the planning, running and analysing of experiments.

coefficient values are unknown. In order to make experimental design feasible and useful, it should be possible to evaluate f for any given $x \in \mathcal{X}$ or at least to obtain a — possibly noisy — measurement of $f(x)$. Such a measurement or approximation can be generated, e.g. by performing an experiment or running a computer–intensive large–scale Monte Carlo simulation. In both cases, the evaluation for a given x is costly. Therefore, one might be interested in approximating $f(x)$ for some x without evaluating f at x, i.e. using solely the information from evaluating f on a small set $\{x_1, \ldots, x_n\}$ of points. One method for 'interpolating' f for other elements of $\mathcal{X}$ consists in fitting a regression 'line' for the given data $\{(x_i, f(x_i)) \mid i = 1, \ldots, n\}$ and to use the forecast of this regression as an approximation of f for further x. For a given n, the contribution of experimental design consists in selecting those points x_i that provide the maximum amount of information about f. Put in other words, for a given number of function evaluations n, experimental design chooses the x_i in such a way so as to obtain a good approximation to $f(x)$ by means of regression analysis or other interpolation methods.

11.1.2   Where is experimental design used?

The discussion of experimental design and analysis in textbooks is mostly concentrated on applications in sciences and engineering.[2] Further applications can be found in social sciences[3], medicine and pharmaceutical.[4]

An elementary example from chemical engineering is provided by Fang and Wang (1994, pp. 200ff.). These authors consider the yield of a chemical production process which may depend on three variables, namely temperature (A), time (B) and concentration of alkali (C), where A can take the values 80, 85 and 90°C, B, 90, 120, and 150 m, and C, 5, 6 and 7%, respectively. Let $A_1, A_2, \ldots, C_2, C_3$ denote these values. Then, the variables are called *factors* and the possible values A_1, A_2, A_3 are called *levels* of factor A. The aim consists in finding a combination of levels which delivers the highest yield. However, as experiments are costly, the number of experiments should be minimized. Some typical approaches to this problem are as follows:

1. Consider all combinations of levels. For the example, there are $3 \times 3 = 27$ combinations. However, for a larger number of factors and/or levels, this number grows very fast. Consequently, the method is very expensive.
2. Consider only the impact of one factor while holding all other factors constant. This method is based on the assumption that there are no

---

[2] See Garcia-Diaz and Phillips (1995). Such literature often describes principles of experimental design without referring explicitly to the inherent optimization problem.
[3] An introduction to experimental design and analysis for the purposes in sociological and psychological studies is given by Brown and Melamed (1990; 1993).
[4] See Toutenburg (1995) for an introduction to experimental design for applications in the pharmaceutical industry, as well as for clinical research in medicine and dentistry.

interactions, i.e. the influence of the different factors is uncorrelated. Clearly, this assumption reduces the number of experiments, but is not very sensible in the case of economic data.

3. Use subsets of all possible combinations, e.g. a so–called *orthogonal design*, which has the following properties: each level of each factor appears the same number of times and each combination of all levels of any two factors also appears the same number of times.

The last method already indicates a starting point for optimization in the context of experimental design. In Section 11.1.4, this argument will be made more explicit.

Although the example is taken from the engineering literature, and despite the concentration of experimental design applications in some research areas, it can be applied whenever a situation as described in the previous section is given. Often an explicit treatment of experimental design is even indicated in such cases. In particular, in economics and statistics the situation of an unknown functional relationship is typical. Unfortunately, experimental design can only be used if the experimenter can influence the choice of the x_i. Therefore, macroeconometric analysis, in general, is not suitable for this approach, since most data are given exogenously, i.e. the x_i are fixed. In contrast, most applications in experimental economics and simulation–based statistics may be enhanced by using experimental design methods. Experimental design is also used in computing. The tuning exercise for threshold accepting discussed above in Section 8.4 is one example of this.

A typical indicator for a situation when the application of ideas from experimental design might be indicated, is the use of a grid of parameter values. Sometimes, such a parameter grid serves only the purpose of illustration, when f is easy to calculate for all x anyway. However, more often it is used in situations when the quality of an estimator or the outcome of an experiment depends on some parameters of the data–generating process in a highly complex manner, in particular if the number of observations is finite. Then, the impact of these parameters is analysed in a simulation experiment. Since it is not possible to obtain results for any parameter vector, only a few such vectors are analysed. Typically, these are chosen from a rectangular grid: for each of d parameters, a finite number of values T_k, $k = 1, \ldots, d$ is considered. Then, for each combination of these parameters, the experiment is run. If $k = 2$, this corresponds to choosing points on a rectangular grid in the two–dimensional space spanned by the values of the two parameters. In particular, if there exist interaction effects between different parameters, the use of such a rectangular grid for modelling the dependency of the outcomes on specific parameter constellations is, in general, not optimal. In fact, choosing the right points is a typical optimization problem. Consequently, an explicit treatment of this optimization aspect inherent in experimental design could improve the results.

11.1.3   Experimental design in economics and statistics

Experimental methods have gained a lot of interest in economics recently. They allow us to test hypotheses on the decision making and behaviour of individuals, while still controlling other influences. Therefore, experimental economics lends itself to the direct application of design theory and optimum design, since the experiments are conducted in a laboratory. The numerical values of many relevant parameters can be set by the experimenter. Despite this definite advantage compared to econometric studies, which have to rely on data sampled by others, the experimental design set–up — in the statistical meaning defined above — is sometimes treated rather ad hoc instead of being based on an explicit analysis. In fact, experimenters in economics spend a lot of effort in order to control many aspects which may influence the interpretation of the results, but seem less to care about the — in general — multivariate design of parameter settings.[5] To my knowledge, the explicit use of design methods is not yet widespread in experimental economics, in contrast to other experimental sciences, in particular chemistry or engineering. It might be expected that with the further development of this research field, experimental design methods will gain interest. Then, optimized or optimal designs are required. See also Section 11.6.1 for possible applications in experimental economics.

Of course, experimental design is not restricted to experimental settings in economics. If the researcher can influence the sample design of a survey or other data gathering, ideas from optimal design may also be useful.

Econometric and statistic analyses provide a much broader area for applications of optimal design. Here, distributions of estimators and other statistics may depend on a large number of parameters, e.g. number of observations, number of regressors and other parameters, lag order, rank of cointegration space, trend specification and other nuisance parameters. For a relevant fraction of these applications, the estimators or statistics are easily computed for any set of parameter values. Then, there is no scope for an experimental design set–up. However, if each evaluation of such a statistic is complex, e.g. because it requires the numerical evaluation of a higher–dimensional integral, it is necessary to derive a maximum of information from a small number of replications. Such a situation is often met during the development of new estimation and forecasting techniques in econometrics, when parameter values or distributional parameters have to be approximated, which themselves depend on some parameters of the estimation problem. In such a situation, simulations are usually performed over a multivariate grid

---

[5] See, for example, the discussion of experimental design in Kagel and Roth (1995). This is mentioned in most of the contributions to this volume, but only under its more general meaning. The design is not evaluated under the statistical aspects described in this present chapter.

of values for those parameters which are considered to be the most influential ones. Ericsson and Marquez (1998) are among the few to note explicitly that they make use of an experimental design in this case. They use a multivariate grid, i.e. a full factorial design (pp. C238 and C245). Given the results for the points of an experimental design such as a multivariate grid, response surfaces can be estimated in order to approximate the dependency of the outcomes on the parameters of the problem. Obviously, this is a special case of an experimental design problem, since the researcher is free to choose for which parameters an explicit simulation is performed. Again, the quality of the results obtained from such a response surface analysis depends on the experimental design used for its estimation. Some examples are discussed in Section 11.6.2.

Similar applications can be found in economic theory, when effects cannot be derived independently from parameter values of the model. Then, as in the case of estimation problems the model is simulated for different parameter constellations in order to gain some insight into the dependency of the modelled effect on specific parameter values. This literature also relies primarily on multivariate grids as experimental design.[6] Again, an explicit treatment of the design aspect of such problems might improve the quality of the results.

A further potential area of application of methods related to experimental design is the simulation of models or estimators. In this context, a mapping from a distribution to some statistic of interest is approximated by using some drawings from the distribution and calculating the corresponding estimate. If the number of such drawings is limited by computational resources, the quality of the simulation might be improved by using so–called quasi–Monte Carlo methods, which are derived from experimental design methods, instead of pseudo–random number generators. An outline of these methods is provided in Section 11.6.3.

11.1.4 Experimental design as an optimization problem

The description of experimental design as a method for extracting a maximum of information with a small number of evaluations of some function $f(x)$ underlines that, basically, it is an optimization problem. In Section 11.2.1, some measures are introduced, which are used for a judgment on the quality

---

[6] For example, Fitzenberger and Franz (1999) study the employment effects of decentralized versus industry–level bargaining in an insider–outsider setting. Their highly complex model depends among others on five parameters. These authors are interested in parameter situations when the employment effects of decentralized bargaining are more favourable than industry–level bargaining, and vice versa. Therefore, they simulate their model, which cannot be solved analytically, over a small grid of parameter values and find different responses within this small subset of the parameter space.

of specific designs, i.e. point sets $\{\mathbf{x}_1, \ldots, \mathbf{x}_n\}$. Then, for a given set of possible evaluation points $\{\mathbf{x}_i\}$, either discrete or continuous, the optimization problem can be stated explicitly.

In Section 11.2.2, it is shown that ad hoc methods and the standard approaches of experimental design theory do not necessarily solve these problems to optimality. Hence, the use of optimization heuristics seems indicated. Sections 11.3 and 11.4 discuss the implementation of threshold accepting and the results for uniform and mixture level designs. A ruin and recreate variant of threshold accepting is also discussed in Section 11.5. Finally, in Section 11.6 an outlook is given to applications in experimental design, response surface analysis, and quasi–Monte Carlo simulation.

11.2 Problem and Complexity of Optimal Experimental Design

11.2.1 The objective

How can the objective of optimal experimental design be described? In fact, the problem as given in equation (11.1) is ill–presented as it is not possible to draw any conclusion about a general function f from a finite number of observations for given $\mathbf{x}_1, \ldots, \mathbf{x}_n$. Hence, some a priori restrictions are necessary. However, in practice, including the examples provided in Subsections 11.6.1 and 11.6.2, it is, in general, not possible to assume a distinctive class of functions, for example linear functions, from the beginning. Therefore, the process of finding a maximum of information about f is sequential by nature. Hence, in a first step $\mathbf{x}_1, \ldots, \mathbf{x}_n$ have to be chosen given only sparse or zero information about possible forms of f. Then, fitting different models (different f) to the data obtained by the experiments, some reasonable, i.e. data–based, restrictions might be imposed. We will consider such a second step of the analysis before turning back to the general problem at the end of this section. For this purpose the linear regression context is used as an example. Of course, any other regression model could be used as well.

The problem from equation (11.1) can be redefined as follows: Choose $\mathbf{x}_1, \ldots, \mathbf{x}_n \in \mathcal{X}$ such that the coefficient vector β in

$$f(\mathbf{x}) = \beta'\mathbf{x} + \varepsilon \qquad (11.2)$$

can be estimated 'as good as possible' under the usual assumptions about the distribution of the error terms ε. Using matrix notation $X = (\mathbf{x}_1, \ldots, \mathbf{x}_n)'$ and $Y = (f(\mathbf{x}_1), \ldots, f(\mathbf{x}_n))'$, in the classical linear model, the optimal estimator for β is given by $(X'X)^{-1}X'Y$. Its dispersion matrix is $\sigma^2(X'X)^{-1}$, where σ^2 denotes the variance of the error process ε. Obtaining a good estimate is equivalent to an estimator minimizing this dispersion or, alternatively, maximizing the corresponding information matrix $(X'X)$ (Pukelsheim, 1993, pp. 24f.), if the error process is assumed to be homoscedastic.

This optimization problem cannot be solved as stated, as the information matrix $(X'X)$ is not scalar and no complete ordering exists on the set of all such matrices. Hence, some information functions, i.e. functions mapping the set of all non–negative definite matrices to $\mathbb{R}$ have to be used.[7] Furthermore, as pointed out, e.g. in Atkinson and Donev (1992, Ch. 5), not only the variances of the estimated coefficients, but also the variance of the model forecasts in the linear regression model depend upon the experimental design, i.e. the choice of $x_1, \ldots, x_n$. Hence, if one has some freedom in setting the design it is reasonable to look for designs with small values for both variances. Unfortunately, not only in this example, but in general it is not possible to obtain good values for all properties, because the goals might be conflicting, e.g. high power and correct size of a statistical test. Consequently, the search for good or optimal designs has to be based on only one or a few important properties.

Turning back to the aim of maximizing the information matrix, one choice for an information matrix consists in choosing the determinant $\det(X'X)$. A design, which maximizes the information content measured by this functional, is called a D–optimum design (Pukelsheim, 1993, pp. 135ff.). This corresponds to a small confidence region for the estimated parameters.[8] Some other related measures are based on the trace of the inverse, the smallest eigenvalue, or the trace of the information matrix giving rise to the A–, E– and T–criterion, respectively (Pukelsheim, 1993, pp. 135ff.). It is beyond the scope of this present chapter to study the plenitude of criteria applied to experimental design problems. It must suffice to say that each, or at least most of them, are of some importance for specific problem classes, i.e. for specific constraints on the functional form of f and the set $\mathcal{X}$.

However, at least in the early stages of experiments it is often difficult to provide any reasonable a priori restrictions for the functional form of f. Instead, one will start with an extrapolative analysis of the data gathered by the experiments and formulate a more specific model based on these observations. Of course, it is not sensible to use one of the aforementioned criteria in this case. A more suitable approach to be followed, if no a priori information is available, consists of completely randomized designs (Brown and Melamed, 1993, pp. 93ff.). This corresponds to choosing $x_1, \ldots, x_n$ at random in the set $\mathcal{X}$, which for convenience is often normalized to the unit cube $C^d = [0, 1[^d$. However, one is not interested in any random selection, but in a covering of $[0, 1[^d$ which should be 'uniform' in order to extract a maximum of information out of the experimental results.

The notion of uniformity is closely related to discrepancy theory. In fact,

---

[7] See Pukelsheim (1993, pp. 114ff.) for some details and conditions for such information functions. In particular, they need to be positively homogenous, superadditive, non–negative, non–constant, and upper semicontinuous.

[8] Atkinson and Donev (1992, pp. 42ff.) discuss some other related criteria, and, on pp. 106ff., present a general discussion of criteria of optimality.

the discrepancy of a set of points is interpreted as its deviation from the uniform distribution (Beck and Sós, 1995, p. 1407). Unfortunately, there also exists a multitude of discrepancy measures used for different purposes, since uniformly distributed sets and sequences are not only important in the areas of application mentioned in the introduction to this chapter, but play a fundamental role in some other fields of mathematics as well (Beck and Sós, 1995, p. 1407).

The original concept of discrepancy as a measure of uniformity was introduced by Weyl (1916). The discrepancy or — in order to mark the difference from other concepts of discrepancy — star–discrepancy $D(\mathcal{P})$ of a set of points $\mathcal{P} = \{\mathbf{x}_1, \ldots, \mathbf{x}_n\} \in [0, 1[^d$ is defined by

$$D(\mathcal{P}) = \sup_{\mathbf{x} \in [0,1[^d} | F_n(\mathbf{x}) - F(\mathbf{x}) |, \qquad (11.3)$$

where $F_n(\mathbf{x})$ denotes the empirical distribution of $\{\mathbf{x}_1, \ldots, \mathbf{x}_n\}$ and $F(\mathbf{x})$ the uniform distribution on $[0, 1[^d$. Obviously, the star–discrepancy is a measure of the representation of $\mathcal{P}$ with respect to the uniform distribution $F(\mathbf{x})$. Consider the test of the hypothesis H_0: 'the underlying distribution of $\mathcal{P}$ is $F(\mathbf{x})$'. Then, $D(\mathcal{P})$ is nothing else than the well known Kolmogorov–Smirnow statistic for the goodness–of–fit test of $F(\mathbf{x})$ (Fang and Wang, 1994, p. 15).

Besides this coincidence with a statistical concept, the star–discrepancy — like some of the other concepts mentioned in this present chapter — exhibits an important property in the context of multivariate numerical integration. Let f be any function on the d–dimensional unit cube. If f has bounded variation $V(f)$ in the sense of Hardy and Krause (Niederreiter, 1992, p. 19), which is the case, e.g. if all dth derivatives exist and are bounded by some L over the unit cube,[9] Koksma and Hlawka (Niederreiter, 1992, p. 19) established an inequality for the numerical integration of f over the unit cube by using a point set $\mathcal{P}$ (Fang and Wang, 1994, pp. 62ff.). Let $I(f)$ denote the integral of f over the unit cube. Then, the following inequality holds:

$$\left| I(f) - \frac{1}{n} \sum_{i=1}^{n} f(\mathbf{x}_i) \right| \leq V(f)D(\mathcal{P}). \qquad (11.4)$$

This Koksma–Hlawka inequality states that the quality of a numerical approximation of the integral of f depends crucially on the discrepancy of the set of points used for the approximation. The more uniformly distributed the point set $\mathcal{P}$, then the better is the approximation quality.

Although providing an intuitive, and in other contexts well known concept, the star–discrepancy exhibits at least two drawbacks for its general application in experimental design. First, it is hard to compute itself, and secondly, it is not invariant to reasonable transformations of the parameter space. The argument

---

[9] In this case, $V(f)$ is bounded by $2^d L$.

of the high complexity of calculating the star–discrepancy for a given set of point is made by L'Ecuyer and Hellekalek (1998, p. 230) who state that 'no efficient algorithm is available for computing' the star–discrepancy for large n and moderate dimension d.[10] However, an implementation of the threshold accepting heuristic proposed by Winker and Fang (1997) provides good lower bounds for the star–discrepancy even if $d > 10$ and $n > 1000$.[11] Thus, the heuristic optimization approach put forward in this book might also help to tackle the first problem with the star–discrepancy.

An alternative to approximations of the star–discrepancy is given by measures of discrepancy based on different metrics in equation (11.3). Replacing the supremum norm by the usual Euclidean norm leads to the L_2–discrepancy D_2 (Fang and Wang, 1994, p.34), which is much easier to calculate. Warnock (1972) provides an analytical formula for calculating the L_2–discrepancy, i.e.

$$(D_2(\mathcal{P}))^2 = 3^{-d} - \frac{2^{1-d}}{n} \sum_{i=1}^{n} \prod_{j=1}^{d} (1 - x_{ij}^2) \tag{11.5}$$

$$+ \frac{1}{n^2} \sum_{i=1}^{n} \sum_{j=1}^{n} \prod_{k=1}^{d} [1 - \max(x_{ik}, x_{jk})],$$

where $\mathbf{x}_i = (x_{i1}, \ldots, x_{id})$.[12] Unfortunately, the L_2–discrepancy does not satisfy the Koksma–Hlawka inequality[13] and exhibits some further disadvantages, as pointed out by Hickernell (1998). In particular, the projection of designs on lower–dimensional subspaces may be of poor quality, and shares the drawback of the star–discrepancy of not being invariant to natural transformations of the parameter space. In order to overcome these disadvantages, Hickernell (1998) proposed three new measures of uniformity which are related to the Euclidean norm, namely the symmetric L_2–discrepancy (SL_2), the centred L_2–discrepancy (CL_2) and the modified L_2–discrepancy (ML_2). All of these three discrepancies satisfy a Koksma–

---

[10] The complexity of the algorithm proposed by Bundschuh and Zhu (1993) grows exponentially in the dimension d. For example, in order to calculate the discrepancy of a U–type design with $d = 5$ and $n = 122$, it takes 1,226.64 seconds of CPU–time on an IBM RS 6000/3AT workstation (SPECfp95: 7.28).

[11] Beck and Chen. (1987) also provide some upper and lower bounds for different measures of discrepancy, including the star–discrepancy. Their results are based on mathematical theory, but leave many unsolved problems open, e.g. to provide a proof for the hypothesis that the star–discrepancy of n points in the d–dimensional unit cube is larger than $C(d)(\log n)^{d-1}$ for some constant $C(d)$ depending only on the dimension d. Only the case where $d = 2$ has been solved (p. 283).

[12] Heinrich (1995) proposes an even more efficient method for the calculation of the L_2–discrepancy.

[13] A similar error bound results only for differentiable functions, if the supremum of the absolute values of the first derivatives exists.

Hlawka type inequality, i.e. each discrepancy has its corresponding measure of variation replacing $V(f)$ in equation (11.4). Recent evidence in Fang, Ma and Winker (2000) indicates that, in particular, considering the centred L_2–discrepancy might result in designs with required features such as orthogonality. Therefore, besides the star– and L_2–discrepancy, the centred L_2–discrepancy (CL_2) is used as an objective for optimizing designs in this chapter. A more in–depth analysis of the different concepts can be found in Hickernell (1998) and Fang, Lin, Winker, and Zhang (2000).

For the centred L_2–discrepancy, Hickernell (1998) provides an analytical expression similar to equation (11.5) as follows:

$$(CL_2(\mathcal{P}))^2 = \left(\frac{13}{12}\right)^d - \frac{2}{n}\sum_{i=1}^{n}\prod_{j=1}^{d}\left(1+\frac{1}{2}|x_{ij}-0.5|-\frac{1}{2}|x_{ij}-0.5|^2\right) \quad (11.6)$$

$$+ \frac{1}{n^2}\sum_{i=1}^{n}\sum_{j=1}^{n}\prod_{k=1}^{d}\left[1+\frac{1}{2}|x_{ik}-0.5|+\frac{1}{2}|x_{jk}-0.5|-\frac{1}{2}|x_{ik}-x_{jk}|\right],$$

which is also easy to evaluate, at least as long as n and d are of moderate size. While both for the star–discrepancy and the L_2–discrepancy, the origin plays a specific role, the centred L_2–discrepancy puts all vertices of the unit cube into the same situation. The modification applied to equation (11.5) in order to obtain equation (11.6) makes the measure of discrepancy invariant under reflections of $\mathcal{P}$ about any hyperplane defined by $x_j = 0.5$.

11.2.2 Solution approaches

Given one of the objective functions introduced in the previous section, the optimization problem of experimental design can be stated explicitly. For given n, it consists in finding n points in the unit cube which approximate the uniform distribution as close as possible. Thereby, the approximation quality is measured by one of the discrepancy measures introduced in Section 11.2.1.

The problem is easy to solve in dimension one. The best approximation of the univariate uniform distribution on the unit interval $[0, 1[$ is given by a set of equidistant points (Fang and Wang, 1994, pp. 16f.). Unfortunately, this result does not allow for a straightforward generalization in higher dimension. In fact, a natural generalization are equidistant grids. However, these often used designs, in general, are far from being optimal in higher dimension (Fang and Wang, 1994, p. 19). In the experimental design literature this approach is known as *full factorial designs*,[14] i.e. for all combinations of all levels of all factors, at least one experiment is performed. Obviously, as the number of levels and factors grows large, this approach becomes infeasible, and consequently, only a subset of experiments can be performed. This subset

---

[14] See Frigon and Mathews (1997, pp. 137ff.) and Atkinson and Donev (1992, pp. 62ff.).

should be chosen such that the information content of the experiments is maximized. This leads to so–called fractional factorial designs (Frigon and Mathews, 1997, p. 163). Closely related is the analysis of orthogonal arrays.[15]

The problem of optimal design is easily set up in higher dimensions. Let $\mathcal{H}_n^d$ denote the space of all subsets of size n of the d–dimensional unit cube. Then, the optimal design problem for a given discrepancy concept D is given by

$$\min_{\mathcal{P} \in \mathcal{H}_n^d} D(\mathcal{P}) . \qquad (11.7)$$

Although looking quite innocent as do most optimization problems in their formal representation, this optimization problem seems to be beyond any standard algorithm. While for very simple problem instances — like the one dimensional case — an analytical solution is possible, its complexity requires the use of numerical methods in most cases. Atkinson and Donev (1992, p. 166) point out that even for small problem instances the direct search over $\mathcal{H}_n^d$ is not practicable. Fang, Ma and Winker (2000) obtain solutions for the optimal design problem by using CL_2 as the discrepancy measure in dimension two for up to five points with a simplex algorithm. However, for $n > 5$ the method breaks down.

Two different approaches have been followed to cope with this highly complex problem. The first approach leaves the domain of deterministic modelling and replaces the set $\mathcal{P}$ by some random sample. This gives rise to the so–called Monte Carlo methods in numerical integration. Unfortunately, the expected error bounds of Monte Carlo methods, in general, are large compared to those which can be obtained by deterministic approaches (Fang and Wang, 1994, p. 19).[16] Therefore, the second approach keeps the explicit optimization framework of equation (11.7), but imposes some a priori restrictions on the search space $\mathcal{H}_n^d$ in order to obtain an integer optimization problem.

While in many practical design problems the design region is, at least in some dimensions, continuous, the efficiency loss in using a discrete approximation is small when compared to the improvement in the calculation of (near) optimum designs. In general, if n points are to be selected for an experimental design, a maximum of n different values is admitted for each factor. Scaling all factors to the unit interval, this corresponds to restricting $\mathcal{H}_n^d$ to the set of point sets with n elements, where each element is placed on the equidistant grid with n segments in each dimension, as shown in the

---

[15] See Bose and Manvel (1984, Ch. 7), Frigon and Mathews (1997, p. 140f.), Dean and Voss (1999, p. 506f.), and the comprehensive book by Hedayat, Sloane and Stufken (1999). See also Section 11.4.4.

[16] An explicit comparison of the expected CL_2 for random designs and designs obtained by number theoretic methods is provided in Fang, Ma and Winker (2000). See also the discussion in Section 11.6.3.

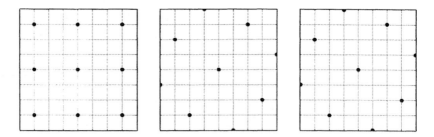

Figure 11.1   Designs for nine points in dimension two.

left and middle panels of Figure 11.1. The resulting designs are called *U–type designs*. A formal definition is provided in Section 11.3. While the equidistant grid points in the left panel are set ad hoc, the point sets in the other two panels are the result of explicit optimization.

For the points in the middle panel, the search for an optimum design was restricted to the grid points. Then, optimization is equivalent to the selection of n points out of a finite, but in general huge, list of candidate points $\mathcal{X}$. The resulting integer optimization problem belongs to the class of highly complex problems which are intractable by standard algorithms (Pukelsheim, 1993, p. 26). Nevertheless, several algorithms have been proposed for generating good designs, e.g. by Atkinson and Donev (1992, pp. 170ff.) or Trinca and Gilmour (2000, pp. 32ff.). These algorithms consist of sequential building schemes, exchange steps and final adjustment steps in different combinations. However, none of these algorithms can be guaranteed to provide a good approximation to the globally optimum design. The reason is that, in general, the algorithms are attempting to find the maximum of a surface with many local extremes. 'Many of the algorithms cannot be guaranteed to find anything more than a local optimum' (Atkinson and Donev, 1992, p. 168). Li and Fang (1995) propose an algorithm for finding uniform designs on a grid in the two–dimensional case. Although their approach might be generalized to multivariate settings, it is expected that the time complexity of this algorithm grows exponentially in the dimension. In fact, due to its high complexity, Li and Fang (1995) present results only for $n \leq 23$. Nevertheless, these exact results can be used as a bench–mark for other algorithms, including the threshold accepting heuristic. The points in the middle panel are obtained by the threshold accepting implementation described in this chapter. The result is identical to the one obtained by the method described in Li and Fang (1995). After having obtained an optimal or optimized U–type design, further improvements might be strived for by relaxing the restrictions of such

Initialization	Choose the first column
Step 1	Find the second column with four 1s, 2s, 3s and 4s such that the design formed by these two columns has the minimum discrepancy value over all two–dimensional designs of this type. Let $t = 2$
Step 2	For the fixed first t columns find the specific $(t + 1)$ column with four 1s, 2s, 3s and 4s such that the design formed by all $t + 1$ columns has the minimum discrepancy value over all $t + 1$ dimensional designs under the constraint that the first t columns are fixed. Let $t \equiv t + 1$
Step 3	If $t = 6$, the procedure is finished and the design obtained in the previous step is the final solution; otherwise go back to **Step 2**

Figure 11.2 Pseudo code for forward procedure.

designs. Donev and Atkinson (1988) have proposed carrying out some final adjustment. Starting from a design on the grid, small changes of its elements are performed in order to improve the value of the objective function. In Fang, Ma and Winker (2000), an explicit optimization step is added after obtaining an optimal U–type design. The results are shown in the right panel of Figure 11.1. The differences to the optimal U–type design, however, are quite small. Therefore, this idea is not implemented in the threshold accepting implementation of this chapter.

Fang, Lin, Winker, and Zhang (2000) propose an alternative construction heuristic — the so–called 'forward procedure' — for generating low–discrepancy designs, which are used as a bench–mark for their threshold accepting implementation.[17] For ease of illustration, the functioning of the algorithm is described only for the problem of finding a low–discrepancy design with 16 points in dimension five, where each factor has four levels. The pseudo code for this example is provided in Figure 11.2. Without loss of generality, the first column can be chosen to be $(1\,1\,1\,1\,2\,2\,2\,2\,3\,3\,3\,3\,4\,4\,4\,4)'$.

Along this description for a specific problem instance, the algorithm is easily extended to the general case. The algorithm provides a local–minimum solution for a specific neighbourhood concept linked to adding

---

[17] The idea is similar to the one used by Atkinson and Donev (1992, pp. 171ff.) for obtaining D–optimum continuous designs.

individual columns. In applications, the algorithm turned out to provide good approximations to uniform designs for many problem instances. However, although being constructive and generating optimal outcomes in each step – given the results of previous steps – the algorithm does not guarantee us ending with an optimal design.

Besides the above mentioned sequential construction algorithms, two further approaches exist for generating good U–type designs. The first relies on results from mathematical number theory and algebraic geometry and is heavily used in the context of quasi–Monte Carlo methods. In particular, if a high number of points can be used in an experiment or numerical integration, these methods are very efficient. However, for smaller instances in dimension up to $d = 10$ and $n \leq 100$, they are not optimal, since the implementation of threshold accepting provides U–type designs of lower discrepancy. The second approach consists in using optimization heuristics either on the optimum design problem in its original version or for finding U–type designs with low discrepancy. Early implementations of simulated annealing to the optimum design problem are reported by Bochachevsky, Johnson and Stein (1986) and Haines (1987). An implementation of threshold accepting to the optimum U–type design problem is presented in the next section, while the results are reported in Section 11.4.

11.3 Threshold Accepting Implementation

11.3.1 The optimization problem

The threshold accepting implementation presented in this chapter applies to the problem of finding low–discrepancy point sets on a given grid as shown in Figure 11.1. We start by providing a formal definition of U–type designs (Fang and Wang, 1994, Ch. 5), which correspond to designs on a given grid:

Definition 11.3.1 *A* **U–type design** $U_{n,d;q_1,\ldots,q_d}$ *is an $n \times d$ matrix $U = (u_{ij})$ with entries $1, 2, \ldots, q_j$ in the jth column and satisfying the condition that each entry in each column occurs the same number of times.*

The induced matrix $X = (x_{ij})$ of U is defined by

$$x_{ij} = (u_{ij} - 0.5)/q_j, \quad i = 1, \ldots, n, \; j = 1, \ldots, d.$$

There is a one–to–one correspondence between a U–type design $U_{n,d;q_1,\ldots,q_d}$ and its induced matrix X so that they can be used interchangeably. In the preceding section, only the case $q_j = n$ for all $j = 1, \ldots, d$ has been discussed. If some q_j are real divisors of n, this corresponds to experiments when some factors exhibit a smaller number of possible levels than others. In an econometric context, the value of a dummy variable may appear as a possible factor. Then, this factor allows only for two levels, while some continuous autoregressive parameter is approximated in the experimental design by a larger number of possible levels. Given the definition of a U–type design and a measure of discrepancy D as introduced in Section 11.2.1, an optimum design can be defined.

Definition 11.3.2 *A U–type design $U_{n,d;q_1,...,q_d}$ is called a* **(U–)uniform design** *under D, where D is a measure of uniformity over the unit cube $C^d = [0,1[^d$, if its induced matrix U has the minimum D–value over all possible U–type designs $U_{n,d;q_1,...,q_d}$. Such a uniform design is denoted as $U_n(q_1 \times \ldots \times q_d)$.*

Obviously, $r_j = n/q_j$ should be a positive integer. When some q_j are the same, the notation $U_n(q_1^{t_1} \times \ldots \times q_m^{t_m})$ is used, where $t_1 + \ldots + t_m = d$. When $q_j = n$ for all j, a corresponding uniform design is denoted as $U_n(n^d)$.

Now, the optimization problem of finding uniform designs under some measure of discrepancy D can be formally defined. Let $\mathcal{U} = \mathcal{U}(n, d; q_1, \ldots, q_d)$ denote the set of all U–type designs for the given parameter values $(n, d; q_1, \ldots, q_d)$. Then, the optimization problem to be solved is given by

$$\min_{U \in \mathcal{U}} D(U). \tag{11.8}$$

As the discrepancy of a set of points does not depend on the labelling of the points, without any loss of generality it can be assumed that the first column of any U–type design is given by $(1, 2, \ldots, n)'$. However, for the second column there are already $n! - 1$ possible choices, $n! - 2$ choices for the third column, and so on. Thus, if d and n are large, the number of U–type designs, i.e. the search space $\mathcal{U}$, is very large, and a simple enumeration algorithm is infeasible. As imposing the restriction on the first column does not seem to improve the performance of the threshold accepting implementation, it is not used in the present implementation.

A further complication for an implementation of threshold accepting to the optimum design problem is given by the computation of discrepancy measures. As pointed out in Winker and Fang (1997), the computation of the star–discrepancy for a single design is already a highly complex task and might require the use of optimization heuristics as soon as d and n grow beyond some moderate values. Therefore, when using this discrepancy measure either only small designs can be obtained, whereby the star–discrepancy can be calculated exactly by using the algorithm provided by Bundschuh and Zhu (1993), or some approximation of the star–discrepancy, as proposed by Winker and Fang (1997), has to be used. In this chapter, only the results for small instances are provided for the star–discrepancy by using the algorithm of Bundschuh and Zhu (1993).[18]

---

[18] The presented results are based on Winker and Fang (1998), who use the C–code provided by V. Chin and J.X. Pan for the exact calculation of the star–discrepancy.

As pointed out in Section 11.2.1, the star–discrepancy has its merits and shortcomings in experimental design. In particular, using different discrepancy measures increases the possibilities of differentiating between designs, i.e. for some problem instances there exist many different designs exhibiting the same star–discrepancy, but different values for other discrepancy measures. Therefore, Section 11.4 also presents results based on other discrepancy measures, in particular the already introduced centred L_2–discrepancy.

11.3.2 Local structure

Using the one–to–one correspondence between U–type designs and their induced matrix, the search space $\mathcal{U}$ can be identified with a finite subset of the d–dimensional unit cube C^d. Therefore, following the guideline provided in Section 9.3, the projections of some ε–spheres with regard to the Euclidian metric in $\mathbb{R}^d$ on $\mathcal{U}$ would be a natural choice for defining local neighbourhoods. Any two elements of $\mathcal{U}$ are considered as being neighbours, if and only if their Euclidian distance is smaller than ε. Unfortunately, this definition of neighbourhoods is not very operational for the optimal design problem. As pointed out by Winker and Fang (1997), at least two problems arise. First, for large d, one would have to choose ε to be very large (> 0.5) to make sure that almost every ε–neighbourhood contains at least two elements of $\mathcal{U}$.[19] The second problem relates to the computational cost of obtaining elements in such an ε–neighbourhood. In principle, given some $U \in \mathcal{U}$, for any $\tilde{U} \in \mathcal{U}$ it has to be checked whether $\|U - \tilde{U}\| < \varepsilon$. As the cardinality of $\mathcal{U}$ is large, the concept becomes infeasible from a computational point of view.

As the problem of finding U–uniform designs shares some similarities in its local structure with the problem of calculating or approximating, respectively, the star–discrepancy, it is not too surprising that the neighbourhood concept introduced by Winker and Fang (1997) can also be used for the uniform design problem. For a given U–type design $U \in \mathcal{U}$, the neighbourhood $\mathcal{N}(U)$ is defined by a generalization of the Hamming distance concept. The design $\tilde{U}$ is in $\mathcal{N}(U)$ if it fulfils the following conditions. The design matrices may differ in, at most, $k \leq d$ columns. Within a given column, the difference is restricted to the exchange of two elements. Let the first of these two elements be given. Then, the second element is restricted to be one of the $l \leq n$ next neighbours of the first one with regard to the natural ordering of all elements in this column. The whole neighbourhood concept can be formally defined as a maximum order norm (Winker and Fang, 1997). Here, the example of a typical exchange step induced by this concept might be sufficient. Figure 11.3 shows an example for a U–type design $U \in \mathcal{U}(5, 3; 5^3)$. The first matrix shows the design with a selected column (box) and element (**2**). The second matrix shows

---

[19] See Figure 5.1 above for a graphical presentation of this problem.

$$
\begin{pmatrix} 1 & 2 & 3 \\ 2 & 1 & 5 \\ 3 & 5 & 1 \\ 4 & 3 & 2 \\ 5 & 4 & 4 \end{pmatrix}
\rightarrow
\begin{pmatrix} 1 & 2 & 3 \\ 2 & 1 & 5 \\ 3 & 5 & 1 \\ 4 & 3 & 2 \\ 5 & 4 & 4 \end{pmatrix}
\rightarrow
\begin{pmatrix} 1 & 2 & 3 \\ 2 & 1 & 5 \\ 3 & 5 & 1 \\ 4 & 3 & 2 \\ 5 & 4 & 4 \end{pmatrix}
\rightarrow
\begin{pmatrix} 1 & 3 & 3 \\ 2 & 1 & 5 \\ 3 & 5 & 1 \\ 4 & 2 & 2 \\ 5 & 4 & 4 \end{pmatrix}
$$

Figure 11.3 An exchange step for the U–uniform design problem.

the two next neighbours of this element according to the natural ordering, i.e. **1** and **3**. Finally, one of these neighbours, **3** in the example, is exchanged with the original element and the design provided by the far right matrix in Figure 11.3 results as an element of $\mathcal{N}(U)$.

Two advantages of this neighbourhood concept are obvious. As the underlying maximum order norm results in the proceeding described in Figure 11.3, an element of $\mathcal{N}(U)$ can be generated at low computational costs. In fact, for generating such an element only the orderings of the entries in each column must be available. It is possible to store these orderings by using memory, growing only at a rate $n \times d$ in the dimensions of the problem. In each iteration step, the orderings are updated at linear costs. Furthermore, the size of the neighbourhoods depends solely on the parameters k and l. Consequently, each neighbourhood has the same size, which can be made small compared to the projected ε–spheres discussed above.

Given the practical advantages of the proposed neighbourhood concept, a simulation similar to the one proposed in Section 9.3 is used to find out whether some local structure of the objective function D is introduced by using this neighbourhood concept. To this end, a large number I (5000 for the following figures) of elements U in $\mathcal{U}$ is randomly generated. For each element U, another element $\tilde{U}$ out of its predefined neighbourhood $\mathcal{N}(U)$ is also randomly chosen. For all resulting pairs of U–type designs $(U, \tilde{U})$ the difference of their discrepancy D is calculated. Then, a frequency plot of the resulting 'local deviations' provides some evidence on the local structure of the discrete optimization problem. Figure 11.4 shows such plots for $\mathcal{U}(20, 5; 20^5)$ and the CL_2 discrepancy.

The figure shows histograms resulting from the trivial neighbourhood concept (all designs are neighbours of all other designs) and for the constructive neighbourhood concept introduced above for $k = 4$, $l = 10$ and $k = 2$, $l = 10$, respectively. All plots show the histograms for the same part of the x–axis. However, while for the really local neighbourhood concepts almost all local deviations fall into the range between -0.2 and 0.2, the local deviations resulting from the trivial topology range over a much larger interval. Both local neighbourhood structures result in a concentration of local deviations around zero. For $k = 2$ and $l = 10$, the chance of

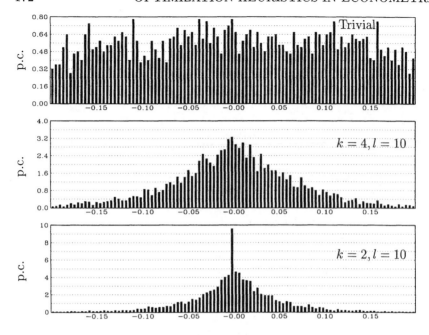

Figure 11.4   Histograms of local deviations (CL_2).

performing exchanges which result either in an identical design or in an isomorphic design with equal discrepancy is quite high. Consequently, some mass of the distribution is concentrated at zero. The consequences of this effect are discussed below in the context of the threshold sequences derived from the same simulation experiment. Here, it may suffice to say that a necessary requirement for a good performance of any threshold accepting implementation is a neighbourhood structure resulting in local deviation plots similar to the middle and lower panel of Figure 11.4.

By using the data generated for the local deviation plots, threshold sequences can be obtained as proposed in Section 9.4. The rationale for this approach is that the calculated local deviations correspond to thresholds which have to be crossed if the algorithm performs some exchange step from a 'better' design to some less uniform one. In order to obtain a threshold sequence, the absolute values of the simulated local deviations are sorted in decreasing order. The upper panel of Figure 11.5 shows the resulting sequences.

This plot indicates that at the very beginning of all three sequences a few very large values are included. The results of the tuning experiments undertaken in Chapter 8 indicate that in such a situation the performance of the algorithm can be improved by using only a lower fraction α of this sequence. The vertical line in the plot corresponds to $\alpha = 0.8$. For the results presented in Section 11.4, values for α ranging from 0.7 to 0.9 have been

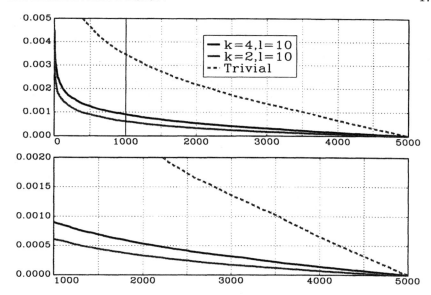

Figure 11.5 Threshold sequences from local deviation simulations.

used. The lower panel of Figure 11.5 concentrates on this lower quantile of the generated sequence. In particular, for the two sequences resulting from the local neighbourhood definitions ($k = 4$, $l = 10$ and $k = 2$, $l = 10$), the steps between two consecutive thresholds T of this sequence become smaller as the sequence approaches zero. In due course, performing a fixed number of iterations for each threshold value implies that the total number of iterations performed for a given interval of threshold values increases as the thresholds become smaller. This feature of a threshold sequence seems to increase the efficiency of an implementation.

A further argument in favour of using a threshold sequence constructed from the data as described above is provided by Li and Fang (1995). These authors observe that there is a lower border to the minimal difference between the discrepancy for two U–type designs in the two–dimensional case. The endogenous generation of threshold values automatically takes into account certain features of the problem such as the fact that it will never use a threshold value smaller than the smallest step to be expected, with the sole exception of the value zero at the very end of the iterations.

A last detail of the implementation is concerned with reducing the number of necessary evaluations of the star–discrepancy for a U–type design.[20] To this end, the observation is used that the discrepancy function achieves its maximum on a finite subset of $[0, 1[^d$. Now, when moving from a current U–

---

[20] This detail is not used for other measures of discrepancy, since their calculation imposes a much smaller computational load.

Initialization	Choose randomly any U–type design $U^c \in \mathcal{U}$ and set the threshold parameter T to its start value T_0
Step 1	Choose randomly any U–type design U^n close to U^c, i.e. $U^n \in \mathcal{N}(U^c)$ and calculate $\Delta D = D(U^n) - D(U^c)$
Step 2	If $\Delta D \leq T$, set $U^c = U^n$; otherwise leave U^c unchanged
Step 3	Repeat **Step 1** and **Step 2** a fixed number of times
Step 4	If the given threshold sequence is not yet exhausted, take the next threshold value and repeat **Step 1** to **Step 3**

Figure 11.6   Threshold accepting pseudo code for uniform design problem.

type design U^c to a new U^n the maximum of the discrepancy function for U^n might be achieved close to the old one. Hence, before applying the algorithm of Bundschuh and Zhu (1993), it is tested whether the discrepancy function for the new U–type design grows by more than the current threshold near the old maximum. In this case, the new solution is not accepted anyway, and a superfluous evaluation can be avoided.

Given the local neighbourhood structure introduced above and some measure of discrepancy D, the threshold accepting implementation for optimizing U–type designs can be summarized by the pseudo code provided in Figure 11.6.

The algorithm provides the design with the best value of the discrepancy measure D obtained during all iterations. Given the convergence results from Chapter 7 this design eventually, i.e. with the number of iterations going to infinity, becomes a globally optimal design provided that the neighbourhood structure and threshold sequence are well chosen. The performance for a finite number of iterations also depends on the choice of a specific local structure (k and l), the predefined sequence of threshold values T_i or α, if the threshold sequence is obtained from the empirical jump distribution as described above, and the total number of iterations.

11.4   Results for Uniform and Mixture Level Designs

This present section summarizes some results of the threshold accepting implementation on U–type designs. First, in Section 11.4.1, the results for the uniform design problem using the star–discrepancy are reported. Due to the high computational costs involved in calculating this discrepancy measure, the application is restricted to rather small instances. Therefore, in Section 11.4.2

alternative measures of discrepancy are considered. The results indicate that designs obtained by minimizing some of these measures also exhibit a small star–discrepancy. In Section 11.4.3, the analysis is extended to cover mixture level designs, i.e. maximizing over $\mathcal{U} = \mathcal{U}(n, d; q_1, \ldots, q_d)$, where the q_i might be different and, in particular, do not have to be equal to n, as for the designs considered in the first two sections. Finally, Section 11.4.4 covers the relationship between low–discrepancy designs and orthogonal designs.

11.4.1 Uniform designs for the star–discrepancy

Given the high computational burden of calculating the star–discrepancy and the high number of iterations a threshold accepting implementation requires in order to achieve high–quality results, the implementation of threshold accepting for uniform designs under the star–discrepancy is run only for small instances ($d = 2, 3$ and $n \leq 44$, $d = 4, 5$ and $n \leq 36$ and $d = 6, 7$ and $n \leq 20$). For this range of values, the star–discrepancy can be calculated in reasonable time by using the algorithm proposed by Bundschuh and Zhu (1993). Furthermore, the preliminary testing of the new candidate solutions discussed before reduces the number of complete evaluations of the star–discrepancy considerably.

For each pair (d, n), 25 different values of the parameter α for the endogenous generation of the threshold sequence are used, and for each value of α, 10 different seeds for the random number generator were selected. Consequently, a total number of 250 runs is performed for each combination of d and n. Due to the high computational cost associated with evaluating the star–discrepancy, only 10 000 iterations are used for each run. A comparison with results for 100 000 iterations for some of the instances documented in Winker and Fang (1998) indicates no improvements for $d = 2$ and only minor improvements for some larger instances in higher dimensions when increasing the number of iterations. This finding strongly indicates the high quality of the generated designs. Tables 11.1–11.3 summarize the results obtained. The columns with the heading 'best' provide the discrepancy of the best design found for all 250 runs,[21] while the columns headed 'mean' report the mean over all runs.

Table 11.1 provides the results for designs in dimension two and three. For $d = 2$ and $n \leq 23$, the results can be compared with the exact results obtained by Li and Fang (1995) using a refined enumeration method. The best designs found by the threshold accepting implementation are equal to these globally optimum designs for all $n \leq 23$.[22] Although the mean is higher than the

---

[21] For $d = 7$ and $n \geq 18$, only 45 different runs are considered due to limited computer resources.

[22] For larger n, when the threshold accepting implementation may fail to provide the global optimum, the design obtained by threshold accepting could be used as input to

Table 11.1   Discrepancy of optimized U–type designs ($d = 2, 3$)

	$d = 2$		$d = 3$			$d = 2$		$d = 3$	
n	Best	Mean	Best	Mean	n	Best	Mean	Best	Mean
5	.250 00	.250 08	.329 00	.335 89	25	.062 80	.068 37	.090 00	.096 86
6	.201 39	.234 50	.296 88	.303 88	26	.061 76	.064 75	.089 62	.093 51
7	.178 57	.207 88	.248 91	.269 54	27	.058 64	.063 08	.086 25	.091 26
8	.160 16	.198 84	.229 25	.250 60	28	.055 80	.061 76	.084 28	.088 95
9	.145 06	.146 77	.211 42	.212 64	29	.054 99	.060 20	.081 79	.086 59
10	.137 50	.139 18	.186 63	.195 78	30	.053 06	.057 54	.079 96	.084 61
11	.130 17	.132 46	.177 40	.182 95	31	.051 77	.056 17	.077 78	.082 56
12	.119 79	.125 08	.163 85	.169 67	32	.050 54	.059 25	.075 47	.080 76
13	.113 91	.117 05	.153 79	.161 25	33	.049 36	.054 14	.073 83	.078 22
14	.105 87	.107 63	.145 73	.150 64	34	.047 79	.053 67	.072 50	.077 15
15	.098 89	.102 87	.136 04	.143 76	35	.047 14	.050 44	.070 69	.075 16
16	.094 73	.099 74	.128 69	.135 70	36	.046 10	.050 64	.069 39	.073 94
17	.087 37	.090 96	.124 54	.129 66	37	.045 11	.048 60	.067 12	.072 25
18	.085 65	.088 12	.118 85	.124 16	38	.043 11	.048 04	.067 17	.070 84
19	.081 02	.083 91	.111 15	.118 70	39	.042 90	.047 20	.065 60	.069 57
20	.075 63	.079 35	.109 77	.114 65	40	.042 34	.045 19	.063 23	.067 91
21	.075 40	.077 26	.104 59	.110 07	41	.040 90	.044 58	.063 21	.067 32
22	.070 76	.074 32	.101 93	.106 26	42	.039 54	.045 09	.061 54	.065 89
23	.067 58	.072 78	.095 97	.102 61	43	.039 35	.043 44	.060 86	.064 42
24	.066 41	.069 16	.093 99	.099 59	44	.038 61	.044 01	.058 54	.063 47

optimum for all instances, indicating that threshold accepting fails to obtain the global optimum for some values of the tuning parameters or initializations, it has a similar order of magnitude.

So far, no algorithm is known which solves the uniform design problem for dimensions larger than two in reasonable time. Therefore, the performance of the threshold accepting implementation for $d \geq 3$ can be compared only with the — in terms of star–discrepancy — best designs obtained by number theoretic methods, which do not necessarily result in optimal designs. Good lattice point (GLP) sets, incomplete Latin squares and U–type designs based on orthogonal designs are among these methods. Fang and Hickernell (1995) provide results for $d \leq 5$ and $n \leq 30$, which are used as a bench–mark for the designs obtained by threshold accepting. Again, a comparison indicates that the designs obtained by threshold accepting are considerably better than the best designs obtained by those number theoretic methods. The gains in terms of the star–discrepancy range between a few and more than 20% for only 10 000 iterations of the threshold accepting heuristic (Winker and Fang, 1998). Such a gain is non–trivial, e.g. the best design found by the threshold accepting

---

the algorithm of Li and Fang (1995) and drastically increase its performance.

Table 11.2 Discrepancy of optimized U–type designs $(d = 4, 5)$

	$d = 4$		$d = 5$			$d = 4$		$d = 5$	
n	Best	Mean	Best	Mean	n	Best	Mean	Best	Mean
5	.403 10	.403 10	.442 79	.442 79	21	.135 76	.142 42	.163 98	.172 06
6	.360 68	.360 68	.400 07	.400 08	22	.131 24	.137 54	.160 90	.166 46
7	.314 89	.322 00	.362 86	.363 49	23	.129 21	.133 79	.156 43	.162 91
8	.276 60	.291 11	.331 05	.332 77	24	.123 27	.130 00	.150 79	.157 91
9	.257 57	.258 09	.303 87	.304 11	25	.121 22	.126 28	.147 56	.154 09
10	.243 03	.243 99	.273 78	.280 74	26	.116 95	.122 75	.144 31	.149 92
11	.217 53	.225 57	.260 38	.262 85	27	.114 85	.120 14	.140 73	.146 59
12	.203 15	.210 49	.242 81	.250 48	28	.112 38	.117 29	.137 83	.142 93
13	.193 66	.199 57	.227 40	.237 04	29	.109 28	.114 56	.134 85	.140 46
14	.181 97	.188 41	.217 29	.225 45	30	.105 60	.111 22	.131 79	.137 34
15	.172 50	.178 06	.209 43	.214 54	31	.105 13	.108 87	.127 23	.134 45
16	.164 45	.171 22	.197 65	.206 58	32	.101 19	.106 57	.125 54	.132 35
17	.159 66	.165 03	.191 85	.199 02	33	.098 48	.104 12	.123 60	.128 84
18	.150 53	.158 10	.182 40	.190 69	34	.098 52	.102 91	.119 38	.126 77
19	.146 03	.152 34	.174 53	.184 31	35	.094 73	.100 64	.119 37	.124 34
20	.141 64	.147 19	.169 66	.178 29	36	.094 84	.098 61	.116 96	.122 14

heuristic for $d = 3$ and $n = 22$ has a star–discrepancy of 0.101 93. In order to obtain an improvement of 20%, $n = 30$ points are required. Consequently, using suboptimal designs in practice might require more experiments in order to obtain the same quality of results. In experimental economics as well as in engineering, performing 22 or 30 experiments makes a difference.

The computation time for the tries with 10 000 iterations ranges from 0.13 CPU–seconds for $d = 2$ and $n = 5$ to more than 117 CPU–seconds for $d = 5$ and $n = 25$ on an IBM RS 6000/43P-260 workstation (SPECfp95: 30.1). Despite of the high computational cost stemming mainly from the evaluation of the objective function D, the time consumption of the threshold accepting implementation is some orders of magnitude smaller than for the (deterministic) algorithm of Li and Fang (1995). This difference will further increase in favour of the heuristic for larger problem instances.

To summarize the findings for the uniform design problem under the star–discrepancy, it can be stated that the threshold accepting implementation provides a good approximation to uniform designs for small problem instances. In fact, the known optimal results provided by Li and Fang (1995) are reproduced without excessive tuning of the optimization parameters. The performance of the algorithm for larger instances is mainly limited by the huge computational burden of an exact calculation of the star–discrepancy. There exist two possibilities for reducing this burden. First, one could use a threshold accepting approximation of the star–discrepancy as presented by Winker and Fang (1997). Secondly, alternative measures of discrepancy could be employed

Table 11.3 Discrepancy of optimized U–type designs ($d = 6, 7$)

	$d = 6$		$d = 7$	
n	Best	Mean	Best	Mean
7	.398 17	.398 23	.439 05	.431 91
8	.366 97	.370 08	.396 92	.401 41
9	.336 38	.345 34	.371 52	.372 42
10	.314 44	.319 95	.352 44	.353 23
11	.289 87	.297 07	.323 68	.338 01
12	.272 69	.281 93	.304 04	.318 65
13	.260 89	.269 47	.288 49	.303 09
14	.251 17	.258 04	.276 37	.286 83
15	.237 60	.247 83	.265 88	.274 94
16	.229 07	.237 07	.256 45	.264 42
17	.219 50	.227 80	.249 38	.256 52
18	.213 46	.220 08	.235 88	.247 31
19	.204 21	.213 23	.233 28	.239 55
20	.198 49	.207 02	.220 88	.231 61

which are easier to evaluate for a given design. This is the approach followed in the next sections.

11.4.2 Uniform designs for alternative discrepancy measures

The implementation of threshold accepting described above can be used directly for generating designs which minimize different discrepancy measures. It is only the 'early rejection' step during the evaluation of the objective function that becomes superfluous in this setting and is thus omitted. Besides the discrepancy measures based on the Euclidean norm, which have been introduced in Section 11.2.1, i.e. D_2, SL_2, CL_2 and ML_2, the notion of D–optimality lends itself also for direct optimization by using the threshold accepting algorithm. A D–optimal design minimizes the generalized variance of the corresponding regression model, i.e. maximizes the determinant of the information matrix (Atkinson and Donev, 1992, p. 42). Therefore, in order to obtain D–optimal designs or approximations thereof, the threshold accepting is applied to the negative of the determinant of the information matrix corresponding to the design matrix, which is denoted by D_O.

The advantage of all these alternative measures compared to the star–discrepancy is that they are much easier to calculate. Furthermore, the modified L_2–discrepancies are invariant to some natural transformations of the parameter space. Nevertheless, the error bounds for simulation and numerical integration may differ depending on the chosen measure of discrepancy. Therefore, it is interesting to compare the U–type designs obtained by optimizing different objective functions.

Table 11.4 reports the best results out of 250 tries for $d = 2, \ldots, 6$, $n = 25$ and 10 000 iterations for each of the six measures of uniformity D, D_2, SL_2, CL_2, ML_2 and D_O. It should be taken into account that due to the high complexity of calculating the discrepancy for a given set of points, a run with 10 000 iterations for D takes more CPU–time than 50 runs with 10 000 iterations each for D_2.[23]

The results shown in Table 11.4 demonstrate that the threshold accepting implementation works reasonably well for all objective functions, as, in general, the optimal values in each block are observed on the diagonal. This means that in order to obtain a design with low D it is best to minimize D, while in order to obtain a design with low CL_2, one rather should minimize CL_2 directly. However, it turns out that the designs obtained by minimizing D_2 or the modified L_2–discrepancies also result in low values for D and high values for D_O, which indicates some robustness of these designs.[24] While maximizing D_O increases the value of D_O only by a small amount, it results in designs with low uniformity in terms of D or the L_2–discrepancy measures. Consequently, designs obtained by maximization of D_O appear to be less robust.

Further evidence on the good performance of the threshold accepting implementation for searching low–discrepancy designs is reported in Fang, Ma and Winker (2000). There, expectation and variance of the centred L_2–discrepancy (CL_2) is calculated for random designs and Latin hypercube designs. These values are compared with the CL_2 of designs obtained through optimization by using threshold accepting. It is found that the CL_2 values for the latter are smaller than the expectation for Latin hypercube designs by a multiple of the standard deviation ranging from one to about eight as the dimension d increases, i.e. in this sense the designs obtained by the threshold accepting implementation are at least significantly better than those obtained by standard methods, if not optimal.

11.4.3 Mixture level designs

The results presented in the preceding section represent only a small fraction of designs obtained by threshold accepting for different measures of discrepancy. Further designs are provided in the references given in the introduction to this present chapter, or can be found at www.math.hkbu.edu.hk/UniformDesign. At this location, not only designs with a number of levels equal to the number of experimental units are shown, but also mixture level designs, i.e. with some $q_i < n$. It turns out that the threshold accepting implementation also provides

---

[23] For $d = 2$, the number of iterations was increased to 200 000. However, the results did not change very much.

[24] Robustness to changes of criteria is considered as an important feature in the experimental design literature; see, e.g. Trinca and Gilmour (2000, p. 27).

Table 11.4 Uniformity of best U–type designs ($n = 25$)

Objective	D	D_2	SL_2	CL_2	ML_2	D_O
			$d = 2$			
D	0.0628	0.3698	0.005 92	0.000 57	0.000 64	69
D_2	0.0660	0.3026	0.004 84	0.000 50	0.000 57	69
SL_2	0.0660	0.3026	0.004 84	0.000 50	0.000 57	69
CL_2	0.0652	0.3026	0.004 84	0.000 50	0.000 57	69
ML_2	0.0660	0.3036	0.004 84	0.000 50	0.000 57	69
D_O	0.1228	0.9145	0.014 63	0.001 11	0.001 18	69
			$d = 3$			
D	0.0939	0.4698	0.050 03	0.002 31	0.002 53	563
D_2	0.0972	0.3733	0.035 27	0.001 85	0.002 15	566
SL_2	0.1086	0.4179	0.027 34	0.001 55	0.001 94	576
CL_2	0.1159	0.4768	0.030 71	0.001 45	0.001 91	576
ML_2	0.1133	0.4394	0.030 34	0.001 49	0.001 88	575
D_O	0.1994	0.5023	0.083 91	0.003 76	0.004 98	576
			$d = 4$			
D	0.1230	0.6128	0.168 25	0.006 07	0.008 81	4435
D_2	0.1292	0.3430	0.142 50	0.005 89	0.006 72	4172
SL_2	0.1566	0.4449	0.111 54	0.003 99	0.005 97	4783
CL_2	0.1694	0.4698	0.117 73	0.003 42	0.005 34	4787
ML_2	0.1368	0.4166	0.125 33	0.003 68	0.005 16	4781
D_O	0.1910	0.7389	0.202 02	0.006 52	0.009 84	4792
			$d = 5$			
D	0.1502	0.4085	0.589 90	0.014 47	0.022 30	31136
D_2	0.1756	0.2645	0.477 05	0.012 79	0.018 71	29352
SL_2	0.1961	0.3607	0.366 67	0.008 62	0.014 96	39502
CL_2	0.2320	0.4234	0.406 89	0.007 14	0.013 67	39746
ML_2	0.1648	0.3397	0.424 77	0.007 90	0.012 65	39433
D_O	0.2238	0.4945	0.606 29	0.014 33	0.025 07	39866
			$d = 6$			
D	0.1748	0.2567	1.372 17	0.023 87	0.047 81	233952
D_2	0.2048	0.1722	1.309 95	0.021 65	0.041 20	221344
SL_2	0.2180	0.2495	1.042 87	0.017 52	0.035 94	307251
CL_2	0.2812	0.2637	1.115 94	0.013 05	0.030 26	327528
ML_2	0.2109	0.2305	1.131 21	0.014 71	0.028 31	322303
D_O	0.2446	0.4105	1.616 55	0.024 28	0.054 76	331662

high–quality results for this purpose. However, again it is difficult to find a bench–mark to better for larger instances. Therefore, these results are not commented upon further within this chapter.

11.4.4 Uniformity and orthogonality

During the research on low–discrepancy U–type designs, some results obtained by the threshold accepting implementation exhibited a remarkable feature — orthogonality. For a formal definition of orthogonality, first the special class of U–type designs called *orthogonal arrays* has to be introduced. An $n \times d$ matrix X with entries from $\{1, \ldots, q\}$ is said to be an *orthogonal array with d levels, strength t and index* λ (for some t in the range $0 \le t \le d$) if every $n \times t$ submatrix of X contains each t–tuple constructed from $\{1, \ldots, q\}$ exactly λ times as a row. The notation $OA(n, d, q, t)$ is often used in this context. The parameter λ is implicitly defined as n/q^t (Hedayat et al., 1999, p. 3f.).

These designs exhibit attractive features for many statistical applications in an experimental set–up.[25] Orthogonal arrays of strength two are also called *orthogonal designs* (denoted by OD). If the labels of the levels of an orthogonal design are chosen symmetric about 0, for example, -1, 0, and 1 for $q = 3$, it is easy to see that the columns of the design are orthogonal to each other, i.e. $X'X$ is equal to some multiple of the $d \times d$ identity matrix (Fang, Lin, Winker, and Zhang, 2000). Although necessary, this condition of column orthogonality (CO) is not sufficient for design orthogonality (DO).[26]

Chan, Fang and Winker (1998) argue that orthogonal designs can be obtained by striving for low–discrepancy designs. Ma and Fang (2000) prove this conjecture to be true for the two–dimensional case and provide further support for this hypothesis. The results presented in Fang, Lin, Winker, and Zhang (2000) indicate that this idea can be used to obtain orthogonal designs by using the threshold accepting implementation with the CL_2 discrepancy measure. Although the threshold accepting implementation provides low discrepancy or even uniform designs for small instances, as soon as the dimension of the problem increases, it becomes less likely to obtain uniform designs unless the number of iterations is increased drastically. This seems to be less harmful when experimental design is applied for the purposes mentioned in the introduction to this chapter, but obviously becomes annoying when used for obtaining orthogonal designs, as orthogonality is a zero–one

---

[25] A few methods are described in Chapter 11 of Hedayat et al. (1999).

[26] A deeper analysis of orthogonal arrays is contained in Raghavarao (1971, Chapters 1 and 2). This author also presents some other methods for constructing orthogonal designs. Hedayat et al. (1999) provide a comprehensive overview on orthogonal arrays, applications and different approaches for construction, e.g. from codes, difference schemes, and Hadamard matrices. Furthermore, extensive tables of orthogonal arrays are included.

OPTIMIZATION HEURISTICS IN ECONOMETRICS

Table 11.5 Orthogonal designs

		$d = 2$		$d = 3$		$d = 4$	
n	q	CL_2	OD	CL_2	OD	CL_2	OD
9	3	.019 98	DO	.033 03	DO	.049 36	DO
16	4	.011 38	DO	.018 92	DO	.028 44	DO
18	3	.019 98	DO	.032 50	DO	.047 36	DO
25	5	.007 21	DO	.011 85	DO	.017 70	DO
27	3	.019 98	DO	.032 32	DO	.046 55	DO
32	4	.011 38	DO	.018 68	DO	.027 44	DO
36	3	.019 98	DO	.032 35	DO	.046 68	DO
36	6	.005 03	DO	.008 28	DO	.012 47	CO
		$d = 5$		$d = 6$		$d = 7$	
9	3	.077 28	NCO	.111 73	NCO	.157 31	NCO
16	4	.041 72	DO	.060 35	CO	.087 19	NCO
18	3	.065 25	DO	.086 90	DO	.113 59	DO
25	5	.025 20	DO	.036 90	NCO	.051 61	NCO
27	3	.063 55	DO	.083 73	DO	.109 00	NCO
32	4	.038 52	NCO	.052 70	NCO	.070 40	NCO
36	3	.063 33	DO	.082 84	DO	.106 28	NCO
36	6	.018 18	NCO	.025 82	NCO	.036 19	NCO

feature. Nevertheless, it is possible to obtain orthogonal designs by using the threshold accepting implementation including designs not reported elsewhere.

Obviously, orthogonal designs can only exist if q^2 is a divisor of n. Thus, Table 11.5 summarizes only results for such mixture level designs. In fact, results not presented in the table indicate that minimizing the CL_2 discrepancy for other instances often still leads to column orthogonal designs. For each problem instance described by dimension d, number of levels q and number of experiments n, the table provides the lowest CL_2–discrepancy obtained by several runs of the threshold accepting implementation with up to 1 000 000 iterations. Furthermore, the entries in the second column for each problem instance indicate whether this best design in terms of CL_2–discrepancy is also design orthogonal (DO), at least column orthogonal (CO) or not even column orthogonal (NCO).

The results presented in Table 11.5 show that minimizing the CL_2 discrepancy leads to orthogonal designs for $d \leq 4$, $n \leq 6$ and $q = 3, \ldots, 6$. As the dimension increases beyond $d = 4$, the number of orthogonal designs generated by using the threshold accepting implementation decreases. However, even in dimension seven, an orthogonal design is still generated.

Different reasons might be responsible for the failure of the threshold accepting implementation to deliver orthogonal arrays in higher dimensions. First, although the condition that q^2 is a divisor of n is necessary, it might not be sufficient. Thus, for given d, n and q an orthogonal design might not exist. Clearly, the threshold accepting implementation might provide the global optimum for the CL_2–discrepancy in this situation without resulting in an orthogonal design. Secondly, the conjecture that minimum CL_2–discrepancy corresponds to orthogonality might be erroneous. However, so far no counter example has been found. Furthermore, several designs with minimum CL_2– discrepancy might exist and the threshold accepting implementation provides only one of these, which is not necessarily the orthogonal one. The opposite case can also be observed, i.e. for some instances (e.g. $d = 6$, $n = 18$, $q = 3$) there exist several orthogonal designs with different values of CL_2. Then, optimization of CL_2 cannot produce all orthogonal designs, but might contribute some so far unknown ones. Thirdly, for larger problem instances, the threshold accepting implementation approximates the global optimum, but cannot guarantee to meet it with a finite and, in general, small number of iterations. In fact, for the instances where the algorithm failed to produce an orthogonal design, an increase in the number of iterations still leads to improvements of the objective function CL_2. Consequently, the designs reported in Table 11.5 are not optimal for these instances. In fact, for $d = 6$, $n = 36$ and $q = 6$, a further increase of the number of iterations performed by the threshold accepting implementation eventually lead to an orthogonal design.

11.5 Ruin and Recreate

As already pointed out in the context of orthogonality, the results for the straightforward threshold accepting implementation presented in the previous sections indicate that this method is quite successful for searching low–discrepancy or uniform designs compared to designs obtained by other approaches, including methods based on deep mathematical backgrounds. Unfortunately, as soon as the problem instances become larger, i.e. the number of factors and levels grows, the threshold accepting implementation, although still improving on existing solutions, cannot provide uniform designs, i.e. globally optimal designs, with certainty or at least high probability unless the number of iterations grows at a very high rate.[27]

Therefore, it seems worth trying to see whether results can be improved by using additional heuristic concepts. The problem of finding low–discrepancy designs, which might eventually become orthogonal, lends itself directly to an application of the ruin and recreate principle already outlined in Section 5.2.6. The main idea of this implementation can be described by using Figure 11.7. Within the design matrix shown on the left side, a submatrix — marked

---

[27] This means that the number of iterations has to grow, e.g. by a cubic rate in the problem size. Note, however, that such a high rate is still small compared to the increase of the size of the search space, which grows exponentially.

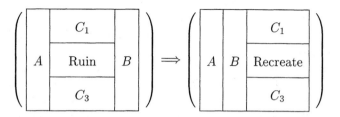

Figure 11.7 A ruin and recreate step.

ruin — is randomly selected. Of course, in the actual implementation this submatrix may overlap from the bottom to the top or from the right side to the left side. Then, it is assumed that all information within this submatrix is lost. The recreation step consists of two parts. First, the remaining parts of the matrix are reorganized as shown in the right side of Figure 11.7. Secondly, the elements of the ruined submatrix have to be filled in sequentially in some reasonable way.

Since the class of admissible design matrices constrains the elements within a column, the recreation step is performed column by column by using the elements in the ruined part of the column for its recreation. The choice of the permutation in this subcolumn follows the forward procedure mentioned in Section 11.3, i.e. all possible permutations of the elements in the ruined part of the column are combined with the already completed columns from submatrices A and B and already finished recreation steps. Then, the permutation resulting in the lowest discrepancy for this lower–dimensional design is selected. Of course, the critique of the forward procedure due to its sequential approach is also relevant in this case. However, only a submatrix is treated this way, and the recreation step is only one step in a larger optimization algorithm. Consequently, the risk of getting stuck in a local minimum due to the sequential approach is smaller in this case.

The difference between such a ruin and recreate step and a simple threshold accepting exchange step is twofold. First, the *ruin* area, in general, is larger than typical changes in a threshold accepting step, when two, three or maybe four elements are exchanged. In a ruin and recreate step for this application, five to eight elements in up to six columns are reshuffled in one single step. Secondly, while the exchanges take place randomly in a standard local exchange try, they are systematic in the *recreate* step.

After having described the ruin and recreate step for a given design matrix, the complete threshold accepting implementation with ruin and recreate can be given. The overall procedure is the same as for threshold accepting, except that some local exchange steps are replaced by a ruin and recreate step. Obviously, typical threshold values differ for a ruin and recreate step. In fact, due to its locally constructive approach it is more likely to move downward than a standard local exchange step. Therefore, different threshold values can be used for standard exchange steps and ruin and recreate steps. Of course, it is also possible to use a pure ruin and recreate approach within the threshold accepting framework, i.e. all exchange tries are of the ruin and recreate type. For the optimum design problem discussed in this chapter, however, such a pure version proved to be less competitive.

So far, a ruin and recreate implementation could not improve on the results obtained by the straightforward threshold accepting algorithm. At least two reasons might be responsible for this failure. First, the tuning of the combined algorithm is more complex than for the simple threshold accepting algorithm. In particular, not much is known about the choice of threshold sequences under this setting. Tuning experiments along the lines pointed out in Chapter 8 might help to improve the performance considerably. Secondly, the recreate step is of crucial importance for the performance of the ruin and recreate algorithm. The application of the forward procedure to a subdesign might not yet represent an optimal choice in this context. Thus, further research is required on this aspect of the ruin and recreate heuristic in order to make it a suitable instrument for applications in the context of experimental design.

11.6   Applications

From the possible applications of experimental design mentioned in Section 11.1.3, only three are considered in a little bit more detail at the end of this chapter. Experimental economics is the research area in economics which probably lends itself most directly to the application of optimal design methods. Section 11.6.1 outlines a few examples and indicates to what extent optimal design aspects are currently used in this context. Section 11.6.2 refers to response surface analysis, which is widely used in statistics and econometrics. The use of experimental design for this purpose seems to be more widespread. Nevertheless, the performance of these applications might be increased by using optimized designs instead of some standard approaches. Finally, Section 11.6.3 refers to quasi–Monte Carlo simulation, which takes up the ideas of optimized design when performing higher–dimensional simulation, e.g. for macroeconometric models or the approximation of integrals.

11.6.1   Experimental economics

Experimental economics is that field of economics where the use of experimental design seems most natural. In fact, the discussion of experimental design plays an important role in this literature (Kagel and Roth, 1995). However, a broader concept of experimental design than that discussed in the previous sections is used here. Selten (1998) indicates the importance that a specific set-up of an experiment might have. This author provides several examples from the literature where different settings lead to quite different outcomes, although the features of the modelled economic situation remained completely unchanged. Thus, it appears to be very important in experimental economics to analyse the complete experimental setting very carefully.

Nevertheless, experiments in economics are also used to derive quantitative relationships by varying the numerical value of some variables and considering the outcome of the experiments by means of regression analysis. This aspect is treated in some detail in Friedman and Sunder (1994, Ch. 3). In particular, they argue that a full factorial design has certain merits, but the number of required trials increases quickly as the number of factors increase. Therefore, they propose using fractional factorial designs. In this setting, the use of optimized designs instead might help to improve the efficiency of such experiments, which are quite expensive.

11.6.2   Response surface analysis

Response surface methodology is also closely linked to experimental design.[28] This approach is used in situations when one is interested in the relationship between the values of some variables as input and the expected response as output. The response is a quantitative continuous variable, and the mean or expected response is a smooth but unknown function of the levels of the factors, which are, at least partially, reel–valued and controllable. The (mean) response can be plotted as a function of the levels of s parameters, thus resulting in a surface of dimension $s + 1$, i.e. the *response surface*.

In statistics and econometrics, such situations are given, e.g. when the distribution of some test statistic or estimator depends on nuisance parameters and/or the sample size. In many cases, for a given set of parameters and a given sample size critical values can be obtained only by simulations which are sometimes very time consuming. Hence, the simulations are carried out only once for a finite, often rather small, set of parameter values and sample sizes. Then, in applied research, when parameter values and sample size will in general be different from one of the few points in the simulation study, some available point close to the given parameter constellation is chosen or some sort of interpolation, which might be based on regression analysis, is performed. Given the results of the previous sections, it might not be too surprising that the conventionally used full factorial designs (with a very limited number of levels for each factor) are not necessarily optimal for this purpose. In fact, often a full factorial design is infeasible due to a high number of factors and the limited resources available for performing simulations or experiments. In both cases, optimized designs might prove to be very useful for improving the

---

[28] See Pukelsheim (1993, pp. 381ff.), Garcia-Diaz and Phillips (1995, Ch. 14), or Dean and Voss (1999, pp. 547ff.).

information content of such a response surface.[29]

It is beyond the scope of this present book to present a response surface analysis for a specific econometric problem and analyse the impact of choosing optimized results. Just to give an idea about the kind of problems which are analysed in this context, an example from the literature on unit root tests is outlined here. Response surface analysis has been used by i Silvestre, i Rosselló and Ortuno (1999) to obtain approximate critical values for the Dickey–Fuller unit root test with structural breaks for any sample size and break date. It is known that the Dickey–Fuller test is inconsistent against structural breaks in the trend. Taking into account segmented trends in the testing procedure, finite–sample time series behaviour becomes important, although only a limited set of critical values is available under this setting.

Therefore, critical values are obtained by Monte Carlo simulation.[30] For three different models (level shift, slope shift and both types of shift), simulations are performed for different values of sample size and the break fraction. While for the break fraction a regular grid with values $\{0.1, 0.2, \ldots, 0.9\}$ is chosen, the considered 31 sample sizes comprise values from 30 to 1250 with increasing distances, which seems a reasonable approach as for the larger sample sizes the critical values should approach their asymptotic values. For each sample size, break fraction and model, 50 000 replications of the data–generating process are performed and the quantiles of the empirical distribution test statistic are stored. Finally, for each quantile, these values are modelled as a polynomial of order four in the break fraction with coefficients depending on the sample size T by T^{-1} and T^{-2}. This specification is motivated by the fact that effects of the sample size should diminish as the sample size grows. As a last step, model selection is performed by setting those coefficients equal to zero, which show a t–statistic less than unity.

Given the fact that the response surface is modelled highly nonlinear in both explaining variables, it is not evident whether the performed simulations are those which — given a fixed total number of simulations — obtain the maximum amount of information and, therefore result in the best fitting response surface. In fact, it might be expected that using optimized designs for selecting the combinations of sample sizes and break fractions, for which a simulation has to be performed, could either improve the results or reduce computational costs associated with such a Monte Carlo experiment. As

---

[29] For example, Trinca and Gilmour (2000) present an algorithm for obtaining optimized designs for a certain type of response surface analysis. Since this algorithm is essentially a standard local search algorithm ('The swap that gives an improvement in the value of the design criterion will be accepted', p. 32), it might be expected that a threshold accepting implementation could improve the performance for this application.

[30] The quasi–Monte Carlo approach presented in Section 11.6.3 might improve the efficiency of this step.

experiments of this type will become more relevant in econometrics, a careful analysis of the experimental design set–up might be useful. Finally, the model selection step of the analysis could also use optimization heuristics (as shown in Chapter 12).

As a replication of the response surface analysis presented in i Silvestre et al. (1999) is beyond the scope of this present chapter, the results of a simulation with artificial data may provide further evidence for the potential of using optimized designs in the context of response surface analysis. For this purpose, mappings of the two dimensional unit cube to $\mathbb{R}$ of the following form are considered:

$$
\begin{aligned}
f(x,y) \quad = \quad & \alpha_1 + \alpha_2 x + \alpha_3 y + \alpha_4 xy + \alpha_5 x^2 + \alpha_6 y^2 \qquad\qquad (11.9)\\
& + \alpha_7 x^3 + \alpha_8 x^2 y + \alpha_9 xy^2 + \alpha_{10} y^3 \\
& + \alpha_{11} \frac{1}{2-x} + \alpha_{12} \frac{1}{2-y} + \alpha_{13} \frac{1}{2-xy} \\
& + \alpha_{14} \log(x+1) + \alpha_{15} \log(y+1) + \alpha_{16} \sin(10xy) + \alpha_{17} \cos(5x)
\end{aligned}
$$

For 1000 replications, the parameter values α_i are generated randomly, with $\alpha_1, \ldots, \alpha_{10}$ from a standard normal distribution and $\alpha_{11}, \ldots, \alpha_{17}$ from the unit distribution on $[-0.5, 0.5]$. Then, both for the regular grid and the optimal U–type design presented in Figure 11.1, the function is evaluated. Based on these nine experimental results, a quadratic function is fitted based on ordinary least squares. Finally, the maximum and mean differences between the fitted response surfaces and the known function values (equation (11.9)) are calculated on a 101×101 grid on the two–dimensional unit plane. Of course, none of the designs is better for all instances and all purposes. However, the response surfaces fitted based on the uniform design reduce the maximum difference in mean by more than 5%. On the other hand, the response surfaces fitted on the regular grid show a better mean performance, with the mean deviation from the true function values being smaller by about 7% in mean. Since the uniform design is obtained through optimization with regard to the star–discrepancy, it is more adequate to compare maximum deviations. Otherwise, a U–type design optimized with regard to some other measure of discrepancy, e.g. some L_2–based measure, is more appropriate. Figures 11.8 and 11.9 show two of the 1000 fitted response surfaces, together with the true function values.

Figure 11.8 shows an instance when the response surface fitted from the regular grid reduced the maximum difference between the estimated and true response by about 60% compared to the uniform design. The left–hand panel shows the true function values $f(x,y)$, the second panel the estimated response surface for the regular grid and the right–hand panel the response surface for the uniform design. Given the low number of nine design points and the restricted functional form of the estimated response surfaces, it is not surprising that both surfaces miss some of the specific features of

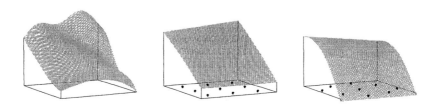

Figure 11.8 Response surfaces for the model represented by equation (11.9) I.

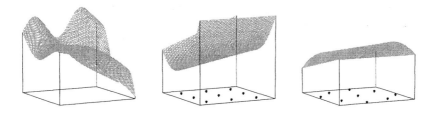

Figure 11.9 Response surfaces for the model represented by equation (11.9) II.

$f(x, y)$. Nevertheless, the overall approximation appears reasonable for both designs. The worse performance for the uniform design stems from the stronger curvature towards the x–axis, which is avoided by the regular grid due to the three design points close to this axis.

The second sample instance shown in Figure 11.9 gives a quite different picture. Both estimated response surfaces fail more or less dramatically in reproducing basic features of the complex function $f(x, y)$. In particular, the strong decrease of $f(x, y)$ towards $x = 1$ and $y = 0$ corresponds to a marked increase of the response surface fitted for the regular grid design and to an only minor decrease for the uniform design. Nevertheless, the uniform design reduces the maximum deviation by about 65% compared to the regular grid design.

It is not possible to draw any general conclusions from this small–scale simulation experiment. The results depend not only on the chosen experimental design, but also on the functional form of the function which

Figure 11.10   Response surfaces for the Rosenbrock function.

has to be approximated, and of the function which is used for approximation. Therefore, a second experiment is conducted by using rational functions of degree 3 and excluding singularities in the considered two–dimensional unit cube. For this functional class, the quality of the approximation by a second–order polynomial appears to be almost independent of the chosen design. Finally, Figure 11.10 shows a notorious example from the optimization literature, i.e. the Rosenbrock function of two dimensions (Goffe, 1997, p. 3). This is defined as

$$f(x, y) = 100(y - x^2)^2 + (1 - x)^2 .$$

The middle panel of this figure shows that the approximation based on the regular grid is mainly driven by the small hill on the right side of the real function landscape shown in the left panel. The approximation based on the uniform design catches the overall features of the function a little better.

For real applications, the functional form of the true response is not known and different functional forms are tested for the approximation. Consequently, the experimental design should be robust to changes of specific assumptions. The comparison of the different measures of discrepancy shown in Table 11.4 indicated some robustness of the optimized U–type designs. Therefore, they may be recommended for the purposes of response surface analysis — at least as a bench–mark for ad hoc approaches and designs tailored for specific settings.

Beyond the estimation of response surfaces, in some applications, in particular in engineering or pharmaceutical, one is mainly interested in the parameter combination leading to the maximum response and the response surface in a vicinity of this maximum. Then, of course, a local search heuristic lends itself for the prospecting of the response surface in order to find the maximum.[31] However, once the maximum is found, one may consider the description of the response surface in its neighbourhood as a general response surface problem given some range to the parameter values. Thus, the same arguments apply as for applications in econometrics, where often an approximation to the response surface is required over a larger region of parameter values.

11.6.3 Quasi–Monte Carlo methods

Quasi–Monte Carlo methods are also based on low–discrepancy point sets. However, the typical application differs from those mentioned in the previous section. In fact, these methods are most commonly applied in a simulation context. In contrast to the response surface analysis, the objective function, in general, is known and more or less easy to evaluate at a given point. However, if the objective function itself is not of much interest, but some statistics of its distribution over a set of parameter values are required, a high–dimensional integration problem results. Such problems come up, e.g. when simulating confidence bands for dynamic nonlinear macroeconometric models (Franz, Göggelmann, Schellhorn and Winker, 2000) or evaluating financial markets instruments.[32] Further applications of quasi–Monte Carlo methods could be found in simulation–based estimators (Gourieroux, Monfort and Renault, 1993).

The numerical problem behind all of these applications consists in evaluating

$$E(f) = \int_B f(\mathbf{x})d\mathbf{x} \, ,$$

where f is some function defined on the multidimensional set B. By using a suitable change of variables and, if necessary, replacing f by f times an indicator function, it can be assumed that B equals the d–dimensional unit cube C^d (Niederreiter, 1992, p. 6). Then, a Monte Carlo approximation is given by

$$\int_{C^d} f(\mathbf{x})d\mathbf{x} \approx \frac{1}{n} \sum_{i=1}^n f(\mathbf{x}_i) \, , \tag{11.10}$$

where the $\mathbf{x}_i$ are chosen randomly in C^d. Niederreiter (1992, p. 7) mentions three deficiencies of this Monte Carlo method as follows:

1. There are only probabilistic error bounds.
2. The regularity of the integrand is not reflected.
3. Generating random samples is difficult.

---

[31] This approach is chosen by Dean and Voss (1999, pp. 548ff.).
[32] See Caflisch and Morokoff (1996), Morokoff and Caflisch (1998), Acworth, Broadie and Glasserman (1998), and Tezuka (1998).

Instead of using random samples in equation (11.10), one could consider using uniformly distributed points (Beck and Sós, 1995, p. 1442). Then, the error bounds become deterministic, as deterministic point sets are used, and the problem of generating actual random samples is avoided. Furthermore, the Koksma–Hlawka inequality (equation (11.4)) provides a guideline for a good choice of the integration nodes. The goal of quasi–Monte Carlo methods, which have been studied and applied since the 1950s, consists in providing good point sets. Most of these methods are based on a rigorously developed mathematical theory (Niederreiter, 1992). However, the example of uniform design discussed in this chapter proves that explicit optimization might be helpful. Thus, it is proposed that optimization algorithms be used for generating good point sets, not only of small size as required in an experimental setting or for response surface analysis, but also of medium size as they are used for simulating macroeconometric models. Then, it can be decided whether such a direct optimization approach can lead to further improvements over the big gains provided by the available quasi–Monte Carlo methods as compared to the standard Monte Carlo framework.[33]

---

[33] For a comparison of Monte Carlo and quasi–Monte Carlo methods for the simulation of time series models, see Winker (1999b) and Li and Winker (2000).

12

Identification of Multivariate Lag Structures

12.1 Introduction

Modelling economic data is a demanding task. It requires knowledge about the data sources, problems related to the definition and generation of the data, a lot of statistical background and a theoretical framework. Consequently, a detailed treatment of the whole modelling procedure is far beyond the reach of this present chapter. Instead, the simplifying assumption is made that the data and theoretical framework are already available of the necessary quality for further analysis.

If the theoretical framework lends itself directly for the derivation of an econometric model, one is well provided for and can continue with the statistical tools to fit parameter values based on the available data. However, economic theory is not always so well determined. In general, it fails both to provide a clear statement on the relevant variables and on the functional relationship between these variables. Rather does it provide a set of potentially relevant variables and a statement on a suitable model class.

Then, the choice of one specific model and a set of influential variables is left to the econometrician or statistician based on statistical features of the data. However, discussions on bottom–up and general–to–specific strategies indicate that there is no consensus approach to this final and decisive step of modelling economic data. Therefore, the results of empirical research may not only depend on the right data and the specific theory, but as well on subjective elements in the final step of model selection. In fact, statistical inference on the estimation results depends on the model selection strategy.[1]

This chapter analyses the contribution that an optimization heuristic such as threshold accepting can make to the model selection problem. For this purpose, it is not necessary to present a complete discussion of econometric modelling strategies, but it suffices to study a specific econometric modelling

---

[1] Cf. Pötscher (1991); for a Bayesian critique see, e.g. Zaman (1984).

problem in detail. The example used in this chapter consists of the selection of the dynamic structure of a vector autoregressive model, i.e. the choice of lags in a multivariate dynamic time series model is analysed.[2]

This problem arises in the context of vector autoregressive (VAR) models or vector error correction models (VECMs). Models of this class are used in nonstructural macroeconometric models and models for financial time series, for Granger causality tests,[3] in self–exciting threshold autoregressive (SETAR) models,[4] or in multivariate (co–)integration analysis, to name just a few areas. Vector autoregressive models are used mechanically as a benchmark for more structured time series models, in particular with regard to their forecasting performance.[5] In the cointegration framework, they are used as structural models where long–run relationships derived from economic theory can be embedded in a short–run dynamic adjustment process which is not specified by theory. Nevertheless, tests of the imposed long–run restrictions along the procedures proposed by Johansen (1988; 1991; 1992) and Johansen and Juselius (1990) are sensitive to a proper treatment of the dynamic part of the VAR model in order to obtain unbiased test statistics for the long–run parameters (Jacobsen, 1995).[6] Asymptotically unbiased estimates can also be obtained by applying a model selection procedure directly to the vector error correction model as discussed in Gonzalo and Pitarakis (1998).[7] Finally, VAR models are used as auxiliary models in indirect estimation of more complex time series models (Martin and Wilkins, 1997). The choice of a 'best–fitting' model seems preferable in this context.[8] Thus, for a strict testing methodology it is preferable not to leave it to the discretion of the econometrician to choose the dynamic part in some ad hoc fashion, but to select it on some a priori defined criterion.

---

[2] Instead of looking at the time dimension of the dynamic model, one could also consider the dimensionality of the process itself, i.e. the number of endogenous variables (Abadir, Hadri and Tzavalis, 1999).

[3] See Pierce and Haugh (1977, p. 289), Sims (1972), and Thornton and Battan (1985). Today and Phillips (1994, p. 284) state: 'Our simulations show the important role played by the choice of lag length in the performance of these tests.'

[4] For SETAR models, not only a single VAR model has to be selected, but at least two — or even more — depending on a data–driven regime indicator, which itself has to be selected from a finite set of possible cases. Consequently, the problem can be solved, in principle, by complete enumeration of all candidate models and selection of the best one according, e.g. to some information criterion (Clements and Krolzig, 1998). However, as soon as the process becomes multivariate and/or the number of regimes is larger than two, enumeration becomes infeasible. Then, an approach similar to the one presented for lag structure identification could be applied.

[5] Cf. Canova (1995) for an overview.

[6] See also Johansen (1995, pp. 21f.).

[7] A similar approach is the posterior information criterion proposed by Chao and Phillips (1999), which takes into account explicitly the different convergence properties of stationary and nonstationary variables involved in a reduced rank regression.

[8] Cf. Gourieroux et al. (1993) and Gallant and Tauchen (1996).

Hence, the problem taken as an example for the implementation of threshold accepting for model selection can be characterized more clearly by making the following two assumptions. First, economic theory has already provided a clear decision on which variables have to be treated as endogenous and influential in the VAR model. However, as each variable may appear with different lags in a dynamic model, and if these lags are treated as separate variables, the list of variables is not completely specified yet. Secondly, the model class is given by linear vector autoregressive models with fixed coefficients. As the number of observations on the variables is finite, it is necessary to limit the number of explanatory variables, i.e. lags of the endogenous variables, in order to keep some degrees of freedom for the estimation of parameters.

Specified this way, the problem of multivariate lag order selection does not represent solely a prototype of the general identification problem,[9] but is of its own interest as it arises in the kind of VAR modelling problems just mentioned. Standard procedures for selecting the lag structure in applications of this type may serve as a bench–mark for the threshold accepting approach. Many different methods have been proposed for this purpose in the literature.[10] A short list of four examples may give an idea about possible approaches without pretending to provide an extensive survey or even to answer the question of which of the proposed methods is best.

First, a lag order is chosen ad hoc, based on some intuitive knowledge of the dynamic part of the model, e.g. four lags for quarterly data. Secondly, a lag order is chosen based on some information criterion like Akaike's final prediction error (FPEn) criterion Akaike (1969; 1971; 1973),[11] the Schwarz or Bayesian information criterion (BIC) (Schwarz, 1978)[12] or the Hannan–Quinn (HQ) estimator (Hannan and Quinn, 1979). A related approach is proposed by Zhang and Wang (1994), who provide a criterion for lag order selection based on the sign of a statistic by comparing the residual variance of the considered model with that of the unconstrained model, i.e. by using all lags up to some a priori determined maximum. Thirdly, a lag order is chosen in order to satisfy some test on multivariate autocorrelation of residuals (Jacobsen, 1995). For all of these first three methods, by selecting a lag order it is implicitly assumed that all lags up to that order are relevant for the model. This seems to be a natural and innocent assumption. However, as

---

[9] Brooks and Burke (1998) present a very similar modelling problem, including higher moments, i.e. GARCH–effects. Model identification is aimed at by using suitable modifications of the lag order selection criteria used in this chapter.

[10] See, for example, the references in Tsurumi and Wago (1991), who propose the use of mean squared errors of forecast for model selection, or Aznar (1989).

[11] For example, the semi–automatic procedure for the univariate case proposed by Kang, Bedworth and Rollier (1982) is based on this criterion.

[12] Phillips (1995) proposes a more general criterion which is asymptotically equivalent to the BIC, if a finite order autoregressive process is considered. Furthermore, this procedure allows for dynamic modifications of the selected model.

soon as the model becomes high–dimensional, this assumption may imply a severe restriction on the number of available degrees of freedom.

The fourth method, proposed in the literature, therefore allows for an unconstrained lag structure, i.e. a lag structure with 'holes'. For the univariate case, McClave (1975) presents an algorithm for unconstrained lag structure identification based on Akaike's FPE criterion. Given a sample from a stationary autoregressive process

$$x_t = \sum_{i=1}^{K} \gamma_i x_{t-i} + \varepsilon_t \,, \tag{12.1}$$

where the ε_t are identically and independently distributed as $\mathcal{N}(0,1)$, an algorithm introduced by Pagano (1972) is used to obtain the Yule–Walker estimates of the γ_i. The further procedure is based on a Cholesky decomposition of the Yule–Walker equations. If some a priori zero constraints can be imposed on the γ_i, a partial updating of the solution is possible, thus reducing the computational effort considerably.

This method for obtaining the estimates for the γ_i is combined with the subset selection algorithm presented by Hocking and Leslie (1967). Their procedure is based on measuring the reduction of the explained variance due to eliminating a subset of variables. If Θ_i denotes the reduction in explained variance due to the elimination of γ_i and the γ_i are relabelled such that

$$\Theta_1 \leq \Theta_2 \leq \ldots \leq \Theta_K \,,$$

then the following property of quadratic forms holds: if the reduction in explained variance due to eliminating any set of variables, for which the maximum subscript is j, is not greater than Θ_{j+1}, then the smallest reduction will result by omitting a set which does not include any variable with subscript greater than j (Hocking and Leslie, 1967, p. 533) Using this property, a search tree with a stopping rule to find the best subset with $p \leq K$ elements can be generated. In a small Monte Carlo experiment reported by Hocking and Leslie (1967), on average 299 of $2^{10} = 1024$ subsets had to be checked to obtain the global optimum. In the examples given by McClave (1975), this algorithm performed well, although the FPE criterion was proven to be asymptotically inconsistent (Shibata, 1976).

For the univariate case and a small set of potentially relevant regressors, even a more naive approach to model identification is feasible given the tremendous increase in available computing power, namely the enumeration of all possible subsets. Thus, e.g. Pesaran and Timmermann (1995) provide an application to financial markets, where all subsets of nine regressors are checked with regard to five different selection criteria. The computing load is further increased by the recursive nature of their application, i.e. the same selection procedure is repeated for different samples.

However, a significantly larger regressor set in the univariate case or, in particular, an extension to the multivariate case, is still not tractable by this naive procedure. So far, no generalization of McClave's procedure to the multivariate case has been presented. Furthermore, even if such a generalization is possible and if it would necessitate the calculation of the selection criterion only for a small part of the possible subsets, as in the one–dimensional examples given by Hocking and Leslie (1967) and McClave (1975), it will be infeasible for most multivariate applications. A bivariate VAR model with a maximum lag length of 24, corresponding to two years of monthly data, allows for more than 10^{14} possible subsets of the lagged endogenous variables. Comparing just a small fraction of these possibilities would take a great deal of time even with the fastest computers. If the dimension increases, the problem becomes practically intractable. Furthermore, the procedure is restricted to criteria based on the sum of squared residuals in order to exploit the property of quadratic forms. However, as pointed out above, for some applications other objective functions might be more appropriate.

There are different ways out of this seeming dead–end. First, some a priori knowledge on the possible structure of the model, or of the kind used for the univariate case described above, might be used. However, a priori assumptions — especially if they cannot be based on economic theory — might be misleading, and even with some a priori restrictions the test of the remaining combinations might still be intractable. Of course, the first three methods mentioned above avoid the high complexity of the selection problem, by imposing the severe 'take all up to the kth lag' restriction.[13] Then, in the two–dimensional example with maximum lag length 24 only 24^2 models have to be compared, but the chance of finding the 'true' model or at least a good approximation just within these 576 out of more than 10^{14} models is extremely small. Webb (1985) relaxes this strong restriction by allowing for different maximum lag lengths for the same variable in different equations. Then, the number of possible models becomes much larger, in particular in a higher–dimensional setting. Therefore, some heuristic search procedure is used to identify a good lag structure based on Akaike's selection criterion.[14]

The implementation of threshold accepting for the automatic identification of multivariate lag structures in vector autoregressive models is also based on explicit selection criteria, e.g. an information criterion. In contrast to the standard procedures, it is not restricted to the 'take all up to the kth lag'

---

[13] The importance of this restriction is not always made explicit. A positive example is given by Clements and Krolzig (1998, p. C51): '... we set the maximum lag length at 5, require the lag orders to be the same across regimes, and do not allow "holes" in the lag distributions'. This contrasts with a similar application of self–exciting threshold autoregressive models to USA business cycle data by Potter (1995), who restricts lags two and three to zero based on Akaike's information criterion.

[14] A similar approach is used in Fackler (1985) for producing five–dimensional VAR models of goods, money, and the credit market.

approach, but — similar to McClave's proposition — allows us to select an arbitrary multivariate lag structure, i.e. a lag structure with 'holes'. As the selection procedure does not exploit any specifics of the selection criteria, some efficiency might be lost. At the same time, much flexibility and ease of implementation is gained. Consequently, the approach can be easily applied to other selection criteria.

This application of threshold accepting has been introduced in Winker (1995). A more detailed and updated description of the approach is reported in Winker (2000a). Winker (2000b) generalizes this method by allowing the included lags to differ between equations. Although parts of the following exposition of the implementation and some of the simulation results are based on this earlier work, it is extended in several directions, thus allowing for a more clear–cut conclusion about the usefulness of the concept.[15] Section 12.2 presents the lag structure identification problem considered as an example for model selection using threshold accepting. Section 12.3 introduces the model selection criteria employed, before the threshold accepting implementation is described in Section 12.4. Results of the different Monte Carlo simulations are presented in Section 12.5. The following section comments on the potential impact of the results of model selection within a vector error correction modelling framework, while Section 12.7 summarizes the findings of this present chapter.

12.2   The Lag Structure Identification Problem

Although the implementation of threshold accepting presented in this chapter is not restricted to specific time series models, a simple vector autoregression model will serve as an example in the present chapter. Let this d–dimensional vector autoregressive process $\{X_t\}$ be given by

$$X_t = \begin{pmatrix} x_{1t} \\ x_{2t} \\ \vdots \\ x_{dt} \end{pmatrix} = \sum_{k=1}^{K} \Gamma_k X_{t-k} + \varepsilon_t, \qquad (12.2)$$

where $\varepsilon_t = (\varepsilon_{1t}, \ldots, \varepsilon_{dt})$ are identically and independently distributed as $\mathcal{N}_d(0, \Sigma)$, where Σ denotes the $d \times d$ covariance matrix of the error distribution, and the matrices

$$\Gamma_k = \{\gamma_{i,j}^k\}_{1 \leq i,j \leq d}$$

provide the autoregression coefficients.

In macroeconometric applications, the number of observations T for a given realization of the process $\{X_t\} = (X_1, \ldots, X_T)$ is often rather small.

---

[15] In this context, I am indebted to D. A. Belsley, T. Büttner, A. H. Pesaran and W. Smolny for constructive comments.

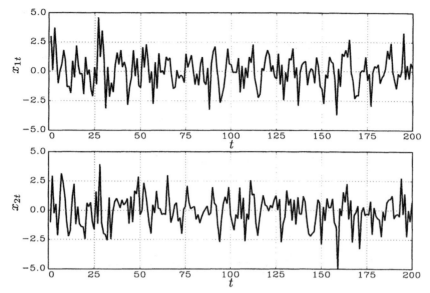

Figure 12.1   Realization of a bivariate VAR process.

Consequently, some reasonable a priori restriction on the maximum lag order $K' \ll T$ has to be imposed. Then, the problem of lag structure identification consists in identifying $K < K'$ along with those $\gamma_{i,j}^k$ that are different from zero in the underlying data–generating process (equation (12.2)).

Figure 12.1 shows the realization of a bivariate stochastic VAR process ($p = 2$) as defined in equation (12.2). The data were obtained by simulation. The process was initialized by setting $X_t \equiv 0$ for $t = 1, \ldots, 20$. For $t > 20$, X_t was calculated according to equation (12.2) by using the coefficients provided in equation (12.4) below in Section 12.5.2 for the matrices Γ_k. Finally, the ε_j were generated independently by using the normal random number generator RNDN of GAUSS 3.01.[16] Only realizations 801 to 1000 for x_{1t} and x_{2t} are shown in the figure.

An actual trivariate economic time series process is plotted in Figure 12.2. This shows the first differences of the German three–month money market rate (r_M), plus a loan (r_L) and a deposit (r_D) rate for the time period 1976.01 to 1997.05.[17] Although the time series do not move perfectly together, a high correlation is obvious. Furthermore, from an economic point of view hypotheses both on long–run relationships and on short–run dynamic adjustment features can be derived (Winker, 1999a). Hence, a VAR modelling

---

[16] The difficulties of using a univariate random number generator for this two–dimensional model are neglected as the figure is for illustration purposes only. For alternative methods, see Section 11.6.3.

[17] All data are taken from the monthly reports of the German Central Bank (Deutsche Bundesbank).

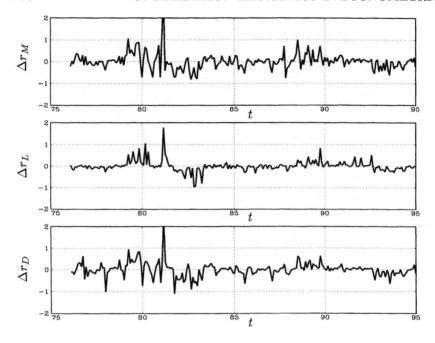

Figure 12.2   A trivariate interest rate process.

approach might be a reasonable first step or bench–mark for studying the underlying dynamic relationships.

When fitting the model represented by equation (12.2) to an observable realization, as in Figures 12.1 and 12.2, none of the estimated coefficients will be zero. Therefore, the model employing all lagged values of X up to the maximum order K' would minimize the sum of squared residuals. However, if the true data–generating process contains only a few non–zero coefficients, this model does not represent a useful identification for most applications, as it contains too many parameters. In the first example, when the true data–generating process is known and contains only a few autoregressive coefficients different from zero, one would rather wish to identify those coefficients. In the second example, when the fitted model will be used for further testing, for example, in a cointegration framework, or for forecasting purposes, the overfitting by including all coefficients is also a very unwelcome feature, since it might reduce the efficiency.

Obviously, when considering the question of finding the 'best' or at least a 'good' model in the class given by equation (12.2), fitting the data as given in Figures 12.1 or 12.2, is ill posed. Any useful concept has to define what a 'good' model should look like. This definition depends on the application under study. If the aim is to identify the data–generating process as well as possible, information criteria can be used to make the trade–off between a good fit of the model and a low number of non–zero coefficients. If the goal is a different one, e.g. to eliminate as much autocorrelation as possible from the residual process without reducing the degrees of freedom too much, a multivariate generalization of the Box–Pierce statistic, i.e. a multivariate portmanteau statistic, might be the proper instrument. The next section will discuss some possible criteria. However, there is no claim made to present an exhaustive overview of all possibly useful selection criteria. Nevertheless, it should be noted that the optimization method introduced in the following sections works well for all selection criteria, as long as the optimization problem can be stated in the form

$$\min_{S \in \mathcal{R}} C(S, X), \qquad (12.3)$$

where C is the selection criterion to be minimized, e.g. one of the information criteria presented in the next section, X is a finite realization of the process $\{X_t\}$ and $\mathcal{R}$ is the set of admitted selection rules. Hereby, a selection rule $S \in \mathcal{R}$ is given by a set of matrices $S_k = \{s_{ij}^k\} \in \{0,1\}^{d \times d}$, $k = 1, \ldots, K$. If $s_{ij}^k = 0$, then the corresponding γ_{ij}^k is constrained to zero for the calculation of $C(S, X)$. Finally, the performance of a threshold accepting implementation depends on the computational load for evaluating $C(S, X)$ as the example of the star–discrepancy versus alternative discrepancy measures in Chapter 11 highlighted. Performance gains might be achievable if a simple updating formula is available for calculating $C(\tilde{S}, X)$, if $\tilde{S}$ is a neighbour of S, given that $C(S, X)$ was already computed in an earlier step.

12.3 Model Selection Criteria

From the huge number of model selection criteria available only information criteria are employed for the threshold accepting implementation in this chapter, while a reference is made to a second class of criteria, which are based on specific properties of the resulting models, e.g. residual autocorrelation. However, as long as the criterion is easy to calculate, the threshold accepting implementation allows us to use any other criterion as well.

Information criteria attempt to approximate the unknown true model as close as possible. To this end, the fit of the model for a given lag structure is combined with a penalty related to the number of included lags, corresponding to the loss in degrees of freedom. In a univariate setting, the model's fit is measured by the sum of squared residuals, while this measure is replaced by the determinant of the estimated residual covariance matrix in the multivariate case. Minimizing such criteria leads to models with high explanatory power (small determinant of the residual covariance) using only few explanatory variables, i.e. lags. The punitive term of these criteria helps to avoid including too many lags and the concomitant overfitting problem.

The second class of selection criteria is based on objectives which are different from a close approximation. Often, the model selection procedure is only a first step of a two– or multistep estimation method, e.g. Johansen's method proposed for the identification of the cointegration space in a multivariate setting (Johansen, 1995). The validity of inference on the second step depends on specific assumptions on the error distribution of the first–step restricted VAR estimates. Here, the objective is not a close approximation, but a model meeting the assumptions of the second step, e.g. white–noise residuals. If the model fails to fulfil this requirement, any inference drawn on the asymptotic critical values can be quite misleading.[18] Consequently, suitable test statistics for white noise residuals may replace the information criteria in the case of Johansen's cointegration framework, while other approaches may require still different first–step features, e.g. criteria related to the forecasting power of the model.

The first information criterion was introduced by Akaike (1969) in the context of univariate autoregressive processes. It is still commonly used and included in most econometric software packages. For a univariate process, Akaike's final prediction error (FPE_o) criterion is given by

$$FPE_o = \left(1 + \frac{k+1}{T}\right) \hat{\sigma}_k \, ,$$

where k is the number of lags included in the estimated model, T the number of observations and $\hat{\sigma}_k$ the estimated residual variance. If the order of the data–generating process is smaller than k, FPE_o becomes unnecessarily large, since additional lags are punished through the term $k + 1$. On the other side, if the true model order is larger than k, $\hat{\sigma}_k$ estimates not only the residual variance, but the bias due to a wrong specification of the model as well. Thus, minimizing FPE_o should result in an estimated model order close to the true one.[19]

In the multivariate setting, the criterion becomes

$$FPE = |\, \hat{\Sigma}_k \,| \left(1 + \frac{2k}{T}\right) \, ,$$

where k again is the number of included lags, $|\, \hat{\Sigma}_k \,|$ the determinant of the estimated sample residual covariance matrix, and T the number of observations. Besides the generalization to multivariate models, the FPE criterion differs from FPE_o by the weight attributed to the number of included lags. In fact, the choice of 2 for this weight in FPE is heuristic and not based on some theoretical argument. Shibata (1976) demonstrates that using this version of FPE leads asymptotically to some overfitting, i.e. FPE overestimates the number of lags.

---

[18] See Jacobsen (1995) and Johansen (1995, pp. 21f.).

[19] Akaike (1973) shows the relationship between the application of information criteria and maximum–likelihood estimation. The latter can be interpreted as a method of asymptotic realization of an optimum estimate with respect to a criterion derived from information theory.

Nevertheless, Akaike (1973) presents some successful applications of the FPE criterion to different practical problems. Meanwhile, this criterion has been used for more than twenty years in many areas of statistical modelling. However, most applications are restricted to the selection of lag subsets, including all lags up to order k. The work by McClave (1975) and Pagano (1972) cited in Section 12.1 seems to be an exception as they provide an application on real subset autoregression. Only recently, in the context of VAR modelling, attempts of relaxing the 'use all lags up to order k' condition can be observed.[20]

In order to overcome the problem of overfitting in the FPE criterion, Schwarz (1978) and Hannan and Quinn (1979), among others, introduce alternative loss functions resulting in asymptotically consistent estimators of the dimension of the model if the process is stationary.[21] These loss functions differ from the FPE criterion solely in the weight attributed to the number of included lags. The constant 2 in this criterion is replaced by $\ln T/2$ in Schwarz's criterion and by $2c \ln \ln T$ with $c > 1$ in Hannan and Quinn (1979), respectively. Quinn (1980) proposes the use of $c = d^2$, where d is the dimension of the process. The constant d^2 is also used in multivariate versions of the other model selection criteria (Jacobsen, 1995, p. 180). Unfortunately, not much is known about the small sample performance of these estimators.[22]

The Hannan–Quinn criterion is given by

$$ \text{HQ} = \ln \mid \hat{\Sigma}_k \mid + \frac{2kc}{T} \ln \ln T , $$

where $\mid \hat{\Sigma}_k \mid$ again is the variance of the residuals in the univariate case and the determinant of the residual covariance matrix in the multivariate case, k the number of included lags, T the number of observations and c a constant greater than 1. For the simulations presented in Section 12.5, the Hannan–Quinn (HQ) estimator with different values for the constant c and Akaike's final prediction error (FPE) criterion are used.

For both criteria, a numerical estimate of $\hat{\Sigma}_k$ has to be calculated based on the available observations. If the same lags appear in all equations, which is the case, for example, for the 'take all up to the kth lag' approach and the restricted version of the threshold accepting algorithm introduced in Winker (1995), ordinary least squares provides a consistent, asymptotically efficient, and computationally tractable estimator. If the lags differ across equations, this no longer holds true. For such instances, the two–step seemingly unrelated regression estimator (SURE) proposed by Zellner (1962) is implemented.

---

[20] See, e.g. Brüggemann and Wolters (1998) and Lütkepohl and Wolters (1998).
[21] See also Sims (1988) for the problems arising in a unit root setting.
[22] Lütkepohl (1985) provides results of an extensive Monte Carlo simulation for bi– and trivariate VAR processes. However, the lag selection rules in his study are restricted to the standard 'take all up to the kth lag' approach.

An example of a model selection criterion, belonging to the second class mentioned above, is the multivariate portmanteau statistic proposed by Jacobsen (1995) for the first step of Johansen's cointegration analysis. This combines a generalization of the univariate Box–Pierce statistic with one of the classical information criteria in order to avoid overfitting, while the portmanteau statistic ensures that the assumptions for the further estimation procedure are met, i.e. no residual autocorrelation. Although it seems reasonable to concentrate on the preconditions for further estimation, the approach presented by Jacobsen (1995) does not yet represent a suitable criterion for automatic model selection, since it does not provide clear indications on the weighting of the two components of the objective function. Furthermore, the computational load for this approach is higher than for the classical information criteria.

Therefore, as pointed out earlier, the simulation experiments documented in this section using automatic lag structure identification through threshold accepting are based solely on classical information criteria.

12.4   Implementation of Threshold Accepting

Given a realization of a VAR process as shown in Figure 12.1 or of a real time series, which according to the Wold representation theorem might be approximated by a VAR process, such as the one depicted in Figure 12.2, how can the underlying lag structure be identified or fitted by using one of the criteria introduced in the previous section? The set of admissible selection rules $\mathcal{R}$, which corresponds to all possible multivariate lag structures of order less or equal to K — possibly after imposing some a priori constraints — is a subset of $\left(\{0,1\}^{d \times d}\right)^{K'}$, where K' is the maximum lag length, which has to be set a priori. In particular, $\mathcal{R}$ is a finite set. Consequently, the optimization problem (equation (12.3)) could be solved by a trivial enumeration method. Unfortunately, this trivial algorithm requires tremendous computing resources, since the finite set $\mathcal{R}$ is very large even for modest values of d and K', if no strong restrictions such as the usual 'take all up to the kth lag' rule are imposed.

The results of Hocking and Leslie (1967), Pagano (1972), and McClave (1975) indicate that, at least for the one–dimensional case, a deeper analysis of the optimization problem may help to reduce the size of the search space considerably. Unfortunately, for the multivariate case no results of this kind are available. Furthermore, even if the search space could be reduced by a factor similar to the one–dimensional case, the problem remains intractable when relying on the trivial enumeration algorithm.

For these reasons, the problem of multivariate lag structure identification is a typical candidate for an optimization heuristic implementation, although no precise results on its mathematical complexity, e.g. by a lower bound for its time complexity, are available. The efficiency of a threshold accepting implementation depends on imposing an adequate local structure and a reasonable choice of the threshold sequence. Thereby, the arguments are similar to the case of optimal experimental design described in Chapter 11 and follow the general outline provided in Chapters 8 and 9.

Hence, first one has to define neighbourhoods $\mathcal{N}(S_0) \subset \mathcal{R}$ for all admitted selection rules $S_0 \in \mathcal{R}$. The set of all possible selection rules of dimension d and maximum lag length K' with no other restrictions imposed is given by $\left(\{0,1\}^{d \times d}\right)^{K'}$. Winker (1995) imposed the restriction that the selected lags have to be equal across equations. Imposing this restriction has the advantage that standard ordinary least squares estimation can be used. However, the restriction is artificial as the two VAR processes used to generate the data for the simulation study in Winker (1995) exhibit different lag structures in different equations. Nevertheless, imposing this restriction did not hinder the implementation to produce good results. Here, a version of the algorithm without this restriction is compared with the original version and the standard 'take all up to the kth lag' approach.

As pointed out in Chapter 9, neighbourhoods may be defined by considering $\mathcal{R}$ as a subset of some $\{0,1\}$ vector space and using the Hamming distance (Hamming, 1950).[23] If, for example, the Hamming distance between two selection rules is less than or equal to two, the rules differ only in two elements. If a Hamming distance of four is admitted, they may differ in up to four entries.

Two slight modifications of this neighbourhood definition are implemented in order to improve the performance. Similar to the approach described in Chapter 11, the first modification consists in restricting the exchange to one of the k next neighbours on the time axis of a given element of S_0. For example, if the lag of order 10 is in S_0 and randomly selected for a modification, then, under the standard Hamming distance definition of neighbourhoods, it may either be just discarded or replaced by any lag of order 1 to K'. Under the modified neighbourhood structure, only lags from, let us say, order 7 to 13 are admitted as replacement. This restriction results from the assumption that changes in the information criteria might be smaller if only lags of similar lag order are exchanged. Consequently, the values of the objective function corresponding to some search path become smoother, which is a necessary requirement for an efficient threshold accepting implementation.

At least if the total number of iterations of the threshold accepting implementation has to be small and given a real economic time series as the input, a further slight improvement can be obtained by biasing the search in favour of selection rules with a small number of non–zero entries. For example, given $K' = 20$, if a lag of order 10 is randomly selected and not yet in

---

[23] Ausiello and Protasi (1995) also use the Hamming distance to define neighbourhoods on $\{0,1\}$ vector spaces.

S_0, then the probability of including it should be less than $1/2$, while the probability of eliminating lag 10 should be larger than $1/2$. This modification of the algorithm does not change the neighbourhoods themselves but the probabilities that individual elements of a neighbourhood are selected in a search step, which routinely is set equal to $1/|\mathcal{N}(\mathbf{x})|$, i.e. equal for all elements of a neighbourhood.

In Chapter 9, a few methods for a heuristic comparison of different neighbourhood structures have been proposed, which have already been used in Section 11.3 for the optimal design problem. One possibility consists in a small–scale simulation of the induced local structure. The results presented in Figure 12.3 are based on simulations of 5000 selection rules $S_{i,0} \in \mathcal{R}, i = 1, \dots, 5000$, which are generated randomly. Then, for each $S_{i,0}$ one element $S_{i,1} \in \mathcal{N}(S_{i,0})$ is selected for the given neighbourhood definition. The value of the objective function C is calculated for both selection rules and the difference $\Delta C_i = C(S_{i,1}) - C(S_{i,0})$ is stored. Finally, histograms of the deviations can be calculated and plotted.[24] Figure 12.3 shows a few examples of such histograms for the HQ information criterion and the first example used in the next section given by the data–generating process described in equation (12.4). The top panel gives the result for the trivial case, where all neighbourhoods are equal to the set of possible selection rules $\mathcal{R}$. As might be expected, no local structure can be found by using this definition. The lower panels give the results for neighbourhoods defined by a Hamming distance (d_H) of four and six, respectively, restricting the exchange steps to one of $k = 10$ next neighbours on the time axis.

While the trivial topology does not lead to any local structure, the other neighbourhood definitions do, as can be detected from the concentration of local deviations around zero. The smaller the neighbourhoods become, then the more concentrated is the distribution of local deviations around zero. However, as pointed out in Chapter 9, defining the neighbourhoods as being too small increases the risk that the algorithm will get stuck in some local minimum of poor quality.

Given this neighbourhood definitions, only the threshold sequence $\{T_i\}_{i=0}^{I}$ is still missing for a fully specified implementation of threshold accepting. This is generated along the lines depicted in Chapter 8 (pp. 127ff.) by a small–scale simulation similar to the one used for Figure 12.3. First, a small number I_0 of selection rules $S_{i,0} \in \mathcal{R}$ is generated randomly for $i = 1, \dots, I_0$. Then, for each $S_{i,0}$ one element $S_{i,1} \in \mathcal{N}(S_{i,0})$ is selected randomly according to the given neighbourhood definition and — if relevant — additional restrictions. The value of the objective function is calculated for both selection rules and the absolute value of the difference $\Delta C_i =| C(S_{i,1}) - C(S_{i,0}) |$ is stored. The generated values of the local deviations contain information about the

---

[24] In the figure, the deviations are multiplied with 100 to keep the axis labelling readable.

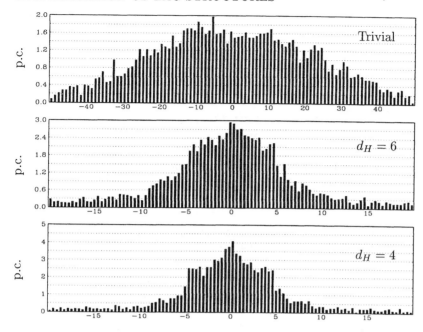

Figure 12.3   Histograms of local deviations.

typical changes in the value of C to be expected when moving from S_0 to an element of its neighbourhood. Consequently, the absolute values of these local deviations ΔC_i, sorted in decreasing order, are used as the threshold sequence. The performance can be improved by using only a fraction α from the end of this sequence. The simulation study of Chapter 8 demonstrates that this performance gain is more important if only a small number of iterations is used. Finally, for each remaining threshold value a fixed number of iterations is performed. As the mass of the distribution of local deviations is concentrated around zero, the algorithm will use up more iterations for small values of the threshold than for very large ones.

Given these ingredients, the threshold accepting implementation works as described in some detail in Chapter 6. It starts with a randomly chosen selection rule S_0. In each iteration step this selection rule is compared with a selection rule in its neighbourhood, which is generated by exchanging some entries. This new selection rule is kept as the current solution if the value of the information criterion is either better or at least not much worse than for the old current selection rule. Thereby, the current value of the threshold sequence sets the limit up to which a worsening is accepted. After an a priori fixed number of iterations is performed for the current threshold value, it is replaced by the next one from the empirical jump distribution described above. Of course, the overall performance of the implementation depends strongly on the total number of iterations performed for one problem instance.

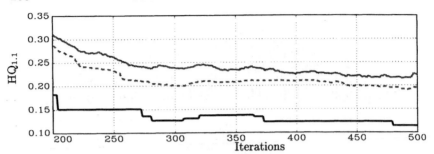

Figure 12.4 Minimization of the HQ information criterion.

Some insights into the functioning of this threshold accepting implementation can be gained from Figure 12.4. This shows the values of the Hannan–Quinn information criterion of the current solutions S_0 and provides information on the values of the candidate solutions S_1 considered during the early stage of an optimization run. Thereby, the constant c in the Hannan and Quinn criterion was set to 1.1.

The lower solid line shows the values of HQ for the current solutions S_0. The hill–climbing behaviour of the algorithm, necessary to escape bad local minima, can be detected between iterations 300 and 400: An increase of the information criterion is accepted in order to achieve the improvement after iteration 360. The grey solid line shows the mean of the last 100 values of the information criterion of the candidate solutions and the grey dashed line their median. As S_1 is always selected in a neighbourhood of S_0, the local structure of the problem leads to a decrease both in the mean and the median as the current solution improves.

Even a perfectly tuned threshold accepting implementation might sometimes fail to identify the 'true' lag structure due to one of the following three reasons. First, the model class can be mis–specified, i.e. either the process cannot be described by a simple VAR model at all, for example if the variables are nonstationary or some of the variables are just the wrong variables, or it cannot be described by a VAR process with maximum lag order K' as it may include a MA–term or very long autoregressive memory. Secondly, both the distribution of the error terms and the identification property of the information criterion are valid only asymptotically, if at all. Hence, for a finite sample the best–fitting model in terms of some model selection criteria does not have to be the 'true' model even if the data are generated exactly according to some VAR model with maximum lag order less or equal to K'. Finally, if the set of admissible selection rules $\mathcal{R}$ is large and the information criterion has many local minima on $\mathcal{R}$ with regard to the chosen neighbourhood definition, which seems to be the typical case in applications, threshold accepting cannot guarantee to find a global minimum with a finite and small number of iterations.

Despite these restrictions, the threshold accepting implementation allows an automatised identification of lag structures according to some predefined objective function, or at least to find a lag structure very close to the best one. Subjective interventions can be avoided in this step of a modelling procedure. Instead, the econometrician might concentrate on the choice of the correct model class.

12.5 Results of Monte Carlo Simulations

This section presents some results of the automatic identification method introduced in the previous sections. The results are based on Monte Carlo simulations for artificially generated data in order to know the 'true model' as a bench–mark case. Section 12.5.1 uses the bivariate VAR process described in Winker (1995) as a bench–mark. However, the restriction that the selected lags have to be equal across equations imposed in the original publication is relaxed. Therefore, the results with and without this restriction can be compared with the results of the traditional 'take all up to the kth lag' approach. Next, in Section 12.5.2, two alternative VAR processes, including one with a dense lag structure, which favours the traditional approach, are used to study the performance of the threshold accepting implementation. Furthermore, in Section 12.5.3 randomly generated structures are used to enlarge the base for the performance comparison. In Section 12.5.4, the method is applied to an actual trivariate process of interest rate changes. Finally, Section 12.5.5 describes the implementation details employed for the money demand example described above in Section 3.2.2.

12.5.1 A bivariate VAR process

The data–generating process in Winker (1995) is given by

$$x_{1t} = 0.12x_{1,t-1} - 0.30x_{1,t-4} + 0.30x_{1,t-12} \qquad (12.4)$$
$$+0.46x_{2,t-1} - 0.20x_{2,t-2} + \varepsilon_{1t}$$

$$x_{2t} = 0.45x_{1,t-1} - 0.4x_{1,t-2}$$
$$+0.3x_{2,t-1} - 0.12x_{2,t-3} + 0.16x_{2,t-12} - 0.17x_{2,t-18} + \varepsilon_{2t},$$

where ε_{1t} and ε_{2t} are independently drawn from a standard normal distribution.[25] Figure 12.1 above shows one realization of this process.

The standard normal random numbers were generated by the random number generator DNRAND of the ESSL–library. This routine is based on

---

[25] As the smallest root of the reverse characteristic polynomial of this bivariate VAR process is of modulus 1.036, the process is stable and hence stationary (Lütkepohl, 1991, Ch. 2).

the Polar method described in Knuth (1981).[26] For this and all following simulations, the VAR process is initialized by setting all starting values for $x_{1,1}, \ldots, x_{1,K'}$ and $x_{2,1}, \ldots, x_{2,K'}$ to zero. Then, the process is simulated for $1000+T$ periods, if a sample of length T is used for lag structure identification. Only observations $t = 1001, \ldots, 1000 + T$ are actually employed for the identification process in order to eliminate the effect of the initial conditions.

By using different initial seed values for the random number generator, several different realizations of the bivariate VAR process are generated. On each of them the threshold accepting implementation is run five times with different initializations of the random number generator used to generate the test solutions. Only for a very small number of iterations, i.e. less than 2000, will the results for the different runs differ with a relevant frequency. Nevertheless, only the results of the best run are used for the summary statistics in this section.

Of course, the results, in general, depend on a large number of parameters referring to the selection criteria on the one hand and to the threshold accepting implementation on the other. For the results presented in this section, all parameters with the exception of six are fixed. The fixed parameters include, e.g. the definition of neighbourhoods. Consequently, the performance of the threshold accepting implementation documented by the results in this section are conditional on these fixed parameters. However, some preliminary tests indicate that their influence is rather limited.

The six parameters analysed in detail in the Monte Carlo simulation are the maximum lag length K', the constant c of the error criterion, the number of iterations I of the threshold accepting implementation on each realization, the number of observations T, the fraction α of the empirical jump distribution, which is used as the threshold sequence, and the restriction on the lag structure over the equations ($r = 0$ if the restriction is imposed, $r = 1$ otherwise). Of course, an exhaustive search of this six–dimensional parameter space is not feasible. Hence, only a small number of different values is used for each parameter, as shown in Table 12.1.

Even this limited set of parameter values allows for a total of 1944 combinations. Hence, an exhaustive simulation is still not feasible, given constraints on available computer resources. Therefore, the impact of the parameters is analysed by using only a small subset of all the combinations. Given that hardly any a priori knowledge on the impact of the different parameters is available, it seems reasonable in the light of the results presented in Chapter 11 to approximate the uniform distribution on the space spanned by the parameters. Therefore, a D–optimal (and, in fact, column orthogonal)

---

[26] Despite the potential shortcomings of such standard congruential random number generators discussed in Chapter 11, the standard method is applied not only for convenience, but also due to the high dimensionality of the problem ($2 \times (1000 + T)$), when quasi–Monte Carlo methods so far fail to improve results considerably.

Table 12.1   Design parameters for threshold accepting implementation analysis

Parameter	No.	Values					
K'	6	12	16	20	24	28	32
c	6	1.01	1.1	1.2	1.3	1.4	1.5
I	3	1000	3000	5000	—	—	—
T	3	100	200	300	—	—	—
r	2	0	1	—	—	—	—
α	3	0.1	0.2	0.3	—	—	—

design with 36 elements is used, which is generated by using the methodology introduced in Chapter 11.[27] Table 12.2 shows the parameters of the 36 different specifications for which a simulation is performed.

For each of these 36 parameter settings, the threshold accepting implementation is run on 20 different realizations of the process represented by equation (12.4). It is obviously not very informative to report all of the results in detail. Summary statistics for each parameter setting are available from this author on request. In particular, these tables include the frequencies that a specific lag is included in the model selected by the algorithm. However, in order to obtain some even more condensed presentation of the results, Tables 12.3 and 12.4 present highly aggregated statistics on lag order selection and lag length.

Before turning to these latter tables, a result on the quality of the approximation obtained by the threshold accepting implementation should be reported. For design (1), i.e. with $K' = 12$ and the maintained restriction that the included lags have to be the same across the equations, it is still possible to obtain the exact result by enumeration. However, it takes about 15 hours computing time on the RS 6000 3CT used for the calculations. In generating 10 different realizations of the process represented by equation (12.4), the threshold accepting implementation with 1000 iterations always produced the global optimum under the constraint $K' = 12$ and equal lag structures across the equations. For all other instances, it is found that the lag structure proposed by the threshold accepting implementation results in a smaller value of the objective function than the 'true' model for a given realization. Thus, if the identified models are different from the 'true DGP', it is rather due to small sample effects than to the performance of the threshold accepting implementation — at least for this smallest problem instance covered in the simulation study.

Table 12.3 shows the frequency that the threshold accepting implementation

[27] Sacks, Welch, Mitchell and Wynn (1989) also advocate the use of experimental design methods for the analysis of computer experiments.

OPTIMIZATION HEURISTICS IN ECONOMETRICS

Table 12.2 Design for threshold accepting implementation analysis

No.	K'	c	Iter.	r	T	α	No.	K'	c	Iter.	r	T	α
(1)	12	1.01	1000	0	100	0.1	(19)	24	1.10	3000	0	300	0.1
(2)	12	1.30	1000	0	100	0.1	(20)	24	1.10	5000	0	100	0.2
(3)	12	1.30	5000	1	100	0.2	(21)	24	1.30	1000	0	200	0.3
(4)	12	1.40	5000	0	300	0.1	(22)	24	1.50	1000	1	300	0.1
(5)	12	1.40	5000	1	100	0.3	(23)	24	1.50	3000	1	200	0.1
(6)	12	1.50	5000	0	300	0.3	(24)	24	1.50	5000	1	100	0.1
(7)	16	1.01	1000	1	200	0.1	(25)	28	1.01	5000	0	200	0.1
(8)	16	1.10	5000	0	300	0.2	(26)	28	1.10	5000	0	300	0.3
(9)	16	1.10	5000	1	200	0.3	(27)	28	1.20	3000	0	100	0.3
(10)	16	1.20	1000	0	300	0.2	(28)	28	1.20	5000	1	300	0.2
(11)	16	1.30	3000	1	200	0.1	(29)	28	1.30	3000	0	100	0.3
(12)	16	1.40	1000	1	100	0.3	(30)	28	1.50	3000	0	200	0.2
(13)	20	1.01	1000	1	300	0.3	(31)	32	1.01	3000	1	200	0.2
(14)	20	1.01	3000	1	100	0.2	(32)	32	1.10	5000	1	200	0.1
(15)	20	1.20	1000	0	200	0.3	(33)	32	1.30	1000	0	100	0.2
(16)	20	1.20	1000	1	300	0.2	(34)	32	1.40	1000	1	300	0.2
(17)	20	1.20	3000	1	200	0.3	(35)	32	1.40	3000	0	100	0.1
(18)	20	1.50	3000	0	300	0.2	(36)	32	1.40	3000	1	200	0.3

included a lag present in the true data generating process (RL) or mistakenly includes another lag (WL). For the instances where the maximum lag length K' was set smaller than the true lag order (18), these frequencies are calculated only with regard to the lags that the algorithm could identify under this restriction. Obviously, the optimal values are 100% and 0%, respectively.

The results indicate that the choice of parameters influences the results quite heavily. The frequency that relevant lags are included in the model selected by the threshold accepting implementation ranges between 46.8 and 85.5%, while the corresponding frequencies for the wrong lags, i.e. lags not present in the true data–generating process but selected by the algorithm, range between 0.3 and 6.2%. Obviously, it is of great interest to identify those parameters which influence the results significantly, and to find good parameter values. A straightforward regression analysis of the results indicates that the relevant variables for the performance on the RL are the constant in the objective function c, the number of available observations T and the dummy indicating whether the restriction on the lags to be equal across the equations is imposed or not (r). Obviously, a large value of c reduces the frequency of identifying a correct lag, while an increasing number of iterations increases it. No significant effect is found for the tuning parameter α, which determines the share of threshold values taken from the empirical step distribution, and for the number of iterations performed by the algorithm.

Table 12.3 Lag selection frequencies, shown as %

No.	RL	WL	No.	RL	WL	No.	RL	WL
(1)	59.4	2.2	(13)	85.5	3.7	(25)	68.3	2.2
(2)	54.4	1.2	(14)	55.0	6.2	(26)	80.0	1.4
(3)	52.5	5.4	(15)	65.0	2.1	(27)	51.1	2.0
(4)	80.0	0.6	(16)	84.1	3.2	(28)	85.0	2.7
(5)	53.0	5.0	(17)	75.9	3.5	(29)	50.6	2.0
(6)	78.1	0.3	(18)	74.4	0.3	(30)	62.8	1.1
(7)	77.0	5.2	(19)	80.0	1.5	(31)	75.9	4.4
(8)	82.5	1.9	(20)	51.7	3.1	(32)	74.5	4.0
(9)	76.5	5.1	(21)	62.2	1.7	(33)	51.1	1.9
(10)	81.9	1.5	(22)	82.7	1.7	(34)	83.2	1.6
(11)	76.0	3.7	(23)	72.7	2.4	(35)	48.3	1.5
(12)	50.5	4.2	(24)	46.8	3.8	(36)	72.3	2.7

This finding indicates that for the considered small–scale problem a small number of iterations is already enough to find a high–quality solution.

For the frequency of mistakenly including a lag not present in the data–generating process (DGP), the same parameters appear to be significant in the regression analysis. Again, a larger number of observations is favourable as this has a negative impact on this frequency. The same holds true for the constant c, in contrast to the finding for the RL. Obviously, these opposite effects reflect the trade–off in the selection criterion between goodness of fit and model parsimony. Imposing the restriction of equal lags across equations ($r = 0$) increases the frequency of including wrong lags, as might have been expected. Finally, again no significant impact of the threshold parameters is found.

Table 12.4 provides some summary information on the lag order K identified by the threshold accepting implementation. The three columns give the shares over all VAR process realizations that a too small, the correct or a too large lag order, respectively, is found. In this case, an optimal result would be 0, 100 and 0%, respectively. However, if the maximum lag order K' is set smaller than the true lag order, all lag orders found by any algorithm will be too small. In the table, only the shares of too small (TS) and too large (TL) lag orders are provided, since the share of correct lag orders is easily imputed.

For identifying the correct lag order, the variable K' is of crucial importance. If K' is chosen to be smaller than the true lag length, any selection algorithm necessarily comes up with a too small lag order. Consequently, for a regression analysis of the impact of parameters on the frequencies of selecting a too short lag order (TS), only designs (13) to (36) are considered. For these designs, the regression analysis indicates a significant effect of c, i.e. a larger value of c reduces the number of included lags, thus making it more likely to end

Table 12.4 Lag order frequencies, shown as %

No.	TS	TL	No.	TS	TL	No.	TS	TL
(1)	100.0	0.0	(13)	5.0	25.0	(25)	35.0	30.0
(2)	100.0	0.0	(14)	20.0	30.0	(26)	20.0	25.0
(3)	100.0	0.0	(15)	45.0	5.0	(27)	50.0	30.0
(4)	100.0	0.0	(16)	10.0	10.0	(28)	5.0	60.0
(5)	100.0	0.0	(17)	30.0	20.0	(29)	50.0	30.0
(6)	100.0	0.0	(18)	40.0	5.0	(30)	40.0	15.0
(7)	100.0	0.0	(19)	20.0	20.0	(31)	0.0	80.0
(8)	100.0	0.0	(20)	45.0	30.0	(32)	5.0	75.0
(9)	100.0	0.0	(21)	40.0	15.0	(33)	50.0	35.0
(10)	100.0	0.0	(22)	5.0	15.0	(34)	5.0	40.0
(11)	100.0	0.0	(23)	25.0	35.0	(35)	50.0	30.0
(12)	100.0	0.0	(24)	15.0	50.0	(36)	10.0	65.0

up with a too short lag structure. Furthermore, a significant impact of K' is found. Thus, even if K' is chosen larger than the true lag length, it has a significant negative impact on selecting a too small order. Increasing the number of observations, other things being equal, reduces the risk of ending up with a too small lag order. No additional effects from the tuning parameters of the threshold accepting implementation appear to be significant in the regression analysis.

While increasing K' reduces the risk of selecting a too small lag order, it significantly increases the risk of selecting a too high one, as the regression results for TL indicate. Increasing the constant c has a significant effect working in the opposite direction. The number of available observations shows the expected negative sign, but is not significant in the regression analysis. While again no significant effect can be found for the threshold accepting tuning parameters, a significant positive effect is found for the dummy variable, thus indicating that the unrestricted model is estimated. Obviously, adding a not relevant lag to only one equation does less harm to the objective function than adding it to both equations, as imposed by the restricted version. Therefore, it is more likely that the selection criterion will result in the inclusion of some high-order irrelevant lag if the unrestricted version is considered.

The findings of this large-scale simulation analysis of the threshold accepting implementation for lag structure identification may be summarized as follows: First, the simulations differ with regard to the sample sizes (*obs*). Differences in the results due to differing sample sizes have to be attributed to the small-sample deficiencies of the information criteria, which are not of central interest in this present chapter. Consequently, for the rest of this chapter only results for sample size 200 are reported. As one might expect, the quality of the identification by threshold accepting increases with sample size.

Secondly, the simulations are repeated for different values of the constant c. Smaller values of the constant increase the number of included lags, while larger values decrease it. Hence, for finite samples, the user faces a trade–off between the risk of overfitting — by choosing a too small value for c — and missing relevant lags — by choosing a too high value. Note, however, that this is not a specific feature of the automatic lag order identification, but stems from the definitions of the information criteria themselves.

Thirdly, the maximum lag length K' is difficult to choose if the true data–generating process is unknown. The simulation results indicate that as long as the maximum lag length is chosen to be higher than the maximum lag length of the data–generating process, the results do not differ much. However, mistakenly choosing $K' < K$ has undesirable effects. Thus, for applications to real data K' should not be chosen too small.

Finally, the tuning parameters of the threshold accepting implementation do not seem to have a relevant impact. However, for larger problem instances, in particular, in higher dimensions, these tuning parameters become more important. For the bivariate models studied in this and the following section, however, 2000 iterations seem to be sufficient.

Table 12.5 summarizes some results obtained from the simulation study for the data–generating process represented by equation (12.4) using the HQ information criterion with the constant c set to 1.1. The threshold accepting implementation is run with 2000 iterations for each try and a fraction $\alpha = 0.3$ of the empirical jump function as the threshold sequence. The maximum lag length K' is set to 20. The VAR process realizations are of length 200. Results for both the restricted and the unrestricted versions of the algorithm are given, as well as the results from the 'take all up to kth lag' approach. The simulation comprises 100 randomly generated VAR process realizations.

The entries in the columns with the headings 'x_1 and x_2' are the frequencies with which $x_{i,t-1}, \ldots, x_{i,t-20}$ are found to be included in the model deemed best by the standard 'take all up to the kth lag' approach. The next two columns indicate the frequencies with which the lags for x_1 and x_2, respectively, are found to be included by the constrained version of the algorithm, that is, the same lags have to appear in both equations. Finally, the last four columns give the results for the unconstrained version of the algorithm in which different lags may appear in different equations. Again, the frequencies are reported. Bold–faced numbers indicate the lags of the 'true' data–generating process.

Table 12.5 exhibits that, due to the sparse structure of the data–generating process used in this simulation, the standard approach never finds the true lag length of 18. It also always fails to detect the annual lag at length 12, while these two lags are detected with frequencies larger than 55 and 90%, respectively, by the automatic selection procedures based on threshold

Table 12.5 Simulation results for process (12.4) obtained by using $HQ_{1.1}$

Method	Take all up to kth lag	Threshold accepting					
		Constrained		Unconstrained			
				Eq. for x_{1t}		Eq. for x_{2t}	
Lag	x_1 and x_2	x_1	x_2	x_1	x_2	x_1	x_2
$t-1$	100	100	100	22	100	100	97
$t-2$	100	98	53	1	69	97	4
$t-3$	41	3	18	4	5	1	42
$t-4$	32	89	1	97	8	8	2
$t-5$	0	3	3	5	3	6	3
$t-6$	0	3	3	4	3	4	2
$t-7$	0	3	5	5	3	6	7
$t-8$	0	1	2	2	4	2	5
$t-9$	0	0	3	4	2	3	6
$t-10$	0	2	0	5	1	5	2
$t-11$	0	2	4	4	9	5	3
$t-12$	0	91	43	97	1	5	64
$t-13$	0	4	1	8	5	7	4
$t-14$	0	1	3	4	2	2	4
$t-15$	0	2	1	2	3	8	1
$t-16$	0	4	4	4	6	5	6
$t-17$	0	1	0	1	3	4	0
$t-18$	0	0	56	3	7	4	74
$t-19$	0	3	1	3	1	4	2
$t-20$	0	0	1	2	6	3	3
Lags	5.46	7.12		11.28			
< True	0	100		100			

accepting.

The last row of the table shows that the information criterion is always smaller for the model identified by threshold accepting than for the true model, even when using the small number of 2000 iterations. Consequently, the fact that the lags of the true model are not always included in the identified model has to be attributed at least partially to the fact that the true model does not fit the generated data best with regard to the given C. This is not too surprising, given the small number of observations (200).

The next to the last row gives the mean number of lags included in the selection rules found by threshold accepting. While the true number of lags in the model represented by equation (12.4) is 9, using the HQ with $c = 1.1$ leads to some underfitting for the constrained version of the algorithm. This can be reduced somewhat by decreasing the value of c. However, increasing the mean number of included lags also increases the probability of identifying some lags not present in the true model. The mean number of included lags of

11.28 obtained for the unconstrained case has to be compared with the true number of 11 lags in the model, because some lags have to be counted twice in this case, as they appear both in the equation for x_{1t} and x_{2t}. Thus, the constant $c = 1.1$ seems to be well chosen in this case.

It should be noted that the number of selection rules in the unconstrained case is not just the double of those in the constrained case, but is, in fact, the square. Consequently, it is easier for the automatic identification procedure to find a model with a lower value of the information criterion than the true model. In fact, this goal is always reached in both cases, but the task of finding an optimal lag structure in the unconstrained case is much more complex than in the constrained case.

The current implementation of the algorithm (with the same small number of 2000 iterations as for the restricted case) still selects most of the true model's lags with high probability, as the bold–faced figures in Table 12.5 exhibit. However, some lags corresponding to small values of the autoregressive coefficients (for example, $x_{1,t-1}$ in the equation for x_{1t}) seem to be not clearly identified given the VAR process realizations.

Table 12.6 shows the effects of a decrease of the constant c to 1.01 (instead of 1.1). Since a smaller value of c in the HQ criterion corresponds to a smaller weight of the loss in degrees of freedom in the information criterion, an increase in the mean number of included lags should result.

In fact, the mean number of lags increases for all three algorithms, although only to a minor extent. The standard approach and the constrained version of the threshold accepting implementation still underestimate the number of lags, while the unrestricted version of the threshold accepting algorithm includes slightly more lags than the true data–generating process. In general, the results do not differ markedly from the results obtained for $c = 1.1$. Thus, for the other processes analysed in the next sections, only $c = 1.1$ is considered.

Alternatively to the HQ criterion, the BIC or FPE criteria can be used. Table 12.7 summarizes the results obtained for the same simulation set–up as before, but in this case by using the FPE criterion with $c = 4$ instead of the HQ criterion. The constant $c = 4$ was chosen in order to obtain a similar mean of the number of included lags for the constrained threshold accepting implementation as in Table 12.5.

The results for the standard approach do not change much when using the FPE criterion. The annual lag at $t - 12$ is included in 2% of all cases instead of 0% for the HQ criterion with $c = 1.1$, while the lag at $t - 18$ is not detected for a single one of the 100 realizations of the VAR process. However, even this small increase in the probability of detecting the annual lag results in a large number of selected lags which do not belong to the true data–generating process. Thus, the standard approach uses a mean number of 6.12 lags, whereas the threshold accepting algorithm (restricted version) achieves much higher identification frequencies with only a slightly higher mean number of 7.75 lags. In fact, the identification frequencies of the

Table 12.6   Simulation results for process (12.4) obtained by using $HQ_{1.01}$

Method	Take all up to kth lag	Threshold accepting					
		Constrained		Unconstrained			
				Eq. for x_{1t}		Eq. for x_{2t}	
Lag	x_1 and x_2	x_1	x_2	x_1	x_2	x_1	x_2
$t-1$	100	100	100	23	100	100	98
$t-2$	100	97	57	1	69	98	5
$t-3$	46	3	21	4	5	2	42
$t-4$	32	90	2	97	9	8	2
$t-5$	0	3	3	7	3	7	4
$t-6$	0	3	3	4	3	4	2
$t-7$	0	3	4	6	4	5	8
$t-8$	0	1	2	2	4	5	5
$t-9$	0	1	3	5	3	4	6
$t-10$	0	3	0	6	1	5	3
$t-11$	0	2	4	5	10	5	6
$t-12$	0	91	43	97	2	6	64
$t-13$	0	5	4	9	5	6	6
$t-14$	0	1	3	6	3	2	2
$t-15$	0	3	1	2	3	8	6
$t-16$	0	4	5	5	6	9	6
$t-17$	0	2	1	2	4	5	2
$t-18$	0	1	60	4	7	5	78
$t-19$	0	4	2	3	1	4	4
$t-20$	0	0	1	2	7	4	7
Lags	5.72	7.36		11.87			
< True	0	100		100			

threshold accepting implementation do not differ considerably from the ones obtained by using the HQ criterion. Hence, the often observed and discussed property of the FPE criterion to include more lags than necessary might be accentuated by the 'take all up to kth lag' approach. In contrast, using the BIC criterion results in a smaller mean number of lags included for all three approaches. The relative performance of the different approaches, however, remains unchanged. Therefore, the following discussion considers only the asymptotically consistent HQ selection criterion.

12.5.2   Further bivariate VAR processes

The first alternative example is generated heuristically from a bivariate process of German interest rates given by the first two series depicted in Figure 12.2. Given these data, a number of relevant lags is selected ad hoc by using a top–to–bottom elimination strategy. Consequently, although the

Table 12.7   Simulation results for process (12.4) obtained by using FPE$_4$

Method	Take all up to kth lag	Threshold accepting					
		Constrained		Unconstrained			
				Eq. for x_{1t}		Eq. for x_{2t}	
Lag	x_1 and x_2	x_1	x_2	x_1	x_2	x_1	x_2
$t-1$	100	100	100	27	100	100	98
$t-2$	100	98	61	1	73	98	6
$t-3$	48	4	22	4	5	2	42
$t-4$	42	92	3	98	8	9	3
$t-5$	2	4	2	6	4	9	6
$t-6$	2	5	2	4	4	5	3
$t-7$	2	4	6	4	3	7	7
$t-8$	2	1	3	2	5	4	5
$t-9$	2	1	3	6	3	3	7
$t-10$	2	3	1	6	1	4	4
$t-11$	2	2	5	4	11	5	3
$t-12$	2	95	47	97	2	8	64
$t-13$	0	5	3	9	5	7	8
$t-14$	0	4	5	7	4	1	3
$t-15$	0	3	2	2	4	7	6
$t-16$	0	5	7	4	8	8	6
$t-17$	0	2	1	4	4	5	3
$t-18$	0	3	59	5	7	7	79
$t-19$	0	5	2	3	1	5	4
$t-20$	0	2	3	3	9	4	6
Lags	6.12	7.75		12.18			
$<$ True	0	100		100			

following data–generating process is artificial, it can be assumed to share some features with the economic time series:

$$x_{1t} = 0.909x_{1,t-1} + 0.105x_{1,t-7} + 0.191x_{1,t-12} - 0.192x_{1,t-13} \qquad (12.5)$$
$$+0.198x_{2,t-1} - 0.069x_{2,t-4} - 0.104x_{2,t-7}$$
$$-0.048x_{2,t-12} + 0.044x_{2,t-17} - 0.041x_{2,t-18} + \varepsilon_{1t}$$

$$x_{2t} = 0.509x_{1,t-1} - 0.304x_{1,t-3} - 0.262x_{1,t-13} + 0.162x_{1,t-19}$$
$$+0.791x_{2,t-1} + 0.218x_{2,t-6} + 0.300x_{2,t-7} + 0.181x_{2,t-9}$$
$$-0.111x_{2,t-11} + 0.197x_{2,t-14} + 0.108x_{2,t-18} - 0.235x_{2,t-19} + \varepsilon_{2t},$$

where ε_{1t} and ε_{2t} are independently drawn from a standard normal distribution. Compared to equation (12.4), this process exhibits a more complicated dynamic structure, including a larger number of lags.

Table 12.8 summarizes some simulation results for this process obtained by

Table 12.8 Simulation results obtained for process (12.5)

Method	Take all up to kth lag	Threshold accepting					
		Constrained		Unconstrained			
				Eq. for x_{1t}		Eq. for x_{2t}	
Lag	x_1 and x_2	x_1	x_2	x_1	x_2	x_1	x_2
$t-1$	100	100	100	100	96	100	100
$t-2$	96	3	2	1	4	4	1
$t-3$	76	71	2	1	4	87	0
$t-4$	4	2	8	2	14	2	0
$t-5$	1	2	6	2	9	2	5
$t-6$	1	0	41	3	5	3	58
$t-7$	1	21	79	18	36	9	79
$t-8$	1	4	6	5	3	0	6
$t-9$	0	4	32	2	1	1	44
$t-10$	0	5	6	2	1	4	3
$t-11$	0	5	12	3	1	7	17
$t-12$	0	43	6	33	7	5	4
$t-13$	0	83	4	30	3	85	4
$t-14$	0	5	52	5	2	5	60
$t-15$	0	2	13	2	2	3	14
$t-16$	0	4	0	8	2	4	3
$t-17$	0	2	3	1	4	2	4
$t-18$	0	8	18	3	6	11	19
$t-19$	0	44	60	1	0	55	67
$t-20$	0	6	5	1	2	10	8
Lags	5.60	8.69		13.20			
$<$ True	0	100		100			

using the same parameter setting as for the results presented in Table 12.5 for the process represented by equation (12.4).

The lag structure of process (12.5) is denser than the first example, including 17 lags without and 22 lags with double counting of lags. Again, the standard approach fails to give a reasonable approximation to the true data–generating process. It includes the irrelevant lag length 2 with a frequency of 96%, while it misses completely the relevant contributions at lag lengths 7, 12 and 13, plus 17 to 19.

Some of the coefficients for neighbouring lags are similar in magnitude but with different signs. Hence, it is not too surprising that the optimized lag structures fail to identify pairs of lags such as $x_{2,t-17}$ and $x_{2,t-18}$ quite often. For some lags (for example, $x_{2,t-14}$) the unconstrained version exhibits a slight advantage.

Again, both the constrained and unconstrained versions of the algorithm always find lag structures having a smaller value of the objective function than the 'true' lag structure of the data–generating process, while the standard approach never does. Furthermore, in this case the constant $c = 1.1$ leads to a clear underfitting of the model. Thus, reducing the constant could improve the results.

The second alternative example is selected in order to allow a fair comparison with the standard 'take all up to the kth lag' approach. This process very much favours the standard approach, as only the first four lags are included in all equations. This lag structure does not have any 'holes', i.e. the lag structure is dense. This data–generating process is described by

$$
\begin{aligned}
x_{1t} \;=\; & 0.40x_{1,t-1} - 0.20x_{1,t-2} - 0.05x_{1,t-3} && (12.6) \\
& + 0.10x_{1,t-4} + 0.20x_{2,t-1} + 0.10x_{2,t-2} \\
& + 0.05x_{2,t-3} - 0.10x_{2,t-4} + \varepsilon_{1t} \\[6pt]
x_{2t} \;=\; & 0.45x_{1,t-1} + 0.10x_{1,t-2} - 0.10x_{1,t-3} \\
& - 0.20x_{1,t-4} + 0.10x_{2,t-1} + 0.20x_{2,t-2} \\
& - 0.10x_{2,t-3} + 0.10x_{2,t-4} + \varepsilon_{2t} \,.
\end{aligned}
$$

Table 12.9 summarizes the findings for this process. In this special case, all three methods select lag structures with a value of the information criterion smaller than the one for the true data–generating process. The information criterion is smaller for the optimization approaches compared to the standard method, as should be expected.

The risk of including lags of too high an order is zero for the standard approach, while still being relevant for the optimization methods. On the other hand, the standard approach fails to find the true lag order of 4 in more than 80% of the cases. This risk is reduced to 12 and 5% for the optimization approaches. However, the latter approaches face difficulties in identifying the contribution of the lags at $t-3$ for the same reason as mentioned above for the previous data–generating process. Finally, although the data–generating process given by equation (12.6) clearly favours the standard approach, unconstrained optimization by using threshold accepting still results in a higher probability of detecting the relevant lags.

Using three ad hoc selected data–generating processes does not allow for far–reaching conclusions about the efficiency of the different methods. Therefore, the next section analyses a large number of randomly generated models. Nevertheless, the data–generating processes introduced so far cover some typical features of economic time series and, therefore, allow for a first tentative conclusion. The optimization approaches always find lag structures with smaller values of the objective function than the standard approach. Furthermore, the models selected by the threshold accepting implementations always result in a smaller value of the information criteria than for the 'true' data–generating process. For sparse structures, they are likely to identify most of the relevant variables. If the true data–generating process exhibits 'holes',

Table 12.9   Simulation results obtained for process (12.6)

Method	Take all up to kth lag	Threshold accepting Constrained		Threshold accepting Unconstrained Eq. for x_{1t}		Eq. for x_{2t}	
Lag	x_1 and x_2	x_1	x_2	x_1	x_2	x_1	x_2
$t-1$	100	100	61	99	61	100	18
$t-2$	69	57	50	61	16	23	54
$t-3$	41	5	15	2	10	3	14
$t-4$	17	75	26	28	19	79	12
$t-5$	0	2	2	1	3	1	1
$t-6$	0	3	2	6	1	2	2
$t-7$	0	3	3	4	1	2	3
$t-8$	0	0	2	3	0	1	4
$t-9$	0	0	0	0	2	0	0
$t-10$	0	1	1	1	2	1	3
$t-11$	0	0	2	1	3	2	1
$t-12$	0	0	1	1	1	1	2
$t-13$	0	4	1	1	1	5	1
$t-14$	0	3	0	2	0	1	3
$t-15$	0	0	1	2	4	0	1
$t-16$	0	3	4	1	1	1	2
$t-17$	0	0	3	1	2	2	1
$t-18$	0	1	2	0	2	1	1
$t-19$	0	1	0	1	0	2	1
$t-20$	0	0	0	0	0	0	2
Lags	4.54	4.34		6.97			
< True	100	100		100			

as in processes (12.4) and (12.5), heuristic optimization methods are clearly superior to the standard method. Furthermore, the last example shows that even for lag structures without 'holes', heuristics may be better suited for identifying the relevant lags.

12.5.3   Randomly generated bivariate structures

Besides the three instances considered as being typical for time series models in econometrics, the automatic model selection by threshold accepting has been tested on a large number of randomly generated bivariate lag structures. The generation of instances for this purpose can be described as follows. First, an expected number of lags k^e is selected. The processes generated with a given k^e should include, in expectation, k^e non–zero parameters. However, the actual number of non–zero entries may range between one and $d^2 K'$. Therefore, as a first step the actual lag number is drawn from a discrete distribution on the set $\{1,\ldots,d^2K'\}$. The probability mass for each $i < k^e$ is set to λi, and for each $i > k^e$ to $\gamma(d^2K' - i + 1)$. Finally, $f(k^e) = \lambda k^e + \gamma(d^2K' - k^e + 1)$.[28] The values of λ and γ are chosen such that

$$\sum_{i=1}^{k^e}(\lambda i)i = \sum_{i=k^e}^{d^2K'} \gamma(d^2K' - i + 1)i = \frac{k^e}{2}. \tag{12.7}$$

Consequently, the density f of the distribution of the number of lags on $\{1,\ldots,d^2K'\}$ is linearly increasing from 1 to $k^e - 1$ and linearly decreasing from $k^e + 1$ to d^2K', with a discrete jump at k^e such that the contribution to the expectation is equal for all lag numbers less or equal to k^e and all lag numbers greater or equal to k^e. The values of λ and γ satisfying the constraints of equation (12.7) are given by

$$\lambda = \frac{3}{(k^e + 1)(2k^e + 1)} \quad \text{and}$$
$$\gamma = \frac{k^e}{\tilde{\gamma}}, \quad \text{where}$$
$$\tilde{\gamma} = (d^2K' + 1)[d^2K'(d^2K' + 1) - k^e(k^e - 1)]$$
$$-1/3d^2K'(d^2K' + 1)(2d^2K' + 1) + 1/3(k^e - 1)k^e(2k^e + 1).$$

Given the actual number of non–zero entries, their position in the lag structure has to be selected. Thereby, column positions, corresponding to variable and equation, respectively, are chosen at random by using a uniform distribution, whereas row positions, corresponding to the time lags, are generated by using the transformation $(1 - \sqrt{x})$ of a standard uniform random variable x. This choice assigns higher probabilities to the inclusion of lags of smaller order, which seems a reasonable assumption for economic time series. The random drawing of column and row positions is iterated until a number of non–zero entries equal to the imposed actual number is reached.

Finally, each non–zero parameter has to be assigned a reel value. Again, the random generation of parameter values is biased in favour of more realistic results. Therefore, given the lag length i of the variable corresponding to the parameter and a standard uniform random variate x, the parameter γ_i is set to

$$\gamma_i = [2x - 1]\sqrt{1 - |2x - 1|}\frac{K'}{K' - 1 + i}.$$

By using this generation procedure, parameter values are biased away from the boundaries at $+1$ and -1. Furthermore, in expectation, parameter values become smaller for higher lag orders.

---

[28] Models with more than $T/4$ lags were excluded to avoid estimation problems due to a lack of degrees of freedom.

Table 12.10   Summarized statistics for randomly generated lag structures $(k^e = 12)$

	Lag identification		Model order		
Method	RL	WL	TS	CL	TL
TA implementation	69.03	4.11	21.90	33.05	45.05
Take all up to kth lag	39.23	18.83	98.60	1.40	0.00

Of course, this layout of the generation mechanism for randomly generated VAR processes is only one possible approach. Nevertheless, it can be assumed that when using a large number of draws, VAR processes with quite different characteristics are generated which, at least partially, share some features with economic time series processes.

Some of the generated VAR processes may exhibit nonstationarity or near singularity of the design matrix. None of those exceptions is tested explicitly. However, if the OLS or SURE estimates fail to provide numerical estimates due to numerical problems, the VAR process is skipped from the sample and replaced by a new drawing.

Tables 12.10 and 12.11 summarize the findings of simulations using randomly generated bivariate lag structures. The results in Table 12.10 are based on 100 randomly generated lag structures with $k^e = 12$, while k^e is set equal to 6 for the 100 lag structures used for Table 12.11. Furthermore, for each lag structure, 20 realizations of the VAR process are generated and the threshold accepting algorithm is run three times on each such realization. As before, only the results obtained from the best run are used. Nevertheless, the mean results presented in these tables are based on 2000 different process realizations. For all processes, the maximum lag length K' is set to 20, the number of observations to 240, the number of iterations for the threshold accepting algorithm to 2000, and the factor α to 0.5. Furthermore, the HQ criterion with $c = 1.1$ is used.

The second and third columns indicate the frequencies at which the threshold accepting implementation and the 'take all up to kth lag' approach, respectively, include a lag present in the true data–generating process (RL) or mistakenly include another lag (WL). The last three columns give the frequencies that the algorithms come up with a model order smaller than (TS), equal to (CL) or larger than (TL) the true model order.

The results indicate that the threshold accepting algorithm identifies relevant lags, i.e. lags of the true data–generating process with a frequency of almost 70% in both settings, whereas the frequency is smaller than 40% for the standard approach if a high number of lags (12) is relevant and drops to less than 31% if the number of relevant lags is small (6). Furthermore, the risk of erroneously including non–relevant lags is much smaller for the threshold

Table 12.11 Summarized statistics for randomly generated lag structures ($k^e = 6$)

Method	Lag identification		Model order		
	RL	WL	TS	CL	TL
TA implementation	68.47	3.29	15.00	30.00	55.00
Take all up to kth lag	30.95	13.53	99.15	0.85	0.00

Table 12.12 Estimated coefficients for the trivariate interest rate process (optimized lag structure)

Equation for	$\Delta r_M(-1)$		$\Delta r_L(-7)$		$\Delta r_D(-3)$	
	Coeff.	std.err.	Coeff.	Std. Err.	Coeff.	Std. Err.
Δr_M	0.3425	0.0606	—	—	—	—
Δr_L	0.1471	0.0308	0.1799	0.0396	0.1471	0.0308
Δr_D	0.3425	0.0456	—	—	—	—

accepting implementation in both cases.

Furthermore, since the standard approach implicitly assumes a dense lag structure, it favours lag structures with a small maximum lag order. Hence, it is not surprising that the true model order is almost always underestimated by the 'take all up to kth lag' approach, while this is the case only in about 22 or 15% of the instances when using the threshold accepting implementation (for $k^e = 12$ and 6, respectively).

12.5.4 A real trivariate process

The complexity of the enumeration approach for the lag structure identification problem grows exponentially with the dimension of the process. Therefore, the evidence for the good performance of the threshold accepting implementation for bivariate processes is augmented by an application to a real trivariate process in this section using the monthly data on German interest rate changes depicted in Figure 12.2. The selection procedure is applied to the 276 observations covering the time span from March 1975 to February 1998. Imposing a maximum lag length of 36 corresponding to three years of monthly data and using the HQ criterion with $c = 4$, the threshold accepting implementation delivered a model with only 5 out of 324 possible lags after 100 000 iterations. The parameter estimates for the coefficients not constrained to zero are reported in Table 12.12.

Applied to the same data, the 'take all up to the kth lag' approach results in a less parsimonious model, including the first lags for all variables, i.e. a total

Table 12.13 Estimated coefficients for the trivariate interest rate process (first lags only)

Equation for	$\Delta r_M(-1)$ Coeff.	std.err.	$\Delta r_L(-1)$ Coeff.	Std. Err.	$\Delta r_D(-1)$ Coeff.	Std. Err.
Δr_M	0.7542	0.1302	0.2178	0.1310	−0.6702	0.1754
Δr_L	0.3887	0.0796	0.1980	0.0800	−0.1563	0.1072
Δr_D	0.7993	0.0971	0.2252	0.0977	−0.5747	0.1308

of 9 estimated coefficients. The estimates for these coefficients are provided in Table 12.13.

While all estimated coefficients are significantly different from zero at conventional levels for the optimized lag structure, some of the coefficients for the standard lag structure are not. Since the models are not nested, it is not possible to test one model against the other. Due to the much more binding constraint for the 'take all up to the kth lag' method, it might be expected that the results presented in Table 12.12 provide a better approximation of the dynamic interaction between the three interest rates. While both models provide similar implications for the short–run impact of changes of the money market rate Δr_M on the loan and deposit rates, respectively, the results differ in other aspects. In fact, under the assumption that only the modelled short–run interactions are relevant, the optimized lag structure provides a different economic interpretation. It does not imply a feedback effect from market interest rates on the money market rate. Furthermore, the changes of the deposit rate do not exhibit persistency, while the changes of the loan rate do. Finally, the optimized lag structure indicates a spillover from changes of the deposit rates on changes of the loan rates, which can be explained with a mark–up pricing model for the loan rate (Winker, 1999a). Of course, the simple nonstructural VAR model of interest rates analysed in Tables 12.12 and 12.13 does not allow for far–reaching economic conclusions. Nevertheless, it should be noted that different methods for lag order selection may result in different economic interpretations of the results!

In order to assess the performance of the threshold accepting implementation for such a trivariate autoregressive model, a further Monte Carlo simulation is performed based on the estimated coefficients in Table 12.12. As for the bivariate case, one hundred realizations of the process are generated and analysed by using the threshold accepting implementation. Again, the obtained identification can be compared with the 'true' model, which of course is only an approximation to the data–generating process of the interest rate changes used above. Table 12.14 summarizes the simulation results for the unconstrained implementation when using the HQ criterion with $c = 4$ and

10 000 iterations of the threshold accepting algorithm.[29]

As for the two–dimensional processes, the threshold accepting implementation always selects a model, which — with regard to the given information criterion — is at least as good as the true data–generating process. Furthermore, the lags appearing in the original model are selected much more frequently than the lags not present in the true model. Consequently, although a higher number of iterations is necessary for this more complex model, the performance of the automatic lag structure selection procedure is still acceptable in a higher dimension. The highly significant first–order lags of the original model are almost always identified by the algorithm, while the spillover effects on changes of the loan rate are identified less frequently. Overall, the threshold accepting algorithm selects the relevant lags with a frequency of 64.8% and irrelevant lags with a frequency of 0.025%.

In contrast, applying the 'take all up to the k–th lag' approach always results in models including only the first lags of all variables. Consequently, relevant lags are found with a frequency of 62.8%, while the frequency of including irrelevant lags is 1.9% and, therefore, larger than for the threshold accepting implementation in this application.

Optimization of the lag structure based on the HQ criterion with $c = 2$, instead of $c = 4$, resulted in a model including 24 lags (not reported), while the standard approach sticks to the first lags as for $c = 4$. Again, all estimated coefficients of the optimized lag structure are highly significant at conventional levels. A simulation study based on this lag structure gave similar results as for the case where $c = 4$. While the threshold accepting implementation finds the relevant lags with a frequency of 46.88%, the standard approach, which still uses only the first lags, does only with a frequency of 20.83%. The risk of including wrong lags is 1.04% for the threshold accepting implementation and 1.33% for the standard approach. However, the economic interpretation of the model with 24 lags is somewhat different from the one with only 5 lags. In particular, the change in the deposit rate exhibits some autocorrelation. Furthermore, the change in the money market rate is also determined by lagged changes of the deposit rate, while no autocorrelation of the changes of the loan market are mirrored in this model. These changes are taken as further evidence for the importance of the model selection step.

12.5.5   The money demand example

For the four–dimensional money demand example taken from Section 3.2.2 above, the threshold accepting implementation described in this present chapter has been extended to allow for the separate inclusion of exogenous variables in each equation. Hence, besides the lagged values of the endogenous

---

[29] Repeating the simulation with only 2000 iterations gave qualitatively similar results.

Table 12.14 Simulation results for the trivariate process

Lag	Eq. for Δr_D			Eq. for Δr_M			Eq. for Δr_L		
	Δr_D	Δr_M	Δr_L	Δr_D	Δr_M	Δr_L	Δr_D	Δr_M	Δr_L
$t-1$	0	**100**	0	0	**96**	0	0	**95**	0
$t-2$	0	0	0	0	0	0	0	0	0
$t-3$	0	0	0	0	0	0	**10**	0	0
$t-4$	0	0	0	0	0	0	0	0	0
$t-5$	0	0	0	0	0	0	0	1	0
$t-6$	0	0	0	0	0	0	0	0	0
$t-7$	0	0	0	0	0	0	0	0	**23**
$t-8$	0	0	0	0	0	0	0	0	0
$t-9$	0	0	0	0	0	0	0	0	0
$t-10$	0	0	0	0	0	0	0	0	0
$t-11$	0	0	0	0	0	0	0	0	0
$t-12$	0	0	0	0	0	0	0	0	0
$t-13$	1	0	0	0	0	0	0	0	0
$t-14$	0	0	0	0	0	0	0	0	0
$t-15$	0	0	0	0	0	0	0	0	0
$t-16$	0	0	0	0	0	0	0	0	0
$t-17$	0	0	0	0	0	0	0	0	0
$t-18$	0	0	0	0	0	0	0	0	0
$t-19$	0	0	0	0	0	0	0	0	0
$t-20$	0	0	0	0	0	0	0	1	0
$t-21$	0	0	0	0	0	0	0	0	0
$t-22$	0	0	0	0	0	0	0	0	0
$t-23$	0	0	0	0	0	0	0	0	0
$t-24$	0	0	0	0	0	0	0	1	0
$t-25$	0	0	0	0	0	0	0	0	0
$t-26$	0	0	0	0	0	0	0	0	0
$t-27$	0	0	0	0	0	0	1	0	0
$t-28$	0	0	0	0	0	0	0	0	0
$t-29$	0	0	0	0	0	0	0	0	0
$t-30$	0	0	0	0	0	0	0	0	0
$t-31$	0	0	0	0	0	0	0	0	0
$t-32$	0	0	0	0	0	0	0	0	0
$t-33$	0	0	0	0	0	0	0	0	1
$t-34$	0	0	0	0	0	0	0	0	0
$t-35$	0	1	0	0	1	0	0	0	0
$t-36$	0	0	0	0	0	0	0	0	0
Lags	3.32								
< True	100								

variables $\Delta m1_t, \Delta p_t, \Delta y_t$, and ΔR_t, where $m1$ is the logarithm of the monetary aggregate M1, p_t denotes the logarithm of the GNP–deflator, y_t the logarithm of the real GNP, and R_t an average long–term interest rate, exogenous variables are included as further explanatory variables. These additional variables comprise an error correction term (ec_{t-1}) taken from Brüggemann and Wolters (1998), lagged changes in import prices (Δpm_{t-4}), seasonal dummies and a shift dummy for the German monetary union.

Model identification is strived for by using the HQ criterion with $c = 1.01$, which resulted in a number of included regressors similar to the model proposed by Brüggemann and Wolters (1998) reported above in Section 3.2.2. The admitted selection rules differ slightly from the pure VAR models, as such a selection rule has to comprise both a matrix $S_k = \{s_{ij}^k\} \in \{0,1\}^{d \times d}$, $k = 1, \dots, K$ indicating the lags chosen for the dynamic modelling and a matrix $R = \{r_{ij}\} \in \{0,1\}^{d \times L}$ for the exogenous regressors. Thereby, $r_{ij} = 1$ indicates that the jth exogenous regressor is included in the ith equation of the system. Obviously, the construction of neighbourhoods and the resulting threshold accepting implementation are quite similar to the pure VAR case.

The optimized identification of the money demand model when using the data from Brüggemann and Wolters (1998) is reported above in Table 3.2 (Section 3.2.2). So far, no Monte Carlo simulation has been performed using this model with exogenous regressors. For this purpose, a slight change has to be added to the simulation of the true data–generating process, as the error correction term cannot be tackled as an exogenous variable in such a simulation setting, but has to be derived from the simulated time paths of the endogenous variables. The setting up of a suitable simulation model is left for future research.

12.6   Second–Step Statistics

The results presented in the previous section indicate that lag structures and lag lengths might differ to a large extent depending on whether one relies on the standard 'take all up to the k–th lag' approach or uses the proposed threshold accepting implementation, respectively. As pointed out in the introduction to this chapter, several econometric tests applied in time series analysis depend heavily on a well–specified dynamic structure. For example, the outcome of Granger causality tests, unit root testing or cointegration analysis might be different when an optimized lag structure is employed instead of some ad hoc rule.[30]

The impact of a specific choice of lag structure identification method on the outcome of Granger causality tests or unit root tests based on the augmented Dickey–Fuller concept can be studied by using further Monte

---

[30] See, e.g. Thornton and Battan (1985, p. 169) and Oke and Lyhagen (1999).

Carlo simulations. The set–up of such simulations is straightforward. Again, realizations of known data–generating processes are used. To each realization, the different methods for lag structure identification are applied and, finally, the relevant test statistic is calculated. Since the true data–generating process is known in such a simulation set–up, the outcomes of the simulations can be compared with the 'true' result. Based on the results presented in the previous section, it might be expected that unconstrained optimization on all lag structures when using the threshold accepting implementation results in better second–step statistics, in particular, if the underlying dynamic structure is complex. However, an assessment of the importance of this improvement will be the subject of future research.

The simulation set–up becomes more complicated for the cointegration analysis in the maximum–likelihood framework proposed by Johansen (1988). The concentration of the likelihood is performed based on the assumption of a lag structure without 'holes'. Of course, asymptotically, this assumption is easily justified. However, for small samples as they arise typically for macroeconometric time series, modelling a dynamic adjustment allowing for 'holes' seems to be preferable. Then, the first step of the estimation procedure cannot be estimated consistently by ordinary least squares as in the original approach. It seems possible to generalize Johansen's approach by allowing for a SURE estimator in this first step. However, different approaches for identifying long–run relationships in a vector error correction framework may lend themselves even easier to the implementation of optimized lag structure identification. For example, Chao and Phillips (1999) propose the application of a so–called posterior information criterion to the simultaneous identification of lag length and cointegration rank. The examples used in their Monte Carlo study include models with 'holes' in the lag structure. However, the length of the true lag structure is either zero or one. Thus, optimization can be performed by enumeration of all alternative lag structures, which becomes infeasible as soon as the lag length increases. Then, this method could be combined with the lag structure identification method based on threshold accepting.[31]

Future Monte Carlo evidence and applications to economic data are necessary in order to assess the importance of a good lag structure selection for such second–step statistics in higher–dimensional dynamic models when given a small number of observations.

12.7   Conclusions

Model *selection* is one of the most important tasks in econometric modelling. This comprises the solution of several optimization problems, which are

---

[31] A related approach is presented by Gonzalo and Pitarakis (1998), who consider specification testing in an error correction framework.

not always made explicit. In this present chapter, one specific optimization problem in econometric modelling has been tackled by using a threshold accepting implementation, namely the identification of multivariate lag structures. This problem becomes intractable by simple enumeration schemes, as soon as the process becomes multivariate and a high lag order may be present. Nevertheless, high–quality solutions are required for applications including large–scale econometric models, Granger causality tests, Johansen's procedure or indirect estimation of even more complicated time series models. Hence, the application of optimization heuristics seems adequate for this type of problem.

The presented implementation is based on the minimization of a given information criterion, although is not restricted to this kind of criterion. Furthermore, it allows us to impose additional a priori constraints on relevant or non–relevant lags. Finally, both a constrained version, which forces the same lags to appear in all equations, and an unconstrained version of the threshold accepting implementation are presented.

The performance of the threshold accepting implementation has been analysed for several bivariate and one trivariate VAR processes. For the first of the bivariate VAR processes, a large–scale Monte Carlo simulation has been performed with regard to those parameters with a potential influence on the results, such as the number of observations, the constant in the information criterion or the number of iterations of the threshold accepting implementation. Based on the results of this simulation, the parameters for the other applications are then chosen. The use of simulated bivariate VAR processes with a priori set coefficients and a large number of randomly generated VAR processes demonstrates the efficiency of this approach. The quality of the results is mainly limited by the available information, i.e. the number of observations. In all instances, the threshold accepting implementation provides a model with a better value of the information criterion than the true data–generating process for any given realization. Furthermore, the identification is always better than that obtained by the standard 'take all up to the kth lag' approach. This advantage becomes most pronounced if the true lag structure includes many 'holes'. This finding is substantiated by the results obtained for a simulated trivariate VAR process. Furthermore, the application to a real trivariate VAR process indicates that the economic interpretation of the results might be different depending on how the identification of multivariate lag structures is strived for.

Therefore, it might be stated that the identification and choice of lag structures can be achieved through the mechanical use of the presented threshold accepting implementation. The econometric model builder is freed from the task of testing hundreds of different lag structures, the resulting subjective factor is avoided, and, because lag structures in general will have many 'holes', the number of degrees of freedom is increased.

Future research will assess the impact of replacing ad hoc methods for lag structure identification by an optimization approach on second–step statistics such as cointegration rank or Granger causality, which depend crucially on the dynamic specification as long as the number of observations is rather limited, which is the typical case for economic time series data. The example presented above in Section 3.2.2 have already provided a first idea about the importance of this aspect.

13

Optimal Aggregation

13.1 Introduction

Aggregation is an essential ingredient to any macroeconomic analysis, both theoretical and empirical. However, after a period of intensive discussion of problems related to the aggregation issue in the 1960s and 1970s, for the last two decades it has rather lived in the shadows. In fact, theoretical analysis avoided discussing the aggregation problem by assuming representative households, firms and so on, while empirical analysis divided into two more or less separate fields. The first one, microeconometrics, avoids aggregation issues by restricting analysis to individual units. Consequently, it is difficult to draw conclusions for economic policy from such studies. On the other side, macroeconometrics takes aggregate data as given as they come from some statistical agencies neglecting the prior aggregation procedure inherent in most economic time series.

A notable recent exception to the tendency of either avoiding or ignoring aggregation issues is the work by Hildenbrand (1994; 1998). Using the increasing amount of data at the individual household level, this author describes an explicit aggregation procedure for an aggregate consumption function based on rather general assumptions about the conduct of individuals, i.e. avoiding unnecessarily strong assumptions on rationality or representative behaviour (McFadden, 1999, p. 99). Through the explicit aggregation, moments of the distribution of relevant variables, for example disposable income, also enter the aggregate consumption function.

The approach towards aggregation problems followed in this present chapter is of a different kind and is related to earlier approaches in the literature, namely the explicit aggregation of individual time series based on their statistical properties. The description of this problem dates back to Theil (1954), Malinvaud (1956) and Fisher (1962), to name just a few of the most influential contributions.

The analysis in this literature starts with the then, and still valid observation, that a widely used method in macroeconometric modelling is to replace the 'true model' by an aggregative one. This aggregated model is obtained by grouping the variables of the more disaggregated model and replacing the variables in each group by sums or weighted averages. In fact, almost every macroeconometric study implicitly or explicitly uses this method. Williams (1978, p. 67) describes input–output analysis as an example. In order to make this feasible, many different firms or even industries have to be grouped together. This aggregation is necessary both for computational reasons — so in diminishing importance with growing computing facilities — and for an economic interpretation of the results. It should be noted that employing aggregate statistics supplied by statistical agencies exactly corresponds to this case, since such aggregate data are the outcome of an explicit aggregation procedure.

The aggregative variables are put into relation with one another in a way that mimics the corresponding relation in the 'true model'. Moreover, the aggregative model is generally treated as if the structural characteristics of the detailed model carry over to it without change, thus enabling one to have — or to believe one has — an understanding of how the economy operates as seen through the model. Geweke (1985, p. 209f.) has pointed out that the distortions introduced by this assumption of perfect aggregation — known as that of the 'representative agent' in current macroeconomic models — may be of the same order of magnitude as the much–studied distortions introduced by ignoring expectation formation.

As becomes clear from the input–output example, often there is no way to avoid this practice, for example for the simple reason that the number of explanatory variables in the disaggregate data set exceeds by far the number of available observations. However, given that aggregation is necessary, it should at least be carried out intelligently. There is no reason to believe that the aggregation procedures of statistical agencies are reasonable or even optimal for all potential applications. Therefore, one has to think about improving aggregation procedures. Two distinct problems arise. The first is that of choosing an aggregative model that *best approximates* the 'true model' when the modes of aggregation are specified in advance; the second, which is mainly analysed in this chapter, is that of choosing the modes of aggregation optimally.[1]

As in the previous chapter for the multivariate lag structure identification problem, an approach towards optimal aggregation first has to introduce some ordering on the set of possible modes of aggregation. In this present chapter, this ordering is derived with explicit reference to an economic model, which also serves as an example for the application of the method. The example stems from trade theory and is related to industrial classification. One supposes, as Samuelson's theory predicts (Samuelson, 1953) , that within a country following fairly liberal trading policies, domestic price movements

---

[1] For the theoretical background to these two aspects of the problem, see Chipman (1976; 1975), respectively.

will closely follow movements in world prices, independently of consumer preferences. Based on the assumption of fixed technical coefficients, a linear–homogeneous multivariate multiple–regression model is postulated with the detailed average import and export prices as exogenous (independent) variables and the detailed average domestic prices of these same groups of commodities as endogenous (dependent) variables.

Obviously, neither an econometric nor an economic analysis of such a model is possible on a very fine level of aggregation. Hence, the objective is to partition the industries into a smaller number of groups at a higher level of aggregation. Comparison of the aggregative endogenous variables with the conditional predictions of these variables from the aggregative model leads to a criterion of mean–square forecast error for a given grouping of the data. Given this objective function — which is denoted by ϕ — one wishes to choose a grouping that minimizes mean–square forecast error. This objective function is derived and analysed in more detail in Section 13.2.

Of course, neither is the aggregation methodology introduced in this chapter restricted to the specific application on industrial classification in an international trade context, nor is a classification system derived from such an approach suitable or even optimal for all potential uses. It refers to only one of such possible uses. Nevertheless, some other uses that come to mind are closely related to this one; for example, one may wish to study the relations between quantities instead of prices. Leamer (1990) studied the mapping from a country's factor endowments to net exports and found that the nine one–digit Standard International Trade Classification (SITC) groupings of the 56 two–digit SITC categories form a far–from–optimal classification. Remarking (p. 157) that the 'calculation costs of a global minimization [...] will [...] be unacceptably high', he settled on a local optimization algorithm.[2]

In this chapter, only the case of optimally partitioning a set of medium–level categories (two– and some three–digit categories of the official classification system) into a specific number of groups is covered. It is obvious, however, that a complete solution of the problem of optimal industrial classification

---

[2] As pointed out in Leamer (1990, p. 157), the number of $m \times m^*$ proper grouping matrices for modestly large m^* is enormous. In fact, the restriction on exactly one non–zero entry per row and at least one per column leads to the following combinatorial expression for the number $P(m, m^*)$ of equivalence classes of $m \times m^*$ proper grouping matrices (considered as unordered sets of m^* column vectors each of order $m \times 1$), i.e. for the number of ways of partitioning m objects into m^* groups (Chipman, 1975, p. 150):

$$P(m, m^*) = \frac{1}{m^*!} \sum_{i=0}^{m^*} (-1)^i \binom{m^*}{i} (m^* - i)^m .$$

For the application to the German price data analysed in this chapter, this amounts to $P(42, 6) = 6.665 \times 10^{29}$. For Leamer's application it is still higher, namely $P(56, 9) = 7.455 \times 10^{47}$.

would entail derivation of a hierarchical classification system at many levels. An approach to this problem has been carried out by Cotterman and Peracchi (1992), who stress the importance of 'consistency', i.e. the requirement that categories once combined should not be broken up at a coarser level of aggregation. However, owing to the tremendous complexity of finding a complete and optimal hierarchical classification system, they content themselves with a sequential procedure which cannot guarantee good or even optimal groupings at any stage of the classification system. Therefore, here the restriction on a single aggregation step is made in order to generate good or optimal groupings.

In Section 13.9 it is shown that the problem of optimal aggregation falls in the category of computationally intractable problems. Hence, the only available alternative to rather simple and suboptimal ad hoc algorithms is the use of global optimization heuristics such as threshold accepting. Consequently, this application differs from the previous ones in so far as threshold accepting is not only used to improve previous results based on other methods, but to enable some real optimization at all. At the same time, the problem of optimal aggregation probably is the most complex one of the examples studied in Part III of this book. At least, the examples employing real data from the German and Swedish classification system exhibit a tremendously high number of potential solutions.

Furthermore, the objective function ϕ considered for this problem includes some matrix inversions. Thus, even when using an optimization heuristic such as threshold accepting, available computing resources impose a constraint. Nevertheless, it is possible to obtain optimized groupings, which are presented and discussed in Section 13.5. Alternatively to the static version of the economic model discussed in Section 13.2, a dynamic version can be considered. The details of the model, changes of the measure of aggregation bias and results of the threshold accepting implementation to the dynamic model, are outlined in Section 13.7.

Parts of this chapter draw on joint work with John S. Chipman, which has been supported by the Deutsche Forschungsgemeinschaft, Long Term Research Project 178, at the University of Konstanz, National Science Foundation grant ES–8607652 at the University of Minnesota, and a senior USA scientist award from the Humboldt Foundation. Some preliminary results are reported in Chipman and Winker (1995a; 1995b).

13.2 Measuring Aggregation Bias

The problem of optimal aggregation which arises, for example, in the model of international price transmission outlined in the introduction to this chapter can be formulated in terms of the multivariate multiple–regression model

$$Y = XB + \mathcal{E} \tag{13.1}$$

where Y is an $n \times m$ matrix of n observations on m endogenous variables, X is an $n \times k$ design matrix of n observations on k exogenous variables, B is a $k \times m$ matrix of unknown regression coefficients to be estimated, and $\mathcal{E}$ is a random $n \times m$ matrix of error terms with zero mean and covariance

$$\mathbf{E}\{(\text{col } \mathcal{E})(\text{col } \mathcal{E})'\} = \Sigma \otimes V \,. \tag{13.2}$$

The above col $\mathcal{E}$ denotes the column vector obtained by stacking the columns of $\mathcal{E}$, Σ is the $m \times m$ simultaneous covariance matrix and V is the $n \times n$ sample covariance matrix. $\mathbf{E}$ denotes the expectation operator. It is assumed that V is positive definite. The more general case with reduced rank (rank $V \leq n$) is treated in Chipman (1975).

The aggregation of several variables to one group by summing up the values of the variables within the group can be described by $k \times k^*$ and $m \times m^*$ (proper) *grouping matrices*, i.e. matrices with exactly one non–zero (in fact, positive) element in each row and at least one non–zero element in each column. Let G and H, respectively, denote such grouping matrices for the exogenous and endogenous variables of the model represented by equation (13.1), respectively. Then, the aggregative model mimicking the true disaggregated one is given by

$$Y^* = X^* B^* + E^* \,, \tag{13.3}$$

where

$$X^* = XG \quad \text{and} \quad Y^* = YH$$

are $n \times k^*$ and $n \times m^*$ matrices of observations on k^* and m^* aggregative exogenous and endogenous variables, respectively. An intuitive representation of the situation, introduced by Malinvaud (1956), is given by the commutative diagram of Figure 13.1.[3]

Hereby, the regression coefficients B at the disaggregated and B^* at the aggregated level are interpreted as linear mappings from the spaces of exogenous variables $\mathcal{X}$ and $\mathcal{X}^*$ to the spaces of endogenous variables $\mathcal{Y}$ and $\mathcal{Y}^*$, respectively. The grouping of variables via grouping matrices G and H can also be interpreted as mappings from the spaces of disaggregated to aggregated variables. Consequently, the regression model at the aggregate level can be compared with a composed mapping $G^{\#} BH$. Another means of comparison is to compare the composed mappings from $\mathcal{X}$ to $\mathcal{Y}^*$ given by BH and GB^*, respectively. Differences between the two mappings are interpreted as aggregation bias.

Three different situations may be distinguished in connection with this model. First, the disaggregated model (13.1) is consistent with the aggregative

---

[3] The meaning of the reverse mapping $G^{\#}$ appearing in the figure will be explained later (see equation (13.7) below).

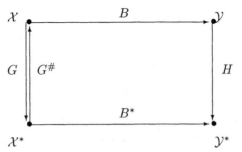

Figure 13.1 Commutative diagram for the aggregation problem.

one (13.3) if

$$XGB^* = \mathbf{E}^*Y^* = \mathbf{E}YH = XBH, \qquad (13.4)$$

where $\mathbf{E}^*$ denotes the expectation operator associated with the aggregative model. In this case, the aggregation bias is zero. However, this situation is hardly fulfilled for economic time series, although Theil (1954) highlights two configurations which may cause such a perfect aggregation. The first case is structural similarity, i.e. the relationships at the disaggregated level directly carry over to the aggregates. The necessary condition for this to happen is known as the 'Hatanaka condition' (Hatanaka, 1952). A simple example of structural similarity is given, if in fact all mappings at the disaggregate level are identical — the case of representative agents. Furthermore, perfect aggregation may result through multicollinearity, if the 'microvariables [are proportional to] the corresponding macrovariables' (Theil, 1954, p. 32). In addition to the pure cases of structural similarity and multicollinearity, there can be many cases of partially restricted structures on a partially restricted domain as described, e.g, by Chipman (1976, pp. 657–665).[4]

Secondly, if perfect aggregation has to be ruled out, the approach of *best approximate aggregation* is to define a suitable measure of aggregation error and choose B^* in such a way as to minimize this error. This approach will also provide the result of perfect aggregation if the conditions are fulfilled, since perfect aggregation corresponds to a minimum of any suitable measure of aggregation bias. Chipman (1975, pp. 125ff.) defines the *mean–square forecast error* of the aggregative model as the matrix

$$F = \mathbf{E}\{(Y^* - X^*B^*)'V^{-1}(Y^* - X^*B^*)|X\} = A + nH'\Sigma H, \qquad (13.5)$$

---

[4] Pesaran, Pierse and Kumar (1989) and Thompson and Lyon (1992) discuss tests for perfect aggregation. In an application to employment demand functions for the UK economy, Pesaran et al. (1989) and Lee, Pesaran and Pierse (1990) clearly reject the hypothesis of perfect aggregation. In fact, these authors observe that relying solely on the aggregate relationship might erroneously indicate a too high long–run real wage elasticity of labour demand.

where

$$A = (BH - GB^*)'X'V^{-1}X(BH - GB^*), \qquad (13.6)$$

which he calls the matrix of 'aggregation bias'. Since the chance of finding $A = 0$ is small for real applications, some ranking of aggregate models based on their aggregation bias measured by F or A, respectively, is required. This is introduced based on the non–negative definiteness of the difference of two matrices. Given the proper grouping matrices G and H, which define the mode of aggregation, minimization of F is then equivalent to minimization of A. Chipman (1976, p. 668) demonstrates that A is minimized with respect to B^* for fixed G and H when

$$B^* = G^{\#}BH, \qquad (13.7)$$

where

$$G^{\#} = (G'X'V^{-1}XG)^{-1}G'X'V^{-1}X, \qquad (13.8)$$

if, as may be expected in practice, the matrix $X^* = XG$ has full rank k^*.[5] In Figure 13.1 one may read off equation (13.7) as the composition of the mapping B^* into the three mappings shown. Furthermore, it can be shown that generalized least–squares estimation of the aggregative model provides the best linear unbiased estimation of the best approximate aggregation.

Thirdly, while best approximate aggregation provides a best aggregate model for given modes of aggregation, *optimal aggregation* treats the optimal selection of grouping mappings G and H. Given a set $\mathcal{G}$ of potential aggregation modes, for each pair $\{G, H\} \in \mathcal{G}$ the aggregate mapping B^* is determined so as to minimize the matrix (13.5) of forecast error, thus resulting in the minimizing matrix

$$F^* = H'B'(I - GG^{\#})'X'V^{-1}X(I - GG^{\#})BH + nH'\Sigma H, \qquad (13.9)$$

where the first term on the right is the minimizing bias matrix, A^*. This expression for the minimizing matrix may then be used to determine G and H optimally. However, the problem of minimizing (13.9) with respect to G and H is ill posed: in general, there will not exist a minimizing F^* matrix. A scalar–valued objective function must therefore be chosen.

The problem of best approximate aggregation remains invariant with respect to replacement of F by $W^{*1/2}FW^{*1/2}$, where W^* is some symmetric positive–definite weighting matrix. Hence, one may replace (13.9) by the criterion function

$$\phi = \alpha + n \operatorname{tr} \{H'\Sigma H W^*\}, \qquad (13.10)$$

where

$$\alpha = \operatorname{tr} \{H'B'(I - GG^{\#})'X'V^{-1}X(I - GG^{\#})BHW^*\}. \qquad (13.11)$$

---

[5] For the general case, see Chipman (1976, p. 668).

For the application presented in this chapter, only two standard weighting functions W^* are considered, although there exists an infinite number of reasonable weighting schemes. First, the Euclidean metric $W^* = I_{m^*}$ is used. This seems a natural choice for the application, since the price data used in Section 13.3 are already measured in a common unit. However, as the data cover a time period of almost 30 years, some explicit correction for heteroscedasticity seems sensible. For the application, the choice of

$$V = \text{diag}\{XX'\} = \text{diag}\left\{\sum_{i=1}^{k} x_{ti}^2\right\} \tag{13.12}$$

is made for this purpose. The second approach is to choose the 'Mahalanobis distance' with no extra correction for heteroscedasticity defined by the choice $W^* = (H'\Sigma H)^{-1}$ and $V = I_n$ (Chipman, 1975). By weighting the variables inversely to their variability, use of the Mahalanobis distance neutralizes the effect that high variability of one variable might otherwise have on the objective function. In fact, it might also partially correct for problems of heteroscedasticity.

So far, the objective function (13.10) has been discussed for pairs of grouping matrices of fixed given dimensions k^* and m^*, respectively. Of course, the choice of these dimensions, i.e. the degree of aggregation may also be the outcome of some optimization (Chipman, 1985). However, it is doubtful whether use of the objective function (13.10) could be justified for comparing grouping matrices of different dimension. Therefore, for the application in this chapter, it will be assumed that the dimensions of the grouping matrices are predetermined, e.g. by the dimensions of some official classification scheme.

Related to the issue of the degree of aggregation are interdependencies between the grouping matrices for the exogenous variables G and for the endogenous variables H. In general, i.e. without any a priori restrictions, a two–step procedure can be used. For each fixed $m \times m^*$ matrix H, an optimization is performed over the set of $k \times k^*$ matrices G, then optimizing over the set of matrices H. Of course, the computational effort can be reduced considerably if this two–step procedure can be avoided. In the application considered in Section 13.3 a natural restriction is introduced, imposing a dependency of G upon H. Consequently, only optimization over the grouping matrices H has to be carried out explicitly.

Since B and Σ are unknown, they have to be replaced by estimates. In the case of an Euclidean metric and correction for heteroscedasticity, the generalized least–squares estimator

$$\tilde{B} = (X'V^{-1}X)^{-1}X'V^{-1}Y \tag{13.13}$$

is used for B, and for Σ the pseudo–maximum–likelihood estimator[6]

$$\hat{\Sigma} = n^{-1}(Y - X\tilde{B})'V^{-1}(Y - X\tilde{B}) = S/n, \qquad (13.14)$$

where

$$S = Y'V^{-1}Y - Y'V^{-1}X(X'V^{-1}X)^{-1}X'V^{-1}Y. \qquad (13.15)$$

These estimators for B and Σ do not depend on the modes of aggregation. Hence, given the data on exogenous and endogenous variables, they can be computed once and for all. Obviously, if aggregation is performed when not enough observations are available to estimate the disaggregated model, this estimation has to use generalized inverses based on singular–value decomposition.[7]

The estimate of ϕ then becomes

$$\tilde{\phi} = \mathrm{tr}\ \{H'\tilde{B}'(I - G\tilde{G}^\#)'X'V^{-1}X(I - G\tilde{G}^\#)\tilde{B}H\} + \mathrm{tr}\ \{H'SH\}, \qquad (13.16)$$

where

$$\tilde{G}^\# = (G'X'V^{-1}XG)^{-1}G'X'V^{-1}X = (X^{*'}V^{-1}X^*)^{-1}X^{*'}V^{-1}X,$$

in accordance with equation (13.8) above.

If the Mahalanobis metric with no correction for heteroscedasticity is used instead, the estimates for B and Σ are just as those given by equations (13.13) and (13.14), respectively, but with V set equal to I_n, and the estimate of ϕ then becomes

$$\hat{\phi} = n\ \mathrm{tr}\ \{H'\tilde{B}'(I - G\tilde{G}^\#)'X'X(I - G\tilde{G}^\#)\tilde{B}H(H'SH)^{-1}\} + nm^*. \qquad (13.17)$$

Dividing through by n one obtains

$$\tilde{\phi} = \tilde{\alpha} + m^* \qquad (13.18)$$

where

$$\tilde{\alpha} = \mathrm{tr}\ \{X(I - G\tilde{G}^\#)\tilde{B}H(H'SH)^{-1}H'\tilde{B}'(I - G\tilde{G}^\#)'X'\}. \qquad (13.19)$$

Due to the assumption that the degree of aggregation is predetermined, m^* is constant in the applications, while $\tilde{\alpha}$ — which is referred to as the 'aggregation bias' — can be used as the objective function in the Mahalanobis case.

The arguments leading to the objective functions (13.16) and (13.19), respectively, which are used in the following sections to obtain optimized

---

[6] If the usual best quadratic unbiased estimator $S/(n - k)$ is used instead, then in the formula for $\hat{\phi}$ below the term m^* would be replaced by $m^*n/(n - k)$.

[7] That is, using the oblique generalized inverse $X^\ddagger = \overset{\circ}{X}{}^\dagger V^{-1/2}$, where $\overset{\circ}{X} = V^{-1/2}X$, in place of $(X'V^{-1}X)^{-1}X'V^{-1}$, where $\overset{\circ}{X}{}^\dagger$ is the Moore–Penrose generalized inverse of $\overset{\circ}{X}$.

groupings, can be summarized as follows. Optimal aggregation based on this objective function selects the grouping matrices G and H that will approximate the conditions for perfect aggregation as closely as possible. For example, if a subset of columns of X are highly collinear, it will tend to aggregate the corresponding variables together; alternatively, if X is well conditioned, it will tend to group variables together so that the corresponding submatrices of B have row sums which are as equal to each other as possible. These conditions are closely related to the intuitive ideas about 'similarity' of commodities and of processes of production in the application to international price spillovers.

13.3 An Application to Price Indices

Given one of the objective functions derived in the previous section, a threshold accepting implementation for the problem of optimal aggregation requires knowledge about the set $\mathcal{G}$ of possible pairs of grouping matrices H and G, a neighbourhood structure on this set and a threshold sequence.

In order to motivate the specific choice for $\mathcal{G}$, the economic model behind the aggregation problem discussed in this section has to be introduced. This model treats the international price spillover, but may also be analysed in the context of factor rentals in an open economy. I am reluctant to use the imprecise word 'globalization', although, however, many people would summarize the following analysis under this heading. The application consists in examining the structural relationship between commodity prices in a country's home markets and the corresponding world prices as represented by the prices (expressed in the country's own currency) of its imports and exports. By the theories of Samuelson (1953) and Shepard (1981), the rentals of the factors of production employed in the country's export and import–competing industries are determined from the external prices by inverting the system of minimum–unit–cost functions dual to the production functions, while the prices of non–traded commodities are determined from the factor rentals directly via the corresponding minimum–unit–cost functions. The resulting composed mapping from external to internal prices can be regarded as a 'generalized Stolper–Samuelson mapping'. If the production technology is of the Leontief fixed–coefficients type, then the minimum–unit–cost functions are linear–homogeneous.

Of course, the assumption of Leontief production technology is not the most convincing one for all industries. A reasonable alternative is to assume a loglinear (Cobb–Douglas) technology, in which case the minimum–unit–cost functions are also of the loglinear type. This would be practicable if the price indices issued by statistical agencies were geometric means. Then, all of the ensuing relationships would be loglinear. Given that price indices are not supplied as geometric, but as weighted arithmetic means, using such a loglinear approach would inhibit a comparison of the outcomes of optimal aggregation with the modes of aggregation implicit in official classification schemes. Therefore, the linear model is maintained.

Then, the aggregation problem can be formulated as follows. X_1 and X_2 denote $n \times m$ matrices of n consecutive monthly observations on import and export price indices of m commodity categories, respectively, and Y denotes the $n \times m$ matrix of internal producer prices for the same commodity categories. Let $X = [X_1, X_2]$ denote the $n \times k$ matrix of observations on the $k = 2m$ independent variables. The regression model is then

$$Y = XB + E = [X_1, X_2] \begin{bmatrix} B_1 \\ B_2 \end{bmatrix} + E, \qquad (13.20)$$

where E is a random $n \times m$ matrix with zero mean and covariance given by equation (13.2), where V is given by equation (13.12). Under the postulated assumptions, the matrix B depends entirely on the production coefficients in the country's industries.

The natural aggregation process is quite simple in this special case. For an $m \times m^*$ grouping matrix H, the $k \times k^*$ grouping matrix G is defined by

$$G = \begin{bmatrix} H & 0 \\ 0 & H \end{bmatrix} = I_2 \otimes H$$

where $k^* = 2m^*$. This choice is based on the intuitive assumption that structural similarity and multicollinearity do not differ much between export, import and domestic prices. Furthermore, grouping all three sets of variables by the same aggregation process allows for a comparison of the result with the mode of aggregation implicit in official classification systems imposing the same constraint. Now, optimal aggregation is reduced to the optimal choice of H. Since the data used in the application in Section 13.5 are price indices multiplied by their weights, the natural weights to use in the grouping matrix H are zero or one, i.e. aggregation within a group is performed by summing up. Therefore, the matrix H has to be chosen out of the class of $m \times m^*$ proper grouping matrices, i.e. matrices with entries in $\{0, 1\}$ with each row having exactly one non–zero entry. Although both dimension and density of the search space $\mathcal{G}$ is reduced considerably by these assumption, as pointed out in Section 13.1, the computational complexity of the resulting optimization problem is still tremendously high.

The impact of aggregation of price indices has been analysed in the context of relative–price variability by Balk (1983) and Goel and Ram (1993). Both authors stress the importance of the level of aggregation on the estimated relative–price variability, but take the modes of aggregation at each aggregation level as given by the official statistics. Therefore, it might be interesting to replicate their results by using optimized modes of aggregation obtained by the threshold accepting implementation described in the following sections.

13.4 Threshold Accepting Implementation

As pointed out in footnote 2 above, the number of $m \times m^*$ proper grouping matrices for a modestly large m^* is enormous. Hence, a simple enumeration algorithm is completely infeasible. In Section 13.9, it is proved that the problem of optimal aggregation in its most general form, i.e. without the restriction on G to be equal to $I_2 \otimes H$, is NP–complete. Consequently, not only is enumeration infeasible, but probably no deterministic algorithm can give an exact solution to this problem with certainty without using computer resources, i.e. computing time or storage capacity, that grow faster than every polynomial in the size of the problem. Therefore, it is not surprising that the few existing empirical approaches to aggregation issues of the kind analysed in this chapter impose severe restrictions in order to obtain some results. For example, the test of perfect aggregation proposed in Pesaran et al. (1989) is applied solely to the case where all categories are aggregated into one aggregate, i.e. $m^* = 1$, while the hierarchical procedure described in Cotterman and Peracchi (1992) implies $m = m^* - 1$, i.e. only the cases of lowest complexity are considered.

Given the high inherent complexity of the optimal aggregation problem, it can hardly be expected that a threshold accepting implementation provides a global optimum for all problems with the small number of iterations which are feasible given the computational load necessary for calculating aggregation bias for a given mode of aggregation. Nevertheless, it can be used to obtain optimized modes of aggregation without imposing very restrictive assumptions. The results can be compared with the outcome of a more restrictive approach or with the official mode of aggregation, which, in general, is not based on some explicit optimization procedure. Furthermore, as long as either m is small or m^* differs not too much from 1 or m, the complexity of the aggregation problem is much smaller than for the example presented in Section 13.5. Therefore, for such a setting the threshold accepting implementation can be expected to produce high–quality solutions. In fact, calculations for a synthetic problem with $m = 10$ and $m^* = 3$ show that the algorithm works well in identifying optimal modes of aggregation in this case.

The implementation of threshold accepting to the optimal aggregation problem for the price index example introduced in Section 13.3 is straightforward. The objective function, which has to be minimized, is given by the aggregation bias measured by $\tilde{\phi}$ or $\tilde{\alpha}$, respectively. The search space consists of the finite, but huge set of proper grouping matrices, i.e. the matrices in $\{0, 1\}^{m \times m^*}$ of full rank, denoted as $\mathcal{H}$ to highlight the difference to the more general case, when the grouping matrix G does not depend on H. Of course, it is possible to introduce additionally a priori restrictions leading to some reduction of the size of the set $\mathcal{H}$. Only for the case when the prior restrictions reduce $\mathcal{H}$ to a very small set, which might loose connectivity for a given neighbourhood concept, the algorithm has to be adapted slightly as discussed in Dueck and Winker (1992).

A neighbourhood structure is imposed on $\mathcal{H}$ by using the concept of Hamming distance introduced above in Section 9.3. The Hamming distance d_H between two grouping matrices $H = (h_{ij})$ and $\tilde{H} = (\tilde{h}_{ij})$ is given by the number of differing entries.[8] As the search space $\mathcal{H}$ consists of proper grouping matrices, two elements with a Hamming distance of 2 can be obtained by moving one commodity from one group to another. Likewise, two grouping matrices with a Hamming distance of 4 might be generated by simultaneously moving two commodities to different groups.[9] For the generation of an element of $\mathcal{N}(H)$, after the exchange of some commodities, it has to be checked whether the resulting matrix $\tilde{H}$ is still a proper grouping matrix, i.e. $\tilde{H} \in \mathcal{H}$, which can be ascertained at low computational cost.

The local behaviour of the objective function with regard to the neighbourhood structure imposed by the Hamming distance can be analysed by using a simulation of local deviations as described above in Section 9.3. Some results of such a simulation for the German data and the setting described in Section 13.5 are provided in Figures 13.2 and 13.3.

Figure 13.2 shows histograms of local relative deviations obtained by the following simulation set–up. For 5000 randomly generated proper grouping matrices H_i, an element in its neighbourhood $\tilde{H}_i \in \mathcal{N}(H_i)$ is chosen. The objective function $\tilde{\phi}$ is calculated both for H_i and $\tilde{H}_i$ and the relative deviation, i.e. $(\tilde{\phi}(\tilde{H}_i) - \tilde{\phi}(H_i))/\tilde{\phi}(H_i)$, is stored. The three panels of Figure 13.2 show the histograms for these relative deviations based on neighbourhoods defined by a Hamming distance (HD) of 4, 8 and 12, respectively.

The upper panel shows the histogram for a Hamming distance of 4, which corresponds to the exchange of a maximum of two elements between groups. The high spike at zero indicates that the probability is high for leaving the grouping matrix unchanged by randomly selecting two commodities and exchanging them between groups, since no restriction is imposed forcing the second commodity to be different from the first one. This effect becomes less pronounced in the second panel, where the Hamming distance for defining

---

[8] E. Ronchetti proposed a modification of this standard concept of Hamming distance by weighting the differing entries with their base–year weights. This modification did not lead to a very different local behaviour. Hence, the application is based on the standard concept.

[9] Note, however, that if two columns of an $m \times m^*$ grouping matrix are interchanged, the Hamming distance between the original and the altered grouping matrix becomes $2m$, even though nothing has changed but the position or 'label' of the group. This anomaly could be avoided by applying the concept of Hamming distance to equivalence classes of grouping matrices (regarded as unordered columns). However, as only small Hamming distances are used in the application, the differences in the resulting topology are not important. Since the generation of a neighboured element is much faster when using the standard notion of Hamming distance, the explicit calculation of equivalence classes is omitted.

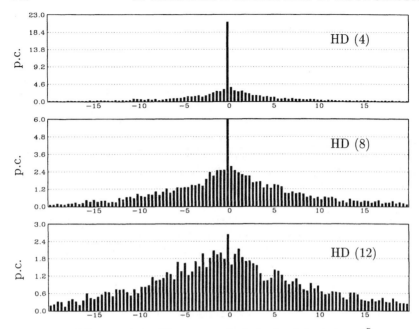

Figure 13.2   Histograms of relative local deviations ($\tilde{\phi}$).

neighbourhoods is increased to 8. The lower panel corresponds to a Hamming distance of 12. The distribution of relative local deviations becomes much less concentrated around zero. Further increasing the admitted distance between two neighboured groupings eventually results in the trivial neighbourhood structure, where every element is a neighbour to all other elements. It has already been pointed out in previous chapters that too large neighbourhoods tend to reduce the efficiency of the threshold accepting algorithm, while too small ones might be too restrictive, as the number of feasible moves in each iteration shrinks. Therefore, for this application, a Hamming distance of 8 has been chosen. However, the performance of the threshold accepting implementation did not change dramatically when turning to a Hamming distance of either 4 or 12.

Following the approach presented above in Section 9.4, the data generated for the local deviation plots are also used to obtain threshold sequences. In order to obtain a threshold sequence, the values of the simulated relative local deviations are sorted in decreasing order. In fact, the maximum of the relative deviation and its inverse are considered for obtaining relative deviations, which are all greater than or equal to one. The upper panel of Figure 13.3 shows the resulting sequences.

Again, only a lower quantile of this sequence is used as the threshold

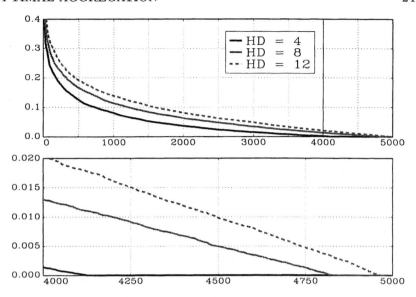

Figure 13.3   Threshold sequences from relative local deviation simulations.

sequence. The size of this quantile is determined by a small–scale simulation similar to the one described earlier in Section 8.4. It turns out that for the optimal aggregation problem with a relative threshold definition and a Hamming distance of 8, only 20–40% of the simulated threshold sequence should be used. This finding is represented in the lower panel of the plot, which shows the 20% lower quantile. Obviously, when using a Hamming distance of 4 instead of 8, the quantile should be chosen larger. Otherwise, the threshold accepting algorithm would almost reduce to a standard local search heuristic.

The definition of the neighbourhood structure and of the method for generating the threshold sequence completely describes the threshold accepting implementation up to the total number of iterations. Unfortunately, as pointed out in the introduction to this present chapter, the calculation of the objective functions $\tilde{\phi}$ or $\tilde{\alpha}$, respectively, requires the inversion of some matrices. Some updating of the objective functions is possible, but the gains in performance appeared to be small for the specific application studied in this chapter. The evaluation of the objective function for one proper grouping matrix H requires a considerable amount of computing time. Therefore, the total number of iterations performed by the threshold accepting implementation, in general, is smaller than for the other applications presented in Part III. On the other side, the search space is much larger. Consequently, it cannot be expected to obtain the global optimum for this application. Nevertheless, it is possible to obtain optimized groupings, which are presented and discussed in Section 13.5.

13.5   Results from Optimal Aggregation

As mentioned briefly in Section 13.3, the data sets used for the application of the threshold accepting heuristic to the optimal aggregation problem consist of monthly observations on import and export price indices (which are formed as weighted averages of prices with fixed weights) and internal producer–price indices (formed the same way). Since the natural way to group them is by forming weighted averages with the given weights, it is most convenient to work with the price indices multiplied by their weights. Then, as pointed out in Section 13.3, aggregation means just summation and the non–zero elements of the grouping matrices are all 'ones'.

Unpublished import and export price–index data of this type, called 'Wertziffern', have been supplied for Germany by the German Federal Statistical Office (Statistisches Bundesamt). They cover a total of 328 months from January 1968 to April 1995.[10] In the case of the internal producer–price index, 'Wertziffern' were not available, and the series used are the published price indices multiplied by their weights. Similar unpublished data for Sweden have been furnished by the Swedish Statistical Office (Statistiska centralbyrån). This data set covers the period 1971–1992. While the data for the years 1971–1979 are calculated on a 1968 base, subsequent data are calculated by using weights from December of the previous year. The resulting series are all with respect to base 1968=100 and multiplied by the 1968 weights (in millions of kronor).

For the German data, the disaggregated level consists of $m = 42$ different commodity categories,[11] whereas the Swedish data contain only $m = 25$ series for different commodity categories. The German Federal Statistical Office provides an official method of grouping the 42 industries into six groups. This official grouping serves as a bench–mark for the results obtained by the threshold accepting implementation.[12] No such official method exists for the Swedish data. Therefore, in Chipman and Winker (1995b) a 'pseudo–official' grouping into $m^* = 6$ groups is generated, which mimics the German one and serves as a bench–mark for the Swedish data. The same 'pseudo–official' grouping, which is essentially a grouping by stage of production like the German official classification, is used in Section 13.5.2.

---

[10] The published data consist of these 'Wertziffern' divided by the weight of the respective commodity category, and then rounded up to one digit after the decimal point. Because of the rounding error, accuracy is lost, especially in the case of the most important (high–weight) commodity groups.

[11] Section 13.10 contains a table of the commodity groups and their weights in imports, exports, and internal production.

[12] The classification system used is the 'Güterverzeichnis für Produktionsstatistiken'. The 42 industries are two–, plus some three–digit categories. A few commodity categories, such as those of electricity, gas, central heating, and water, as well as watercraft and aircraft, are not represented in the import– and export–price–index series, and have therefore been omitted from the producer–price–index series.

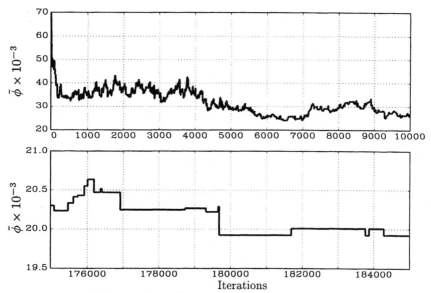

Figure 13.4   The way to an optimized grouping.

13.5.1   The German industrial classification system

This section provides computational results of the threshold accepting implementation for the German industrial classification system. To begin with some computational aspects, the threshold accepting implementation used for this application is coded in FORTRAN by using some ESSL–subroutines for matrix operations. The results presented in this section were obtained on an IBM RS 6000/3AT workstation. The optimized grouping matrix presented in the following has been achieved by 250 000 iterations in about 20 000 CPU–seconds.

The development of the objective function $\tilde{\phi}$ during the optimization procedure is tracked by Figure 13.4. The algorithm starts with a high value of the threshold parameter. Consequently, a new current solution is accepted very often at the beginning (upper panel of the figure). As the optimization proceeds, both the current values of the objective function and the threshold parameter decrease. Hence, it becomes less likely that a grouping matrix in the neighbourhood of the current solution is accepted. The value of $\tilde{\phi}$ changes less often (lower panel of the figure) and, in particular, increases of the objective function become smaller. Nevertheless, both parts of the figure show the typical 'hill climbing' feature of any efficient threshold accepting implementation: In order to achieve an improvement with regard to $\tilde{\phi}$ a temporary worsening has to be admitted. Otherwise, the algorithm risks getting stuck in a suboptimal local minimum.

Before turning to the optimized modes of aggregation provided by the threshold accepting implementation, the bench–mark grouping is introduced.

Table 13.1   The official grouping (Germany)

Agricultural, forestry and fishery products	Capital goods
– Agricultural, forestry and fishery products	– Steelworking products
	– Structural-steel products and rolling stock
Mining products	– Machinery (including farm tractors)
– Coal	– Road vehicles (excluding farm tractors)
– Crude oil and natural gas	– Electrical products
	– Precision and optical goods, clocks, etc.
Basic materials	– Ironware, sheet-metal ware and hardware
– Petroleum products	– Office machinery and data-processing
– Quarrying products	equipment
– Iron and steeel	**Consumer goods**
– Nonferrous metals	– Musical instruments, toys, film, etc.
– Iron, steel and malleable cast iron	– Fine ceramics
products	– Glass and glassware
– Products of drawing and cold-rolling	– Wood products
mills	– Paper and paperboard products
– Chemical 41 (inorganic)	– Printed and duplicated matter
– Chemical 42 (organic)	– Plastic products
– Chemical 43 (fertilizers)	– Leather
– Chemical 44 (plastics)	– Leatherware (including travelware)
– Chemical 45 (chemical fibres)	– Footwear
– Chemical 46 (colouring matter, etc.)	– Textiles
– Chemical 47 (pharmaceuticals)	– Apparel
– Chemical 49 (other)	**Food, beverages and tobacco**
– Sawn timber, plywood, etc.	– Food and beverages
– Wood pulp, cellulose, paper, etc.	– Tobacco products
– Rubber products	

The official grouping, as given by the publications of the German Federal Statistical Office, is presented in Table 13.1. This grouping can be characterized as a grouping by stage of production, or a 'horizontal' grouping. The aggregation bias introduced by this grouping for the linear–homogenous regression model under consideration is measured in terms of the objective function ϕ. The value for the official grouping is about three times as large as the best value for a grouping achieved by the threshold accepting implementation. Hence, as far as the regression problem is concerned, the official grouping is far from being optimal. In fact, the average aggregation bias of some randomly generated grouping matrices is of the same order of magnitude as for the official grouping. Thus, the latter is not worse than a randomly generated one, but is not better either.

Most of the groupings obtained with the optimization procedure contain at least a partial 'vertical grouping', which means that a group with a given commodity tends to contain also products of the preceding and following stages of production, rather than only commodities at the same stage of production as implied by the official grouping. Given real data and the complex interactions between the prices in all commodity groups, one should not expect to find a grouping which is completely obvious or can be given some intuitive interpretation in any detail. In particular, for commodity categories with small weight such as fine ceramics or leatherware,[13] the objective function will not change much if they are moved to some other group. Since threshold accepting cannot be expected to deliver the global optimum of this problem due to its extreme complexity, it is therefore not too surprising that the position of these smaller categories differs for different runs of the optimization algorithm using different initializations. In Table 13.2, the grouping with the best value of $\tilde{\phi}$ detected during several optimization runs is presented. In the following, this grouping is denoted as the 'optimized grouping'.

Before discussing the commodity groups of the optimized grouping in some detail, it should be noted that the numbering of the groups has no meaning in contrast to the official grouping. Neither has the ordering of commodities within groups any meaning for the optimized grouping. The latter just follows the ordering of the official grouping for ease of comparison, while the groups are presented in the ordering given by the optimization algorithm.

Group 1 in the optimized grouping contains a strong cluster around the vertical grouping of crude oil, organic chemicals and plastics, which reappears in several other groupings with low aggregation bias obtained while tuning the threshold accepting implementation. The relationship between the other commodities in this group is less pronounced, although electrical products are an important input to car manufacturing. **Group 2** also provides some convincing vertical groupings such as chemical fibres together with textiles or leather and leatherware together with footwear. Furthermore, sawn timber, plywood, and other worked wood appear in the value–added chain for wood products. The inclusion of rubber products, and glass and glassware would rather fit into a horizontal aggregation scheme, as they represent final consumer goods like wood products, footware or textiles. Finally, the pricing of coal in Germany does not follow world market prices. Thus, it is not too surprising that it is not grouped together with crude oil. However, it is less intuitive to find it in this second group.

Group 3 comprises parts of the value–added chain for structural–steel products, namely iron, steel and malleable cast iron products, as well as quarrying products. No pronounced vertical clustering can be found in **Group 4**, although it comprises commodities at different stages of production. The same holds true for **Group 5**, while the last group (6) again provides two vertical subgroups consisting of inorganic chemicals and fertilizers on the

---

[13] The weights of the commodity categories are given below in Table 13.7 (Section 13.10).

Table 13.2 The optimized grouping (Germany)

Group 1	Group 4
– Agricultural, forestry and fishery products	– Steelworking products
– Crude oil and natural gas	– Machinery (including farm tractors)
– Road vehicles (excluding farm tractors)	– Musical instruments, toys, film, etc.
– Electrical products	– Chemical 46 (colouring matter, etc.)
– Ironware, sheet-metal ware and hardware	– Chemical 49 (other)
– Chemical 42 (organic)	– Office machinery and data-processing equipment
– Chemical 44 (plastics)	– Wood pulp, cellulose, paper, etc.
– Apparel	**Group 5**
– Tobacco products	– Petroleum products
Group 2	– Iron and steel
– Coal	– Nonferrous metals
– Chemical 45 (chemical fibres)	– Precision and optical goods, clocks, etc.
– Glass and glassware	– Chemical 47 (pharmaceuticals)
– Sawn timber, plywood, etc.	– Fine ceramics
– Wood products	– Plastic products
– Rubber products	– Food and beverages
– Leather	**Group 6**
– Leatherware	– Products of drawing and cold-rolling mills
– Footware	– Chemical 41 (inorganic)
– Textiles	– Chemical 43 (fertilizers)
Group 3	– Paper and paperboard products
– Quarrying products	– Printed and duplicated matter
– Iron, steel and malleable cast iron products	
– Structural-steel products and rolling stock	

one side, and paper and paperboard products, together with printed and duplicated matter on the other.

The forecast error $\tilde{\phi}$ for the above optimized grouping amounts to 19923.1 compared with 57678.0 for the official grouping. Unfortunately, due to the high complexity of the aggregation problem, it cannot be expected that this grouping is optimal or at least very close to an optimum solution. However, it is significantly better suited than the official grouping for studying international price spillovers. Furthermore, an extensive tuning of the threshold accepting implementation and a large number of runs of the algorithm allow for the conclusion that a significantly improved result cannot be achieved without using tremendous computing resources. Finally, some features of this best grouping also appear in other optimized groupings with slightly higher values of the objective function. Hence, these features appear to be robust. Some further arguments on the robustness of the presented optimized grouping are

Table 13.3   The pseudo–official grouping (Sweden)

Agricultural, hunting, forestry and	Capital goods
fishery products	– 3841 Ships and boats
– 100 Agricultural, hunting, forestry and	– 384\3841 Other transport equipment
fishery products	– 383 Electrical products
Mining and quarrying products	– 385 Instruments, photographic and
– 200 Mining and quarrying products	optical goods
Basic materials	– 381 Fabricated metal products
– 369 Other non–metallic mineral	– 3825 Office, computing and accounting
products	machinery
– 371 Iron and steel	– 382\3825 Other machinery
– 372 Nonferrous metals	**Consumer goods**
– 351 Industrial chemicals	– 361+362 Fine ceramics, glass and
– 352 Other chemical products	glassware
– 353+354 Petroleum products, lubricating	– 330 Wood products
oils, asphalt and coal products	– 356 Plastic products
– 331 Sawn timber, plywood and	– 323+324 Leather, leatherware and
other worked wood	footwear
– 355 Rubber products	– 321 Textiles
– 340 Pulp, paper, paper products and	– 322 Apparel
printed matter	**Food, beverages and tobacco**
	– 310 Food, beverages and tobacco

summarized below in Section 13.6.

13.5.2   The Swedish industrial classification system

The number of possible grouping matrices for the Swedish data is considerably smaller than for the German case, since only 25 commodity categories can be considered due to constraints in data availability. Nevertheless, the search space still comprises $P(25,6) = 3.7026 \times 10^{16}$ elements, leaving any direct enumeration approach out of reach. Hence, the same threshold accepting implementation as before for the German industrial classification system is used.

Before turning to the optimized groupings, the 'pseudo–official' grouping used as a bench–mark should be introduced. This is given in Table 13.3, where the code numbers refer to the Swedish industrial classification (Svensk standard för näringsgrensindelning (SE–SIC 69)), which is a refinement of the United Nations Standard Industrial Classification of All Economic Activities (ISIC). The six aggregated groups are formed according to the official grouping for the German data, i.e. by stage of production.

For the Euclidian metric', this 'horizontal' grouping yields a value $\tilde{\phi} = 701.95$ for the mean–square forecast error, which as for the German data is

Table 13.4   Optimized grouping (Sweden)

Group 1	Group 3
- 100 Agricultural, hunting, forestry and fishery products	- 352 Other chemical products
- 200 Mining and quarrying products	- 3825 Office, computing and accounting machinery
- 353+354 Petroleum products, lubricating oils, asphalt and coal products	- 330 Wood products
- 3841 Ships and boats	**Group 4**
- 384\3841 Other transport equipment	- 372 Nonferrous metals
- 340 Pulp, paper, paper products and printed matter	- 351 Industrial chemicals
Group 2	- 355 Rubber products
- 371 Iron and steel	**Group 5**
- 385 Instruments, photographic and optical goods	- 369 Other non–metallic mineral products
- 381 Fabricated metal products	- 331 Sawn timber, plywood and other worked wood
- 382\3825 Other machinery	- 383 Electrical products
- 356 Plastic products	**Group 6**
- 323+324 Leather, leatherware and footwear	- 361+362 Fine ceramics, glass and glassware
- 310 Food, beverages and tobacco	- 321 Textiles
	- 322 Apparel

about three times as large as the results for optimized groupings.

The optimized grouping for the Euclidean metric seems to be a very strong attractor in the underlying set of proper grouping matrices. It was obtained several times by using different initializations and different numbers of iterations ranging from 20 000 to 100 000. The value of the objective function $\tilde{\phi}$ for this grouping is only 279.02. The groups of this optimized grouping are presented in Table 13.4.

Group 1 accounted for 31% of Sweden's imports and 40% of its exports in 1968. Crude petroleum accounted for 61% of category 200 imports in 1968 (and much more after the two oil shocks of 1973 and 1979); this is combined in Group 1 with petroleum products. Category 100, which has a substantial forestry component, is combined in this group with paper products which comprised 48% of Group–1 exports. The two transport categories are also combined.

Group 2 accounted for 30% of imports and 32% of exports in 1968. Categories 371, 381, and 382\3825, consisting of iron and steel and their products, accounted for 60% of Group 2's imports and 87% of its exports. While some of the combinations in Groups 3 to 5 are hard to explain, textiles and apparel appear together in Group 6.

It should be noted, however, that the aim of using optimized modes of aggregation does not primarily consist in providing intuitive groupings. This goal is much easier reached by using the official modes of aggregation or similar procedures. The advantage of optimized modes of aggregation consists in the fact that the bias introduced by using these aggregates for further econometric analysis is reduced.

13.6 Robustness

The previous section indicated that despite of the huge computational complexity of the optimal aggregation problem, a threshold accepting implementation can obtain improved modes of aggregation. However, the presented groupings are optimized with regard to a specific objective function — corresponding to a specific aggregate model — and for a given data set. It would be interesting to know how robust the outcomes of this aggregation procedure are with regard to revisions or changes of the data, or a change of the considered aggregate model. The related question of robustness of the achieved groupings is the subject of this present section.

A first remark on the robustness of the real optimal grouping with regard to changes or errors in the data can be derived from a formal argument. The optimal grouping H^*, i.e. the proper grouping matrix, which minimizes $\tilde{\phi}$, is an element of the discrete space $\mathcal{H}$, and $\tilde{\phi}$ can be thought of as a mapping

$$\mathcal{H} \times \mathcal{X} \longrightarrow \mathbb{R},\tag{13.21}$$

where $\mathcal{X}$ denotes the space of data sets (X, Y); $\tilde{\phi}$ is uniformly continuous in the second argument, as $\mathcal{H}$ is finite. Hence, in the general case

$$H \neq \tilde{H} \Longrightarrow \tilde{\phi}(H) \neq \tilde{\phi}(\tilde{H})\tag{13.22}$$

and therefore

$$\delta \equiv \min_{H,\tilde{H}\in\mathcal{H}} |\tilde{\phi}(H) - \tilde{\phi}(\tilde{H})| > 0\,.\tag{13.23}$$

As $\tilde{\phi}$ is continuous in (X, Y), there exists a small positive ε such that any change of the data due to revisions, etc. by less than ε in the Euclidean norm will lead to a deviation in the values of $\tilde{\phi}$ for any $H \in \mathcal{H}$ by less than $\delta/2$. Consequently, the resulting optimal grouping will remain the same for small perturbations or errors in the data.

Unfortunately, two aspects reduce the power of this rather strong finding on the robustness of optimal groupings. First, the threshold accepting implementation does not behave in a completely deterministic manner. In particular, it does not give the global optimum with certainty. Thus, a small change in the data might lead to a different outcome, although the optimal grouping does not change.[14] Secondly, even if the algorithm behaves

---

[14] Given the earlier results of Section 7.4, this problem disappears as soon as the number of iterations of the threshold accepting implementation is allowed to grow with the rate required for the asymptotic convergence result to hold.

Table 13.5   Different measures of aggregation bias for optimized groupings

Objective	Euclidian distance			Mahalanobis distance		
	$\tilde{\phi}$	$\tilde{\phi}_{25}$	$\tilde{\phi}_{50}$	$\tilde{\alpha}$	$\tilde{\alpha}_{25}$	$\tilde{\alpha}_{50}$
$\tilde{\phi}$	19 923	19 317	27 674	39.171	26.126	11.105
$\tilde{\phi}_{25}$	19 727	18 873	27 944	47.486	32.525	15.009
$\tilde{\phi}_{50}$	34 300	33 588	26 534	56.401	39.010	19.505
$\tilde{\alpha}$	34 052	34 243	37 056	16.437	10.601	4.043
$\tilde{\alpha}_{25}$	34 198	31 409	39 707	17.575	9.828	3.900
$\tilde{\alpha}_{50}$	40 292	38 822	36 419	19.294	11.375	2.871
Official grouping	57 678	59 313	63 822	65.653	40.896	17.515

deterministically and always gives the global optimum, the order of magnitude of the admissible perturbations in the data, i.e. ε, remains unknown and real revisions of data might be larger.

A different approach to robustness compares the outcomes of the threshold accepting implementation in terms of the objective function and some central features of the resulting groupings. In this understanding of robustness, an optimized grouping is considered as robust if it is still of high quality compared to an optimized grouping for a slightly different setting of parameters or sample periods. This approach is similar to the one used earlier in Section 11.4 for the U–type designs.

Table 13.5 presents some results for the German data. It contains the values of different objective functions for groupings optimized with regard to one of these criteria. Besides the measures of aggregation bias based on the Euclidian distance $\tilde{\phi}$ and on the Mahalanobis distance $\tilde{\alpha}$, respectively, both measures have also been calculated by using a reduced rank regression for the calculation of the covariance matrix V. Thus, e.g. $\tilde{\phi}_{25}$ stands for the aggregation bias based on the Euclidian distance and imposing a rank reduction of 25 when estimating V.

In the upper left cell of the table, we find the entry for the optimized grouping presented in Table 13.2. The other entries of the first row are estimates of the aggregation bias for this grouping based on the Euclidian or Mahalanobis distance, respectively, and using reduced rank estimates for V. The fourth row presents the corresponding values for the best grouping obtained when minimizing the aggregation bias $\tilde{\alpha}$. Comparing the values for both groupings, it turns out that a grouping optimized with regard to the Euclidian distance $\tilde{\phi}$ results in a much smaller value for this measure than the grouping optimized with regard to $\tilde{\alpha}$. Nevertheless, independent of the used measure of aggregation bias for optimization, almost all optimized groupings exhibit a much smaller aggregation bias than the official grouping (last row) with regard to all measures of aggregation bias.

Given the layout of Table 13.5, the smallest values of aggregation bias are expected for the diagonal entries. However, it turns out that the best grouping obtained with regard to $\tilde{\phi}_{25}$ also exhibits a slightly smaller aggregation bias in terms of $\tilde{\phi}$ than the grouping optimized with regard to this objective function, which has been reported in the previous section. However, the difference in aggregation bias is small, and both groupings are quite similar. In fact, two groups are identical, and the other groups differ only in a few entries. Nevertheless, this finding indicates that a further increase in the number of iterations might result in further slight reductions of the objectives. Thus, the tremendous complexity of the optimal aggregation problem does not permit the belief that one has found an exact global optimum.

Using measures of aggregation bias based on reduced rank regressions has similar effects as small changes of the data. However, it is also possible to perform a direct comparison of optimized groupings obtained after slight perturbation of the data or a small change of the sample, respectively. Unfortunately, such a direct approach is hampered by the stochastic nature of the threshold accepting heuristic. Nevertheless, similar to the comparison presented for different criteria or models, it is possible to calculate the value of the aggregation bias function both for the original grouping and the groupings obtained by optimizing with regard to the slightly modified data. If the values do not change dramatically, this is a further hint for the robustness of the method. Of course, if the heuristic even provides the same optimized grouping, this would be an even stronger robustness indicator.

Finally, robustness may be assessed rather with regard to the tuning of the threshold accepting heuristic than with regard to the resulting groupings. Since, historically, optimal aggregation has been the first problem in econometrics and statistics tackled by threshold accepting, most methods and approaches for tuning and performance testing discussed in this present book have been applied to this problem. In particular, an extensive tuning of optimization parameters has been undertaken. Besides the automatically generated threshold sequence, deterministic linear and nonlinear threshold sequences have also been tested. Furthermore, the number of iterations used for these tuning experiments ranged from 10 000 to 1 000 000. These experiments allow for the conclusion of a negative correlation between the number of iterations and the achieved values of the objective function. Furthermore, using different functional forms for the threshold sequence does not seem to have a strong influence as long as the thresholds are not too small in the beginning. Finally, for the German data, the best groupings obtained for different parameter combinations resulted in values of $\tilde{\phi}$ always below 25 000, and often smaller than 21 000, and all these 'good' grouping matrices shared some patterns, e.g. the tendency to 'vertical grouping', with the optimized grouping presented in the previous section.

13.7   A Dynamic Model

The discussion of robustness in the previous section concentrated on using different measures of aggregation bias. However, it is also possible to change the model examined for this aggregation problem in Section 13.3. In fact, it seems natural to allow the international price adjustment to take some time, for example, changes in world prices of raw materials may take some time until they manifest themselves in the domestic prices of finished products. Then, a dynamic version of the price relationships is more adequate. Such a dynamic version of the model is introduced in this section. It is demonstrated that the aggregation problem itself does not change — only the measure of aggregation bias becomes different and, unfortunately, more complex. Nevertheless, it is possible to use the threshold accepting implementation to also obtain optimized groupings for this dynamic version of the model. Some results are presented and a comparison with the results from the static version is conducted along the lines depicted in the previous section.

A straightforward dynamic formulation consists of a distributed-lag model. Retaining the linearity assumption, this will be of the form

$$y_{tj} = \sum_{i=1}^{k} \sum_{l=0}^{L} z_{t-l,i} \gamma_{ij}(l) + \varepsilon_{tj}, \quad \mathbb{E}\varepsilon_{tj} = 0, \quad \mathbb{E}\varepsilon_{tj}\varepsilon_{t'j'} = \sigma_{jj'} v_{tt'}, \quad (13.24)$$

$$(j = 1, 2, \ldots, m)$$

where $t = 1, 2, \ldots, n$. The expectations are taken to be conditional on the current and lagged values of the explanatory variables $z_{t-l,i}$, and the $v_{tt'}$ are assumed to be known. The periods covered in the sample are $1 - L, 2 - L, \ldots, 0, 1, 2, \ldots, n$, numbering $T = n + L$. However, using $L = 12$ lags to model adjustments, which take place within one year, would result in a blow–up of the design matrix X in equation (13.1) by a factor of $L + 1 = 13$. Instead of solely including import and export prices of the current period, the lagged values of these variables for all commodities also have to be included. Consequently, the estimation of the disaggregated model becomes impossible due to missing degrees of freedom. Even if enough observations were available, the calculation of the aggregation bias for this extended model would require too much computing resources to allow an efficient threshold accepting implementation for the example considered in this chapter.

Chipman (1985, pp. 112ff.) shows that it is possible to reduce the number of estimated coefficients by imposing a spline structure on the lag coefficients. This method is implemented for the optimal aggregation of price indices. Thereby, the contemporaneous lag coefficients $\gamma_{ij}(0)$ are estimated freely, and the remaining L lag coefficients $\gamma_{ij}(l), l = 1, 2, \ldots, L$ are assumed to lie on a natural cubic spline with four knots. Considering the regression equations obtained by imposing the spline structure on the lag coefficients, a transformed system results, which is of order $n \times 3k$ instead of $n \times (L + 1)k$ of the

Table 13.6   Aggregation bias for optimized groupings (dynamic model)

	Euclidian distance $\tilde{\phi}$		Mahalanobis distance $\tilde{\alpha}$	
Objective	static	dynamic	static	dynamic
$\tilde{\phi}$, static	19 923.1	439.43	39.17	370.00
$\tilde{\phi}$, dynamic	39 801.5	270.44	50.50	290.85
$\tilde{\alpha}$, static	34 051.9	675.63	16.44	186.67
$\tilde{\alpha}$, dynamic	44 972.6	364.77	28.54	121.02
Official grouping	57 678.0	898.10	65.65	692.93

original dynamic system, where k denotes the number of commodities in the disaggregated model. The details of the derivation of the restricted dynamic model are somewhat burdensome and are therefore omitted. Such details can be found in Chipman (1985) and Greville (1969), respectively.

Aggregation of the unrestricted dynamic model proceeds as for the static case. The same spline transformation, which has been applied to the disaggregated model, has to be applied to the distributed lags of the aggregated model. This transformation delivers a formula for X and, consequently, V, which can be employed as in the static case for the calculation of measures of aggregation bias. Thus, the sole difference of the dynamic version consists in a higher computational burden for the calculation of these measures[15] and a loss of 12 observations due to the dynamic formulation.

Due to these constraints, only preliminary results for the German data are reported in this section. Table 13.6 summarizes the findings for the official grouping, which is the same as that provided in Table 13.1 for the static case, and the groupings obtained by the threshold accepting implementation minimizing the different objective functions for the static and dynamic model, respectively. It should be noted that due to the higher complexity of the objective functions in the dynamic setting, the number of iterations performed for each run of the threshold accepting implementation for the dynamic model is rather small (50 000). Further increases of this number might improve the results.

The entries on the diagonal of Table 13.6 are the smallest within each row. This indicates that optimizing with regard to the dynamic model results in groupings better suited for the dynamic setting, while optimizing with regard to the static model provides good groupings for the static case, as could have

---

[15] The increase by a factor of three of the matrix dimension results in the computational burden growing by a factor of around 27 due to the matrix inversions involved in calculating aggregation bias.

been expected. Again, a good grouping obtained for the static case is still much better for the dynamic case than the official grouping (439.43 compared to 898.10 for the Euclidian distance, and 186.67 compared to 692.93 for the Mahalanobis distance). Consequently, for the application to German price data, the result of performing optimized aggregation appears to be rather robust when switching from a static to a dynamic model, and vice versa.

13.8   Conclusions

This chapter has provided an application of the threshold accepting heuristic to the aggregation problem of economic time series. This specific application consisted in aggregating commodity categories into groups for the purpose of assessing and forecasting the impact of changes. The following conclusions can be drawn:

1. *Aggregation matters.* The process of aggregation, and the mode of aggregation chosen, can have a substantial impact on the results obtained in econometric research.

2. *Optimal aggregation is not trivial.* Section 13.9 provides a proof that choosing an optimal mode of aggregation is intractable from the point of view of mathematical complexity theory. Consequently, the use of optimization heuristics is indicated. The threshold accepting implementation still allows us to obtain optimized groupings at an intermediate level of aggregation. However, the generation of complete hierarchical classification systems or partitions for very large sets of groups seems to be beyond the scope of this implementation.

3. *Standard methods of aggregation are far from optimal.* The modes of aggregation implied by official classification systems and the groupings provided by statistical agencies in presenting their data may be far from the optimal classification system needed for purposes of econometric estimation and prediction. Consequently, econometric analysis based on these groupings can be misleading.

4. *Threshold accepting offers a way to better groupings.* The reduction in the value of the criterion function — mean–square forecast error or aggregation bias — when using the threshold accepting implementation can be very considerable.

5. *The economic meaning of 'better groupings' is not yet completely obvious.* While the optimized groupings exhibit a tendency for 'vertical' groupings — groupings which take account of input–output relationships between industries — to outperform 'horizontal' groupings, which group commodities by stage of production, certain commodity combinations appearing in optimized groupings cannot be easily explained by intuitive reasoning.

13.9 Appendix: The Computational Complexity of Optimal Aggregation

This appendix summarizes some findings on the computational complexity of Optimal Aggregation. The exposition benefitted from helpful comments by W. Bob and T. Schneeweis and the participants of a mathematical–economic workshop at the University of Konstanz.

13.9.1 Introduction

The discussion of the inherent complexity of the optimization problems presented above in Sections 2.3 and 3.3 provided the rationale for considering a new optimization paradigm in statistics, which was introduced in Chapter 4. In this appendix, the notion of complexity is used in its formal meaning, as developed in the mathematical complexity theory. In particular, in order to categorize a problem as intractable it is not sufficient that no available algorithm can solve it to optimality with computer resources growing 'only' polynomially in input size. In order to be considered as a complex problem in the formal sense, it is required that no algorithm — including any algorithm not known at present — fulfils this task.

A simple example, which nevertheless marks an important contribution to the analysis of computational complexity, may demonstrate the difference. The multiplication of two 2×2 matrices according to

$$\begin{pmatrix} a_{11} & a_{12} \\ a_{21} & a_{22} \end{pmatrix} \begin{pmatrix} b_{11} & b_{12} \\ b_{21} & b_{22} \end{pmatrix} = \begin{pmatrix} a_{11}b_{11} + a_{12}b_{21} & a_{11}b_{12} + a_{12}b_{22} \\ a_{21}b_{11} + a_{21}b_{21} & a_{21}b_{12} + a_{22}b_{22} \end{pmatrix}$$

requires eight scalar multiplications and four additions. Depending on the computer and required accuracy of the calculations, a scalar multiplication may require much more time than an addition. Under such circumstances, it is reasonable to consider the required number of scalar multiplications as representing the complexity of an algorithm for performing matrix multiplication. Then, the method described above has a complexity of n^3 for matrices of size $n \times n$. Provided that no other algorithm is known, the heuristic concept of complexity used in Part I leads to the acceptance of this computational complexity for matrix multiplication.

However, Strassen (1969) proposed an algorithm requiring solely 7 multiplications for the 2×2 case and $n^{\log_2 7}$ for the general case. Now, the a priori assumption about the computational complexity of matrix multiplication must be revised. Nevertheless, it is unclear whether the new algorithm represents the optimum. Consequently, different approaches are necessary in order to obtain lower bounds for the computational complexity of a problem which ensure that no algorithm can do the job faster. In fact, for the problem of matrix multiplication, 7 scalar multiplication are optimal for the 2×2 case. The proof of this conjecture, however, is not trivial, but requires the use of mathematical complexity theory.

When it comes to the theoretical analysis of the complexity of a problem in this strict mathematical sense, interesting cases can be observed. Maybe the most puzzling result of research in this area has been the discovery of certain problems for which all algorithms — known and unknown ones — are inefficient. A related result, which is the basis of the following discussion of the complexity of the problem of optimal aggregation, is the discovery of a class of problems, known as NP–complete, which are supposed to be of high computational complexity. Despite the efforts of many of the most famous researchers in this area, no efficient algorithm has been found so far for any of these problems. It is well known, however, that whenever there exists an efficient algorithm for one of these problems, then there has to be one for all of them. This means that whenever a given problem is in this class of problems it is 'just as hard' as a large number of other problems which are commonly regarded as being difficult.

After a brief introduction to the concept of NP–completeness in Section 13.9.2, the problem of optimal aggregation is discussed within this framework. In Section 13.9.3, it is then shown that at least some specific versions of this problem are NP–complete. Consequently, it is not possible to obtain optimal solutions with a deterministic algorithm when using only a reasonable amount of computing resources.

13.9.2 NP–completeness

A complete introduction to the theory of NP–completeness and 'inherently intractable' problems is clearly beyond the scope of this present section. Instead, a few basic concepts of the theory of computation are introduced in a concise and comprehensive way. For a more detailed and technical description the interested reader is referred to the standard literature.[16]

A 'problem' is given by a question depending on some parameters, which describe the specific instance of the problem. For example, the problem could be to find a shortest tour connecting a given set of points (cities). In this case, an instance is described by the coordinates of all points in the Euclidian space. Consequently, any feasible set of parameters defines an instance of the problem. The size of this instance is defined as the length of some 'natural' encoding of the parameter set.[17]

---

[16] See, for example, Aho, Hopcroft and Ullman (1974), Ferrante and Rackoff (1979), Garey and Johnson (1979), Hopcroft and Ullman (1979), Papadimitriou and Steiglitz (1982), Paul (1978), Reischuk (1990), Wilf (1986) or Cook et al. (1998, pp. 309ff.). See also the presentation in Ausiello et al. (1999, pp. 175ff.).

[17] In particular, the encoding has to be concise, avoiding any unnecessary information or symbols. Then, it can be shown that the computational complexity of a problem does not depend heavily on the chosen encoding.

In the mathematical theory of computation, algorithms for problem solving are connected to theoretical models of computers, which, in fact, are older than the first electronic computer, but still provide a good description of their properties. The most famous among them is known as the 'Turing machine' in honour of the important contributions to the theory of computation by Alan Turing. For the purposes of this present section, one can think of an algorithm as a computer program written in some higher computer language for some real computer. An algorithm is said to 'solve' a specific problem if it provides a solution to the general question for any problem instance.

The type of problems considered in this part of the book are combinatorial optimization problems, i.e. answering the question: 'Which element out of a finite set minimizes a given objective function $f(x)$?'. For these problems, a straightforward algorithm consists in calculating the objective function for all elements of the finite set and in choosing one element corresponding to the minimum value. Unfortunately, the finite search space, in general, becomes too large for this algorithm, even for small problem instances. Then, the enumeration algorithm certainly is not 'efficient'.

In general, one measures the efficiency of an algorithm in terms of the time requirements depending on the instance size. This concept has been introduced earlier in Section 3.3 under the heading 'time complexity'. After having chosen a specific encoding of the possible instances, the time complexity function can be defined for any algorithm which solves the given problem. This gives for each possible instance size the largest amount of time needed by the algorithm to solve a problem instance of that size. Using this concept, in Section 3.3 it has been argued that only algorithms working in polynomial time can be considered as 'efficient', while problems which require algorithms with time complexity functions, which cannot be bounded by polynomials, are considered as intractable.[18] Of course, there do exist non–polynomial time algorithms which work quite well in practice, such as the simplex algorithm for linear programming. However, such examples are rare, not well understood so far and always risk having enormous time requirements for some instances. As the presented measure of efficiency is a worst–case measure, i.e. the maximum amount of computing time needed for any instance of a given size is considered, only polynomial time algorithms will be regarded as efficient.

Certainly, those problems as described by Turing in the 1930s, which are not decidable at all, i.e. there exists no algorithm for solving such problems, have to be considered as intractable in this sense. Furthermore, there exist some 'natural' decidable but provably intractable problems.[19] Most of these problems cannot be solved in polynomial time even with a 'nondeterministic' algorithm. Such an algorithm can be viewed as being composed of two separate steps. Given a problem instance, the first step just 'guesses' a potential

---

[18] See Table 3.3 (Section 3.3) for an intuitive argument for this classification.

[19] Mayr and Meyer (1982) and Winker (1991) provide a formal discussion of one of these.

solution. It is assumed that this step can be performed in parallel for an infinite number of potential solutions. Both the instance and the 'guessed' solution are provided as inputs to the second step. The second step checks whether the provided potential solution is really a solution for the given instance. For example, for a given set of cities for the travelling salesman problem a first–step guess is just one tour connecting all cities. The second step then checks whether the length of this tour is less than a given value. If both the guessing and the checking step are executed in polynomial time, the algorithm is called nondeterministic polynomial (NP). A problem belongs to the class of NP decision problems if it can be solved by such a nondeterministic algorithm in polynomial time. In particular, any problem which may be solved by a deterministic polynomial time algorithm belongs to this class. A still open question, however, is whether there exist problems in the class NP which cannot be solved by a polynomial time algorithm.[20]

The interrelation of problems with respect to their time complexity can be described by reduction mappings. Thereby, a reduction of one problem to another is given by a constructive transformation, which maps any instance of the first into an equivalent instance of the second. If a reduction of one problem to another can be found, then the first problem can be solved whenever there is an algorithm solving the second. If this reduction is performed by an algorithm working in polynomial time in the input size, one then speaks of 'polynomial time reducibility'. This term was introduced by Cook (1971) who also proved that the 'satisfiability' problem, which comes up in formal logic, is in the NP class and that any other problem in the latter can be reduced to it in polynomial time. This means that if the satisfiability problem can be solved with a polynomial time algorithm, so can every problem in the NP class — one just constructs the solving algorithm as a sequence of the reduction algorithm and the solving algorithm for the satisfiability problem. The other way round, if any problem in such a class is intractable, the satisfiability problem must also be intractable. Therefore, the satisfiability problem belongs to a group of 'hardest' problems in the NP class.

A problem is called NP–complete, if every problem in the NP class may be reduced to it with an algorithm working in polynomial time in the input size. As the polynomial time reducibility is transitive, the following result is evident. If a NP–complete problem can be polynomially reduced to another problem in this NP class, the latter also has to be NP–complete.

Meanwhile, literally hundreds of problems have been shown to belong to the class of NP–complete problems.[21] The question, however, of whether a polynomial time algorithm may be found for the NP–complete problems remains open. After many fruitless attempts, it will need a major

---

[20] Cook's hypothesis, which is commonly accepted in the mathematical complexity literature, states that there exist such problems.

[21] For an impressive list of such problems, see the appendix of Garey and Johnson (1979).

breakthrough in mathematics and computer science to decide this. For the moment, a problem in the class of NP–complete problems must be regarded as being intrinsically difficult.

13.9.3   The computational complexity of Optimal Aggregation

In this section, it is shown that the recognition version[22] of *Optimal Aggregation* is NP–complete. At first, the problem of *Optimal Aggregation* is restated in its general version. In two further sections it will be proven that two simpler versions of *Optimal Aggregation*, namely $Group_1$ and $Group_2$ are NP–complete. This will be done by using a polynomial reduction of the NP–complete problem $Knapsack$[23] to the recognition version of the considered problems $Group_1$ and $Group_2$. Finally, it will be shown that the general version of *Optimal Aggregation* is at least as difficult as $Group_2$.

Optimal Aggregation

Given an $n \times m$ matrix Y of observations of the dependent variables, an $n \times k$ design matrix X, a $k \times m$ matrix B of regression coefficients and the $n \times m$ matrix E of residuals, the regression model is given by

$$Y = X B + E. \tag{13.25}$$

The grouping or aggregation of the variables is carried out by regular grouping matrices G and H. In the simplest case each row of G and H contains exactly one 1, with 0 elsewhere. The reduced regression model becomes

$$YH = XG B^* + E^*.$$

The problem of *Optimal Aggregation* is to choose G and H in an optimal way, i.e. in order to obtain a good approximation of the 'true' model (equation (13.25)). In Section 13.2, it has been shown that under some assumptions about the model this is achieved by minimizing the objective function

$$\alpha = \mathrm{tr}\{X(I - G\tilde{G}^{\#})\tilde{B}H(H'SH)^{-1}H'\tilde{B}'(I - G\tilde{G}^{\#})'X'\}, \tag{13.26}$$

where

$$\tilde{G}^{\#} = (G'X'XG)^{-1}G'X'X,$$
$$\tilde{B} = (X'X)^{-1}X'Y,$$
$$S = (Y - X\tilde{B})'(Y - X\tilde{B}).$$

---

[22] For a discussion of the different versions of a combinatorial optimization problem, see Papadimitriou and Steiglitz (1982, p. 345).

[23] See Garey and Johnson (1979, p. 247), Papadimitriou and Steiglitz (1982, p. 374), or the original paper by Karp (1972, p. 94).

In the following, it is shown that the recognition version of this problem is NP–complete. This means that the problem to decide whether there are grouping matrices G and H in such a way that the corresponding α lies in a given range is already very hard. Of course, the problem of finding the optimal groupings, i.e. the matrices minimizing explicitly the objective function α, is at least as difficult as the recognition version.

$Group_1$ is NP–complete

In this section, it is shown that a simpler version of *Optimal Aggregation*, namely $Group_1$, is NP–complete. This will be done by a polynomial reduction of *Knapsack*, which is well-known to be NP–complete. First, the *Knapsack* problem and the problem $Group_1$ considered in this subsection are introduced. Comparing the two problems, a natural nearly one–to–one correspondence between optimally packed knapsacks and optimal groupings can be observed, which leads immediately to the reduction given in the proof of proposition 13.9.1 (see below).

Formally, *Knapsack* is given by an integer $n > 0$, a vector of integers $\bar{x} \in (\mathbb{N}_{>0})^n$ and two positive integers k and k'. The question is whether there exists a $\bar{\delta} \in \{0, 1\}^n$ such that $k' \leq \sum_{i=1}^{n} \delta_i x_i \leq k$. Less formally, one might consider a knapsack of given volume and a set of n boxes of different volumes x_i. How can the volume of the knapsack be used in an optimal way? The question seems to be quite simple, but when experimenting with a number of 20 boxes, it is easy to see that there is no trivial way to find the best possibility. Hence, it is not too astonishing to learn that *Knapsack* had already been proven to be NP–complete as early as 1972 (Karp, 1972, p. 94).

The problem $Group_1$ is defined as follows. For $\tilde{m}, m \in \mathbb{N}_{>0}$ with $\tilde{m} < m$ let the set of feasible grouping matrices $GR(m, \tilde{m})$ be given by

$$GR(m, \tilde{m}) := \{H \in \{0, 1\}^{m \times \tilde{m}} \mid \forall i \in [m] \, \exists! \, j \in [\tilde{m}] \, H_{ij} = 1,$$
$$\forall j \in [\tilde{m}] \, \exists \, i \in [m] \, H_{ij} = 1\} \, ,$$

i.e. the matrices H in $GR(m, \tilde{m})$ are exactly those for the simplest case of *Optimal Aggregation* discussed above. Furthermore, let the set of symmetric, positive definite, regular matrices of rank m be given by:

$$P(m) := \{S \in GL(\mathbb{Q}, m) \mid S = S', S \text{ positive definite } \}$$

The elements S of $P(m)$ correspond to the S in equation (13.26). Then, the grouping problem is to optimize the trace of $(H'SH)^{-1}$ for given $S \in P(m)$.[24]

---

[24] Note that S assumed to be in $P(m)$ implies that $H'SH$ is positive definite and regular (Johnston, 1972, p. 105), i.e. the inverse $(H'SH)^{-1}$ exists.

Consequently, one has to regard the following set of optimal groupings:

$$\widetilde{Group}_1 := \{(H, S, m, \tilde{m}) \mid \tilde{m}, m \in \mathbb{N}, \tilde{m} < m, H \in GR(m, \tilde{m}), S \in P(m),$$
$$\text{tr}\{(H'SH)^{-1}\} = \min_{\tilde{H} \in GR(m,\tilde{m})} \text{tr}(\tilde{H}'S\tilde{H})^{-1}\}.$$

The optimization version of the optimal grouping problem $\widetilde{Group}_1$ corresponds to a recognition problem $Group_1$ with instance $\tilde{m} < m$, both in $\mathbb{N}_{>0}$, $S \in P(m)$ and $K \in \mathbb{Q}$ and the question as to whether there exists a grouping matrix $H \in GR(m, \tilde{m})$ with $\text{tr}\{H(H'SH)^{-1}H'\} \leq K$. It is obvious that the optimization version $\widetilde{Group}_1$ is at least as difficult as the recognition version $Group_1$ (Papadimitriou and Steiglitz, 1982, p. 345). In order to simplify the proof, the following version of $Group_1$ will be considered: the instance is given by $\tilde{m} < m$, $S \in P(m)$ as before, and $K' < K \in \mathbb{Q}$. Now, the question is whether there is a grouping matrix $H \in GR(m, \tilde{m})$ with $K' \leq \text{tr}\{(H'SH)^{-1}\} \leq K$. $Group_1$ is in the NP class because for a given feasible grouping matrix $H \in GR(m, \tilde{m})$ the question whether $K' \leq \text{tr}\{(H'SH)^{-1}\} \leq K$ is easily answered in polynomial time by straightforward matrix computations. Therefore a nondeterministic algorithm would just calculate all the values of $\text{tr}\{(H'SH)^{-1}\}$ and test whether at least one of them falls in the range defined by K' and K.

Proposition 13.9.1 $Group_1$ *is NP–complete.*

An instance of *Knapsack* is given by $n \in \mathbb{N}_{>0}$, $\bar{x} \in (\mathbb{N}_{>0})^n$ and two positive integers k and k'. The question is whether there exists a $\bar{\delta} \in \{0,1\}^n$ such that $k' \leq \sum_{i=1}^n \delta_i x_i \leq k$. Without loss of generality, it might be assumed that $k' \leq k$ and $N := \sum_{i=1}^n x_i \geq C$ for some given constant $C > k$.[25]
Now let us consider the instance for $Group_1$ given by $m := n > 2$, $\tilde{m} := 2$, $S_x := \text{diag}(x_1, \ldots, x_n)$, i.e. a matrix containing non–zero elements only in the diagonal. Obviously S_x is in $P(m)$. To each $\bar{\delta} \in \{0,1\}^n$ corresponds a $H_{\bar{\delta}}$ in $GR(m, \tilde{m})$, defined by:

$$H_{i1} := \delta_i, \quad 1 \leq i \leq n$$
$$H_{i2} := 1 - \delta_i, \quad 1 \leq i \leq n.$$

Some straightforward calculations result in the following:

$$\text{tr}\{(H'_{\bar{\delta}}S_x H_{\bar{\delta}})^{-1}\} = \text{tr}\begin{pmatrix} \sum_i \delta_i x_i & 0 \\ 0 & \sum_i (1 - \delta_i)x_i \end{pmatrix}^{-1}$$
$$= \frac{N}{(\sum_i \delta_i x_i)(N - \sum_i \delta_i x_i)}.$$

---

[25] One just has to add a $x_{n+1} = C$. Obviously, δ_{n+1} has to be zero for all feasible solutions.

It can be seen that the restricted version of $Group_1$ with a diagonal matrix S and only two columns for the grouping matrices H mimics $Knapsack$. A 1 in the first column of H corresponds to a box, which is put into the knapsack, whereas the 1s in the second column correspond to the boxes left outside.

More formally, for $N \geq C := 2k$, $K' := N/(k(N-k))$ and $K := N/(k'(N-k'))$, there is a solution for a given instance of the $Knapsack$ problem if, and only if, there is one for the corresponding instance of $Group_1$ given by S_x, K' and K as defined above. Since the calculation of S_x out of x is polynomial, the reduction of $Knapsack$ to $Group_1$ is polynomial. Therefore, $Group_1$ is NP–complete, which terminates the proof.

$Group_2$ is NP–complete

Let the set of feasible grouping matrices GR and the set of symmetric, positive definite, regular matrices P be given as defined above. Then, the grouping problem considered here is to optimize the trace of $H(H'SH)^{-1}H'$ for a given $S \in P(m)$. As a result, the following set of optimal groupings has to be regarded:

$$\widetilde{Group_2} := \{(H, S, m, \tilde{m}) \,|\, \tilde{m}, m \in \mathbb{N}, \tilde{m} < m, H \in GR(m, \tilde{m}), S \in P(m),$$
$$\text{tr}\{H(H'SH)^{-1}H'\} = \min_{\tilde{H} \in GR(m, \tilde{m})} \text{tr}\tilde{H}(\tilde{H}'S\tilde{H})^{-1}\tilde{H}'\}.$$

As in the last section, the corresponding recognition problem $Group_2$ will be considered with instance $\tilde{m} < m$, $S \in P(m)$ and $K' \leq K \in \mathbb{Q}$. The question will be whether there is a grouping matrix $H \in GR(m, \tilde{m})$ with $K' \leq \text{tr}\{H(H'SH)^{-1}H'\} \leq K$. $Group_2$ is in the NP class for the same reasons as $Group_1$.

Proposition 13.9.2 $Group_2$ *is NP–complete.*

It will be shown that a restricted version of $Knapsack$ is polynomial reducible to $Group_2$. Then, as this restricted version can be shown to remain NP–complete (Garey and Johnson, 1979, pp. 65 and 223), $Group_2$ has to be NP–complete, too.

An instance of the restricted version of $Knapsack$ is given by an even integer n, a vector $\bar{x} \in (\mathbb{N}_{>0})^n$ and two positive integers k' and k. The question is whether there exists a $\bar{\delta} \in \{0,1\}^n$ such that $k' \leq \sum_{i=1}^n \delta_i x_i \leq k$ and $\sum_{i=1}^n \delta_i = n/2$. As for $Group_1$ one might assume that $k' \leq k$ and $N := \sum_{i=1}^n x_i \geq C$ for some given constant $C > k$.

Now consider the instance for $Group_2$ given by $m := n > 2$, $\tilde{m} := 2$ and $S_x := \text{diag}(x_1, \ldots, x_n)$. To each $\bar{\delta} \in \{0,1\}^n$ corresponds a $H_{\bar{\delta}}$ in $GR(m, \tilde{m})$ as defined above. Some straightforward calculations result in the following:

$$\text{tr}\{H_{\bar{\delta}}(H'_{\bar{\delta}}S_x H_{\bar{\delta}})^{-1}H'_{\bar{\delta}}\} = \text{tr}\{H_{\bar{\delta}}\begin{pmatrix} \sum_i \delta_i x_i & 0 \\ 0 & \sum_i(1-\delta_i)x_i \end{pmatrix}^{-1}H'_{\bar{\delta}}\}$$

$$= \frac{\frac{n}{2}(N - \sum_i \delta_i x_i) + (n - \frac{n}{2})(\sum_i \delta_i x_i)}{(\sum_i \delta_i x_i)(N - \sum_i \delta_i x_i)}$$

$$= \frac{\frac{n}{2}N}{(\sum_i \delta_i x_i)(N - \sum_i \delta_i x_i)},$$

and it follows as for $Group_1$ that $Group_2$ is NP–complete, which terminates the proof.

The computational complexity of *Optimal Aggregation*

This section is completed with an argument on the computational complexity of *Optimal Aggregation* itself. Therefore, the real problem of optimal aggregation has to be regarded with the resulting objective function:

$$\alpha = \text{tr}\{X(I - G\tilde{G}^{\#})\tilde{B}H(H'SH)^{-1}H'\tilde{B}'(I - G\tilde{G}^{\#})'X'\}.$$

As the covariance–matrix S is symmetric and positive semidefinite in general, S is in $P(m)$ in almost all cases. In order to see that *Optimal Aggregation* is of high computational complexity it is enough to treat the case where S is diagonal. If the choice of G is independent of the choice of H

$$R := X(I - G\tilde{G}^{\#})\tilde{B}$$

is a constant matrix with respect to H. For the matrices R with equal sum K for the squared elements of each row an easy calculation shows that

$$\text{tr}\{RH(H'SH)^{-1}H'R'\} = K \cdot \text{tr}\{H(H'SH)^{-1}H'\}.$$

Thus, the general problem of *Optimal Aggregation* is at least as difficult as $Group_2$ but still belongs to the NP class in its recognition version. Therefore, *Optimal Aggregation* is also NP–complete.

13.9.4 Conclusions

This appendix has provided a short introduction to the mathematical theory of complexity, and has described a group of problems which are intrinsically hard to solve. The major goal has been to prove that the problem of optimal aggregation discussed in the previous sections belongs to this group of hard problems. This provides a formal justification for the use of a heuristic optimization algorithm such as threshold accepting.

13.10   Appendix: Commodity Groups and Weights

Table 13.7   Weights per thousand for German price indices $(1991 = 100)^{a)}$

Item	Category	Code	Imports	Exports	Prod.$^{b)}$
1	Agricultural, forestry and fishery products	0000	56.07	12.10	‡52.48
	[Electricity, gas, and water]	1000	0.00	0.00	93.08
	[Mining products]	2100	63.93	5.93	54.45
2	Coal and lignite	2110	3.06	2.74	15.20
3	Crude petroleum and natural gas	2120	53.89	†0.69	*38.58
	[Crude petroleum]	2121	*37.55	0.00	0.25
	[Natural gas]	2122	16.34	†0.69	38.33
	[Other mining products]				
	(210 − 211 − 212)	—	†6.98	†3.19	†0.67
	[Iron ore]	2130	3.99	0.00	—
	[Nonferrous metal ore]	2140	2.24	†0.08	—
	[Potash salt]	2150	0.00	†1.68	0.37
	[Rock and pit salt]	2160	†0.13	†0.26	0.22
	[Fluorspar, heavy spar, graphite, etc.]	2170	†0.57	†0.08	—
	[Peat]	2180	†0.05	†0.40	—
4	Petroleum products	2200	28.21	5.94	34.14
	[Nuclear fuels]	2400	1.53	2.35	0.00
5	Quarrying products	2500	9.50	8.89	25.04
6	Iron and steel	2700	28.78	34.42	20.97
7	Nonferrous metals	2800	31.06	19.86	11.82
8	Foundry products	2900	2.92	4.56	8.57
	[Iron, steel and malleable cast iron products]	2910	2.50	4.04	5.22
	[Nonferrous metal foundry products]	2950	†0.42	†0.52	3.35
	[Drawing, cold-rolling and steelworking products]	3000	9.77	14.18	20.45
9	Drawing and cold-rolling products	3010	4.01	5.10	5.88
10	Steelworking products	3020	5.76	9.08	14.57
11	Structural steel products and rolling stock	3100	6.28	12.37	21.50
12	Machinery	3200	71.19	161.81	87.31
13	Road vehicles (excluding farm tractors)	3300	116.96	176.42	95.10
	[Ships and boats]	3400	0.00	0.00	0.13
	[Aircraft]	3500	0.00	0.00	0.00
14	Electrical products	3600	107.98	124.72	90.74
15	Precision and optical goods, clocks and watches	3700	19.01	21.22	9.77
16	Ironware, sheet metal ware and hardware	3800	23.17	29.87	35.86

continued overleaf

Table **13.7** (*continued*)

Item	Category	Code	Imports	Exports	Prod.[b]
17	Musical instruments, sporting goods, etc.	3900	11.73	8.19	4.61
	[Chemicals]	4000	93.50	135.24	70.18
18	Inorganic materials and chemicals	4100	5.43	8.39	4.66
19	Organic materials and chemicals	4200	25.08	31.06	10.91
20	Fertilizers and insecticides	4300	3.97	4.19	1.68
21	Plastics and artificial rubber	4400	20.13	25.58	12.08
22	Chemical fibers	4500	2.20	4.66	2.24
23	Dyes and paints	4600	4.83	13.93	7.79
24	Pharmaceuticals	4700	11.72	17.75	13.10
25	Other chemical products	4900	20.14	29.68	17.72
26	Office machinery and data-processing equipment	5000	41.84	22.63	9.52
27	Fine ceramics	5100	4.42	3.34	2.54
28	Glass	5200	6.57	6.81	6.71
29	Sawn timber, plywood and other worked wood	5300	8.06	3.29	6.99
30	Wood products	5400	14.43	10.26	25.84
31	Wood pulp, cellulose, paper and paperboard	5500	22.97	11.49	7.86
32	Paper products	5600	6.13	9.60	16.59
33	Printed and duplicated matter	5700	4.58	9.19	23.82
34	Plastic products	5800	21.13	28.20	34.68
35	Rubber products	5900	10.39	9.71	8.22
36	Leather	6100	2.23	1.81	0.73
	[Leatherware and footware]	6200	16.25	4.33	3.49
37	Leatherware (including travelware)	6220	3.77	1.28	1.13
38	Footware	6250	12.48	3.05	2.36
39	Textiles	6300	53.64	36.93	20.09
40	Apparel	6400	43.90	15.83	16.14
41	Food and beverages	6800	61.06	45.41	113.63
42	Tobacco products	6900	0.81	3.10	17.43
	[Prefabricated housing]	7000	0.00	0.00	2.00

[a] Square brackets indicate items omitted from the data set, and an asterisk (*) in front of a weight indicates data computed by aggregation. A dagger (†) indicates that published data are not available, while a double dagger (‡) indicates a weight computed from 1991 data. The commodity codes used are those of the Güterverzeichnis für die Produktionsstatistik (GP).

[b] The following abbreviation is used: Prod. denotes weights in the domestic producer price index.

14

Censored Quantile Regression

14.1 Introduction

Censored data are often encountered in statistical modelling. Usually, a model
can be presented as a mapping of a vector of exogenous variables $\mathbf{x}$ to an
endogenous variable y. Then, estimates of model parameters are based on
pairs of observations $(\mathbf{x}_i, y_i)$, $i = 1, \ldots, T$. An observation is called *censored* if
for a given $\mathbf{x}_i$, y_i cannot be observed. However, it is known that y_i lies beyond
some cut–off values, e.g. $y_i \geq yc_i$. Such situations appear in all kind of hazard
rate models, for example in failure time analysis in an engineering context.
Censoring also appears quite often in microeconomic data sets, either as a
result of some economic decisions which imply censoring, e.g. unemployment
duration, or due to institutional features, e.g. data protection or cut–off values
for data gathered for different purposes (income in social security statistics).

There exist estimation procedures for censored data which can be tackled
by standard optimization routines. However, these methods, e.g. hazard rate
estimation, are limited to the analysis of the mean response. Recently, the
interest in modelling such data shifted from this concentration on mean
responses to an analysis of the whole conditional distribution, which is
represented by the quantiles of the conditional distribution (Fitzenberger,
Koenker and Machado, 2000). Then, the estimation problem becomes a
problem of censored quantile regression (CQR), which has been introduced
by Powell (1984). For the case when the censoring points yc_i are known
and fixed, the estimation problem has been analysed and a few applications
exist.[1] Again, the first applications came up in the context of failure time
analysis, both in econometrics and engineering (Womersley (1986); Horowitz
and Neumann (1987)). A different application is presented by Buchinsky
(1994), who models quantiles of the conditional wage distribution for the
USA by using a censored quantile regression approach.

To the extent that more and more microeconomic data sets became available

---

[1] See Fitzenberger (1997a) for a survey, and details of applications in Fitzenberger et al.
(2000).

which often exhibit some censoring, the interest in this modelling approach has increased significantly. In particular, it became apparent that the large heterogeneity of microeconomic data requires a modelling beyond the simple conditional mean analysis in order to extract a maximum of information from the available data. Nevertheless, the experience with applications of censored quantile regression is still limited. One reason for this reluctance to use the adequate modelling framework can be found in the difficulties in computing the censored quantile regressions and the resulting performance problems of the algorithms proposed in the literature.

As will be discussed in Section 14.2, the calculation of a CQR estimate is carried out by minimizing a non–differentiable and nonconvex objective function. An example provided in Section 14.2 will show that this optimization problem may exhibit multiple local optima even in very simple settings, as the example in Section 14.2 will show. Consequently, the CQR estimation problem is a typical instance of a problem which cannot be tackled satisfyingly within the classical optimization paradigm. In fact, a simulation study performed by Fitzenberger (1997a; 1997b) shows that several optimization algorithms proposed for this problem in the literature often fail to give satisfactory results. In order to improve the performance, Fitzenberger introduced the algorithm BRCENS, which is tailored to the CQR problem by taking into account special features of potential solutions — the interpolation property discussed in Section 14.2. In particular, the continuous global estimation or optimization problem becomes a discrete optimization problem by exploiting this property. Previous simulation studies by Fitzenberger (1997a; 1997b) indicate that the results, and hence the applicability, can be improved considerably by using this approach. Unfortunately, if in particular a high degree of censoring is present in the data, BRCENS may also fail to provide high–quality estimation results for many instances.

Given that available algorithms may fail to provide estimates of the CQR estimation problem of reasonable quality, it seems straightforward to consider optimization heuristics for the solution of the large–scale discrete optimization problem with multiple local optima obtained by exploiting the interpolation property. In fact, Pinske (1993) has already proposed the use of simulated annealing in order to improve the results. Unfortunately, no explicit implementation is given and no results are reported. Therefore, it seemed reasonable to challenge the available algorithms for the solution of the CQR estimation problem by a threshold accepting implementation, which has been introduced by Fitzenberger and Winker (1998; 1999). The results presented in the following sections mainly draw on these papers.

While the other methods already proposed for the CQR problem[2] may fail to tackle the problem, threshold accepting, at least asymptotically, will almost

---

[2] See, e.g. Pinske (1993) and Fitzenberger (1997a; 1997b).

certainly find a global optimum, as pointed out earlier in Section 7.3. However, a much stronger argument for the use of threshold accepting in this context consists in the fact that it is able to give better results for many problem instances as compared to all of the other algorithms used for this purpose so far. As the results of a first simulation study presented in Section 14.4 show, threshold accepting can improve the estimation of CQRs considerably. However, this technique uses more computing time than other algorithms.

In Section 14.2, a short formal outline of the CQR estimation problem is presented, which allows the motivation of the use of integer optimization methods. Section 14.3 presents some algorithms used for the CQR estimation problem which serve as a bench–mark for the threshold accepting implementation also introduced in this section. The results of a simulation study comparing the performance of the threshold accepting implementation with the performance of other algorithms are summarized in Section 14.4.

14.2 Interpolation Property

This section starts with a compact introduction to the CQR estimation problem,[3] and derives its representation as an integer programming problem. For this purpose, some notation has to be introduced. For a sample of size T, let the values of the dependent variable be $y = (y_1, \ldots, y_T)$. The values of k explanatory variables are summarized in the $T \times k$ design matrix $X = (\mathbf{x}_1, \ldots, \mathbf{x}_T)'$, where $\mathbf{x}_i = (x_{i,1}, \ldots, x_{i,k})$ denotes the values of the explanatory variables for observation i. Finally, the censoring values, which are assumed to be fixed and known, are denoted by $yc = (yc_1, \ldots, yc_T)$. For this present chapter, only the case of censoring from above is considered, i.e. y_i is reported only if $y_i \leq yc_i$. This is the case appearing in duration analysis and for income data from social security statistics, respectively. However, both the formal analysis and the implementation of the algorithms can be easily generalized to other forms of censoring.

When looking at the quantile $\theta \in]0, 1[$ of the conditional distribution, the CQR estimation problem consists in minimizing the piecewise linear distance function

$$S = \sum_{i=1}^{T} \{\theta I[y_i > g(\mathbf{x}_i'\beta, yc_i)] + (1-\theta) I[y_i < g(\mathbf{x}_i'\beta, yc_i)]\} \cdot \mid y_i - g(\mathbf{x}_i'\beta, yc_i) \mid$$

(14.1)

with regard to the $k \times 1$ parameter vector β. Thereby, censoring is taken into account by the nonlinear response function $g(\mathbf{x}_i'\beta, yc_i) = \min[\mathbf{x}_i'\beta, yc_i]$, while $I[.]$ denotes the indicator function. If $\widehat{\beta_\theta}$ solves this minimization problem, the

---

[3] For a more comprehensive introduction, the interested reader is referred to Fitzenberger (1997a).

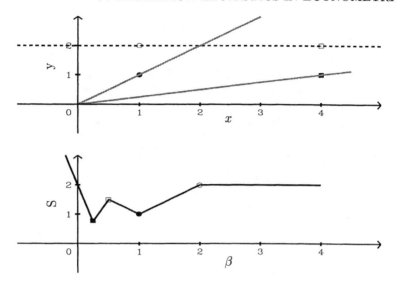

Figure 14.1   Local optima for a simple CQR.

term $x_i'\widehat{\beta_\theta}$ captures the estimated θ–quantile of the underlying uncensored dependent variable conditional on x_i.

The CQR distance function S given by equation (14.1) is piecewise linear. Consequently, the CQR estimation problem does not necessarily have a unique optimal solution. The following example shows that the situation is even worse, as the distance function S may exhibit suboptimal local minima with regard to the standard Euclidean metric. The example can be described by two observations $i = 1, 2$ with

$$x_1 = 1 \quad y_1 = 1 \quad yc_1 = 2$$
$$x_2 = 4 \quad y_2 = 1 \quad yc_2 = 2.$$

These data points are depicted by a solid circle and a square, respectively, in the upper panel of Figure 14.1. The censoring values are represented by the dotted line. For these data, the model $y = \beta x$ is fitted for the median ($\theta = 0.5$) by using censored quantile regression. To these data, the model $y = \beta x$ is fitted by using CQR for the median ($\theta = 0.5$). The regression lines for $\beta = 1$ (through (x_1, y_1)) and $\beta = 0.25$ (through (x_2, y_2)), respectively, are depicted by the grey lines through the origin in the upper panel.

The value of the CQR distance function S for the median ($\theta = 0.5$) is plotted in the lower panel of Figure 14.1 against values for the only parameter β of the model. This plot shows that S is piecewise linear between points corresponding to regression lines which either fit a data point (solid symbols) or a censoring value (open symbols). Furthermore, besides the global optimum for $\beta = 0.25$,

S exhibits a local optimum for $\beta = 1$ and a flat region for $\beta > 2$. Obviously, standard optimization techniques based on (refined) gradient methods are challenged by these features of the problem.

However, at least the piecewise linearity of S can also be considered as an advantage. In fact, analogous to standard quantile regressions without censoring (Koenker and Bassett, 1978), a discretization of the optimization problem can be derived by using this property. This discretization stems from the *Interpolation Property* for CQRs provided by Womersley (1986):

Interpolation Property. If the design matrix X has full rank k, then there exists a global minimizer $\widehat{\beta_\theta}$ of the CQR distance function such that $\widehat{\beta_\theta}$ interpolates at least k data points, i.e. there are k observations $\{(y_{i_1}, \mathbf{x}_{i_1}), \ldots, (y_{i_k}, \mathbf{x}_{i_k})\}$ with

(IP) $\quad y_{i_l} = \mathbf{x}'_{i_l}\widehat{\beta_\theta}$ for $l = 1, \ldots, k$ and the rank of $(\mathbf{x}_{i_1}, \ldots, \mathbf{x}_{i_k})'$ equals k,

using the formulation of Fitzenberger (1997a). It should be noted that some of the interpolated data points in (IP) might be censored observations, i.e. $y_{i_l} = yc_{i_l}$.

Given this result, the CQR estimation problem can be solved by considering only the finite set of all k–tuples of data points satisfying IP. The minimum of the objective function S over this set is also a global minimizer. However, it is not guaranteed that it is a unique global minimizer. Thus, if there is a unique global minimizer it can be obtained by minimization over the finite set of all k–tuples. Otherwise, at least one global minimizer can be obtained through this approach.

14.3   Algorithms for the CQR Estimation Problem

Reducing the CQR estimation problem to an integer programming problem over the search space $\mathcal{X}$, which consists of all k–tuples out of the T data points which satisfy IP, allows for a straightforward implementation of threshold accepting. Some details of this implementation are provided in Section 14.3.2. Furthermore, $\mathcal{X}$ is a finite set. Consequently, at least for small problem instances, i.e. a small number of observations and a low dimension of the parameter vector β, the exact solution can be obtained by enumeration, as described in Section 14.3.1. This exact solution serves as a bench–mark for the comparison of the results of different algorithms in the simulation study summarized in Section 14.4.

Besides the exact enumeration algorithm, which is not feasible for real applications due to its tremendous computational complexity, three further algorithms discussed in the literature on CQRs are included in the simulation study. These algorithms are the iterative linear programming algorithm (ILPA) proposed by Buchinsky (1994), the general interior point algorithm for nonlinear quantile regression problems (NLRQ) developed by Koenker and Park (1996), and the BRCENS algorithm introduced by Fitzenberger (1997a; 1997b). A previous simulation study in Fitzenberger (1997a) compared these three algorithms, since they seem to be the most efficient ones proposed in the literature for the CQR estimation problem. Nevertheless, ILPA is not guaranteed to converge and, both for ILPA and NLRQ, convergence does not even guarantee a local minimum of the CQR problem. In contrast, BRCENS guarantees a local minimum, and exhibits a higher frequency than the two other algorithms to find the global minimum of S. However, as soon as a high degree of censoring is present in the data, the simulation results in Fitzenberger (1997a) indicate that all algorithms perform quite poorly.

14.3.1 Enumeration

Since $\mathcal{X}$ defined by the set of all k–tuples of data points satisfying IP is a finite set, the global minimizer can be found by a deterministic enumeration algorithm,[4] which is denoted by IPOL. This consists of a complete enumeration of all k–tuples of data points with linearly independent regressor vectors. To each such k–tuple corresponds an interpolating regression line and, consequently, an estimator $\hat{\beta}$ and a value for the objective function S in equation (14.1). The enumeration of all k–tuples of data points with linearly independent regressor vectors provides at least one minimizer $\hat{\beta}^*$ of (14.1) over all such k–tuples. The interpolation property (IP) guarantees that this minimizer $\hat{\beta}^*$ is also a global minimizer of the CQR distance function S and, therefore, a correct CQR estimator.

Although IPOL provides the correct CQR estimator with certainty after a finite number of evaluations, it is not a feasible choice for real applications due to its high computational complexity. In fact, given the number of observations T, IPOL involves the evaluation of at most $\binom{T}{k}$ k–tuples. Given a large number of observations T and a comparatively small number of parameters k to be estimated, the computational effort grows approximately with $T^k/k!$. Although this figure is growing only as a polynomial in the number of observations, IPOL becomes impractical, as k may reach values of more than 20 in real applications for several thousand observations. Nevertheless, it is the best choice as a bench–mark for a simulation of small–scale problems.

---

[4] See Pinske (1993) and Fitzenberger (1997b). Hawkins and Olive (1999, p.120) note for a related estimation problem, i.e. the least trimmed sum of absolute deviations regression, that 'in general, there is no completely reliable method other than full enumeration to identify the "best" subset to cover [...], and so computing any of these high breakdown estimators involves a substantial combinatorial problem'.

14.3.2   The threshold accepting implementation

The threshold accepting implementation for the CQR estimation problem also relies on the interpolation property. Therefore, it operates on the set of k–tuples of data points with linearly independent vectors of exogenous variables $(\mathcal{X})$. This set is given local structure by the following construction of neighbourhoods $\mathcal{N}(\mathbf{z})$ for each k–tuple $\mathbf{z}$ of data points with linearly independent vectors of exogenous variables. Leaving out one of the k data points in $\mathbf{z}$ describes a one–dimensional subspace in the dual coefficient space as the orthogonal complement to the space spanned by the $k-1$ remaining data points. A new data point may be searched in both directions along this subspace. By calculating the residuals $(y_i - \mathbf{x}_i'\beta)$ for the parameter vector β corresponding to the initial k–tuple $\mathbf{z}$, only those points and associated directions are considered for which the absolute size of the residual is strictly reduced when moving into one of the search directions. Now, all elements $\tilde{\mathbf{z}}$ of the neighbourhood $\mathcal{N}(\mathbf{z})$ can be described as those k–tuples obtained from $\mathbf{z}$ by leaving out one point and replacing it by another found along one of the admissible search directions.

It should be noted that this definition of local neighbourhoods is much more problem–specific than those used in the previous chapters of Part III. In fact, instead of replacing a left–out data point by one of the data points along the specified search directions, any new data point could be included. This concept would result in much larger neighbourhoods. However, the choice of the smaller neighbourhoods is based on expert knowledge about the local behaviour of the objective function, which is also used by the BRCENS algorithm. Finally, preliminary tests using a more simple neighbourhood definition — like replacing one data point by any other — did not improve the performance.

For the CQR estimation problem, the quotient version of the threshold accepting algorithm is used, i.e. the thresholds indicate up to what percentage a worsening of the objective function is accepted in each iteration. The threshold sequence itself is generated from the data of the problem based on the concept of an empirical jump function described, e.g. in Section 11.3.2. To this end, a large number of k–tuples $\mathbf{z_i} \in \mathcal{X}$ is randomly selected. For each $\mathbf{z_i}$, an element $\tilde{\mathbf{z}}_i \in \mathcal{N}(\mathbf{z_i})$ is also randomly chosen. The relative difference of the values of the objective function S for $\mathbf{z_i}$ and $\tilde{\mathbf{z}}_i$, respectively, is denoted by ΔS_i. Consequently, the set of all ΔS_i approximates the distribution of relative local changes of the objective function with regard to the neighbourhood mapping introduced above. Using the relative differences to the smaller value of the objective function, i.e. $\max\{\Delta S_i, 1/\Delta S_i\}$, sorting these values in decreasing order, and using a fraction at the lower end provides a threshold sequence decreasing to one.

In the implementation used for the simulation study in the next section, the empirical jump function is based on 1000 replications and the values of the lower 5–percentile are used as the threshold sequence. The fact that only very small threshold values have to be used corresponds to the specific choice of local neighbourhoods, which drive the algorithm faster to 'good' parts of the landscape as for the other applications.

Given the data–based threshold sequence and the neighbourhood definition, the performance of the threshold accepting implementation will depend solely on the total number of iterations. Increasing this number improves the quality of the results. However, for the small instances used in the simulation study, the enumeration algorithm IPOL produces exact results with a comparatively small number of evaluations. Consequently, the threshold accepting implementation should not use more computational resources than this exact algorithm. Therefore, the numbers of iterations used for the threshold accepting algorithm range from 5000 to 100 000. The corresponding versions of threshold accepting are denoted by TA_5, TA_{10}, and so on up to TA_{100}. As soon as the problem instances become slightly larger, the computational resources required for IPOL 'explode', and threshold accepting becomes competitive even if it requires some million iterations in order to obtain high–quality estimators.

If the number of iterations is small, it cannot be guaranteed that this implementation ends in a local minimum with regard to the chosen neighbourhood structure. Therefore, a hybrid heuristic is also considered, which combines a threshold accepting optimization run with BRCENS. Starting BRCENS at the final solution of the threshold accepting algorithm makes sure that the final result is a local minimum. This hybrid algorithm is denoted by TA–BRCENS.[5]

14.4 Simulation Set–Up and Results

From previous simulation studies (Fitzenberger, 1997b), it is known that all conventional algorithms may fail to provide the global optimum. Since it is not expected that the threshold accepting implementation will always provide this global optimum, a comparison has to be based on a simulation study. Unfortunately, the conclusions which can be drawn from such a simulation study, are restricted by the simulation design. Strictly speaking, results on relative performance only apply to the specific problem instances. Nevertheless, if one algorithm consistently outperforms the others on a number of differing problem instances, a superior performance might also be expected for problem instances not included in the simulation study. However, any further simulation experiment may challenge this finding.

Keeping these limitations in mind, the comparison of the performance of the different algorithms is based on a finite set of problem instances. In the case

---

[5] See Fitzenberger and Winker (1999) for more details on this hybrid algorithm.

Table 14.1 Data–generating processes for the CQR simulation study[a]

DGP	Censoring values yc_i	True coefficients $(\beta_1, \beta_2, \beta_3, \beta_4)$	Regressor values
(A)	C	$(0,0,0,0)$	$x_{i,2}, x_{i,3} \sim \mathcal{N}(0,1)$
(B)	C	$(0,0,0,0)$	$x_{i,2} = -9.9 + 0.2i$, $x_{i,3} = x_{i,2}^2$
(C)	$C + 0.5$	$(0.5, 0.5, -0.5, 0.5)$	$x_{i,2}, x_{i,3} \sim \mathcal{N}(0,1)$
(D)	$C + 0.5$	$(0.5, 0.5, -0.5, 0.5)$	$x_{i,2} = -9.9 + 0.2i$, $x_{i,3} = x_{i,2}^2$
(E)	$\mathcal{N}(C,1)$	$(0,0,0,0)$	$x_{i,2}, x_{i,3} \sim \mathcal{N}(0,1)$
(F)	$\mathcal{N}(C,1)$	$(0,0,0,0)$	$x_{i,2} = -9.9 + 0.2i$, $x_{i,3} = x_{i,2}^2$
(G)	$\mathcal{N}(C+0.5,1)$	$(0.5, 0.5, -0.5, 0.5)$	$x_{i,2}, x_{i,3} \sim \mathcal{N}(0,1)$
(H)	$\mathcal{N}(C+0.5,1)$	$(0.5, 0.5, -0.5, 0.5)$	$x_{i,2} = -9.9 + 0.2i$, $x_{i,3} = x_{i,2}^2$

[a] C denotes a constant taking values in $\{0, 0.5, 1\}$. The $x_{i,4}$ are always distributed as i.i.d. $\mathcal{N}(0,1)$ as well as the random variables ε_i. Finally, $i = 1, \ldots, T$ numbers the observations.

of the CQR estimation problem, such instances are given by data–generating processes. All considered instances in the simulation study are based on the limited dependent variable model

$$y_i = \min(yc_i, \beta_1 + \sum_{j=2}^{k} \beta_j \cdot x_{i,j} + \varepsilon_i), \qquad (14.2)$$

where k denotes the dimension of the parameter space ranging from two to four. The estimation problem is a censored least absolute deviation regression ($\theta = 0.5$) with one to three regressors and an intercept. The data–generating processes (DGP) (A) – (H) described in Table 14.1 differ in the assumptions about the processes for y_i and the $x_{i,j}$ on the one hand, and the values of the β_j on the other hand. The data–generating processes are selected to cover a broad range of typical features of CQR estimation problems.

The data–generating processes in Table 14.1 are for the four–dimensional case ($k = 4$). The data–generating processes used in dimension two and three are obtained by removing the additional information on β_4 and $x_{i,4}$ or β_3, β_4 and $x_{i,3}, x_{i,4}$, respectively. The eight data–generating processes differ in three aspects. First, the coefficients β_i are all zero in data–generating processes (A,B,E,F) and equal to $(0.5, 0.5, -0.5, 0.5)$ for the other cases. Since the initial configuration of BRCENS is to start with a zero coefficient vector, it could make a difference whether the starting values are close to

Table 14.2   Average share of censored observations (in %)

	$k = 2$			$k = 3$			$k = 4$		
	$C =$			$C =$			$C =$		
DGP	1.0	0.5	0.0	1.0	0.5	0.0	1.0	0.5	0.0
(A)	15.9	30.9	49.9	15.9	31.1	50.1	15.8	30.8	50.0
(B)	15.9	30.9	49.8	15.9	31.1	50.2	15.8	30.8	50.0
(C)	18.7	32.9	49.9	18.7	32.8	49.9	18.4	32.6	32.6
(D)	40.0	45.1	49.9	40.1	45.1	50.0	40.0	45.0	50.1
(E)	24.1	36.4	50.1	24.1	36.2	50.1	23.8	36.0	49.8
(F)	24.1	36.4	50.1	24.1	36.2	50.1	23.8	36.0	49.8
(G)	25.4	37.1	50.0	25.3	37.0	49.9	25.1	36.8	49.7
(H)	40.1	45.0	50.0	40.0	45.0	50.0	40.0	45.0	50.0

the true parameter values.[6] Secondly, while for the data–generating processes (A,C,E,G) all regressors are random variables, the first two regressors are given by deterministic sequence of numbers for (B,D,F,H). These sequences also introduce nonlinearity in designs of dimensions three and four due to $x_{i,3} = x_{i,2}^2$. Thirdly, the processes differ by the assumptions about censoring. While for processes (A,B,C,D) the censoring points are fixed, they are random variables for the last four data–generating processes. Furthermore, the choice of $C \in \{0, 0.5, 1\}$ determines the (expected) degree of censoring. In particular, $C = 0$ represents a situation where on average 50% of the observations are censored, i.e. the exact CQR ($\theta = 0.5$) typically reaches the censored region. Censoring is less pronounced for $C = 0.5$ and $C = 1$, respectively, as shown in Table 14.2, which provides the average shares of censored observations in the random samples for all data–generating processes.

For 48 different situations (given by the eight data–generating processes (A)–(H), three different values for C, and dimension $k = 2, 3$) 1000 random samples with 100 observations each were generated, while due to the computational load of calculating the global optimum with the enumeration algorithm IPOL, in dimension $k = 4$ only 100 random samples of size 100 were generated for each of 24 data–generating processes. The enumeration algorithm IPOL described in Section 14.3.1 is used to obtain the global optimum. It uses at most $\binom{100}{2} = 4950$ evaluations of 2–tuples in dimension two. In dimension three this number increases to 161 700, and in dimension four to 3 921 225. The design always exhibits full rank. The enumeration of all

---

[6] This assumption is confirmed by the detailed results reported in Fitzenberger and Winker (1999).

k–tuples was also used to check whether the exact CQR estimate is unique.[7] Random samples for which the exact CQR estimate proves to be non–unique are dismissed from the further analysis, which occurred typically for less than 1% of all samples drawn.

Given this simulation set–up, all of the algorithms discussed in Section 14.3, namely ILPA, NLRQ, BRCENS, TA$_{iter.}$ and TA$_{iter.}$–BRCENS were run on each generated data set. The following sections summarize some findings on absolute performance (Section 14.4.1), relative performance in terms of the objective function S (Section 14.4.2), and relative performance with regard to the properties of the obtained coefficient estimates (Section 14.4.3). Finally, the relative computing times for the different algorithms are reported in Section 14.4.4.

14.4.1 Absolute performance

Table 14.3 shows the mean empirical frequencies of cases for which the different algorithms achieve the true optimum. For all three dimensions $k = 2, 3, 4$, the first row gives the mean over all 24 data–generating processes, while the second row gives the mean for the eight processes(A)–(H) with $C = 1.0$ corresponding to a low and the third row for $C = 0.0$ corresponding to a high degree of censoring, respectively.

The simulation results indicate that BRCENS dominates the other algorithms suggested in previous research (ILPA and NLRQ). However, this is challenged by the threshold accepting implementation, in particular if a high degree of censoring ($C = 0$) is present. In fact, for the corresponding DGPs the threshold accepting implementation already finds the global optimum more often than the BRCENS algorithm for a small number of 5000 iterations. For a low degree of censoring, threshold accepting is not yet competitive with BRCENS when using such a small number of iterations. However, either by using the hybrid algorithm TA–BRCENS with the same small number of iterations for the threshold accepting part, or by increasing the number of iterations, threshold accepting clearly outperforms all other algorithms with regard to their absolute performance. The frequency of finding the global minimizer tends to 100% when the number of iterations is increased further and censoring is low, and to much higher figures than for the other algorithms when the degree of censoring is high. Finally, it can be stated that the hybrid algorithm TA–BRCENS improves the results of a simple threshold accepting implementation only in the case of a small number of iterations. In real applications, which, in general, include more regressors and observations than the simulated instances, the number of iterations of a threshold accepting implementation will be small compared to the size of the

---

[7] See Fitzenberger and Winker (1999) for the details.

Table 14.3 Mean frequencies (in %) of cases for which algorithms achieved optimum[a]

DGP	Algorithm[b]								
	ILPA	NLRQ	BRC	TA$_5$	TB$_5$	TA$_{20}$	TB$_{20}$	TA$_{100}$[c]	TB$_{100}$[c]
Dimension $k = 2$									
All	78	41	82	72	86	92	95	96	97
$C = 1$	90	54	92	69	87	94	97	99	100
$C = 0$	60	24	66	70	82	87	90	90	92
Dimension $k = 3$									
All	67	39	72	49	65	79	82	93	93
$C = 1$	84	57	87	45	67	83	87	99	99
$C = 0$	45	18	54	48	58	70	73	81	81
Dimension $k = 4$									
All	61	36	69	—	—	—	—	91	91
$C = 1$	82	54	86	—	—	—	—	97	97
$C = 0$	37	15	48	—	—	—	—	81	81

[a] Based on 1000 random samples for $k = 2$ and $k = 3$ and 100 random samples for $k = 4$ and each data–generating process. All numbers are rounded up to integer values.

[b] The following abbreviations are used: BRC for BRCENS, and TB$_{it.}$ for TA$_{iter}$–BRCENS.

[c] Only 50 000 iterations for $k = 2$.

set of possible solutions. Then, the use of the hybrid algorithm TA–BRCENS may still improve the results of a simple threshold accepting implementation, as it guarantees convergence to a local minimum. If the threshold accepting algorithm has already come close to the global minimizer, this local minimum is likely to be the global one.

14.4.2 Relative performance

The dominance of the threshold accepting algorithm with a total number of 20 000 iterations is confirmed by the analysis of relative performance. Table 14.4 is based on a direct pairwise comparison of the algorithms with regard to the objective function. To this end, the left entries in each column of this table reports the mean empirical frequencies of cases when the value of the objective function S obtained by ILPA, NLRQ and BRCENS, respectively, was strictly lower than for TA$_{20}$, while the right entries give the corresponding frequencies when TA betters the other algorithms.

The results of this direct comparison confirm the findings of the previous

Table 14.4 Mean frequencies (in %) of cases for which one algorithm results in smaller value of S than the other algorithm[a]

DGP	ILPA vs. TA_{20}		NLRQ vs. TA_{20}		BRCENS vs. TA_{20}	
Dimension $k = 2$						
All	4	20	4	56	5	17
$C = 1$	5	9	5	44	5	8
$C = 0$	4	36	3	71	7	30
Dimension $k = 3$						
All	11	31	9	56	12	25
$C = 1$	15	15	13	39	15	12
$C = 0$	8	50	6	76	11	41
Dimension $k = 4$						
All	19	36	17	56	21	28
$C = 1$	27	18	24	37	28	13
$C = 0$	12	58	10	75	14	46

[a] Based on 1000 random samples for each data–generating process. All numbers are rounded up to integer values.

section. When using a small number of 20 000 iterations, the threshold accepting implementation is much more likely to result in a smaller value of the objective function S than the other algorithms if the degree of censoring is high ($C = 0$), while for a low degree of censoring ($C = 1$) ILPA and BRCENS outperform TA_{20}. However, the threshold accepting implementation outperforms ILPA, NLRQ and BRCENS consistently if the number of iterations is increased.

14.4.3 Properties of coefficient estimates

Measures of the absolute and relative performances reported in the previous sections indicate that threshold accepting is more likely to obtain the correct CQR estimator and to obtain a smaller value of the objective function, if it fails to provide the global minimizer, than the other algorithms. However, for real applications it is less important to what extent the different algorithms miss the global minimizer of S, than how far the coefficient estimates $\hat{\beta}$ are away from the true parameters if the global minimum is missed. The quality of this approximation can be measured either by the average root–mean–squared deviation (RMSD) or some robust estimates like the 90% quantiles of absolute deviations. Qualitatively, both measures provide similar conclusions. Therefore, in Table 14.5 only the RMSD of the coefficient estimates for β_2

Table 14.5 Average root–mean–squared deviation (RMSD) values of coefficient estimates for β_2 and $C = 0.5$ from true CQR estimates[a]

DGP	Algorithms[b]								
	ILPA	NLRQ	BRC	TA$_5$	TB$_5$	TA$_{20}$	TB$_{20}$	TA$_{100}$[c]	TB$_{100}$[c]
	Dimension $k = 2$								
(A)	.309 99	.260 97	.309 05	.597 55	.597 55	.121 06	.121 06	.139 25	.139 25
(B)	.670 78	.670 95	.671 02	.641 63	.641 63	.733 97	.733 97	.668 11	.668 11
(C)	.161 79	.110 70	.155 77	.087 01	.087 01	.146 16	.146 16	.042 51	.042 51
(D)	.017 66	.012 23	.017 39	.006 35	.006 35	.007 81	.007 80	.007 35	.007 35
(E)	.051 01	.037 20	.048 01	.016 08	.015 57	.012 00	.011 67	.010 74	.010 74
(F)	.007 24	.005 68	.007 31	.002 26	.002 05	.001 37	.001 27	.000 46	.000 45
(G)	.050 53	.037 63	.049 30	.011 38	.009 92	.009 36	.008 72	.003 82	.003 77
(H)	.019 13	.011 84	.018 20	.004 84	.003 71	.003 77	.003 71	.001 63	.001 57
	Dimension $k = 3$								
(A)	1.3500	1.3491	1.3501	1.3487	1.3487	1.3385	1.3385	.88474	.88474
(B)	67.478	67.478	67.478	67.347	67.347	66.496	66.496	67.497	67.497
(C)	.283 95	.220 83	.269 42	.065 45	.065 44	.119 09	.119 09	.061 07	.061 07
(D)	.156 92	.138 88	.141 57	.055 04	.055 04	.035 82	.035 82	.051 83	.051 83
(E)	.053 18	.046 65	.051 28	.023 50	.020 12	.012 10	.011 19	.008 89	.008 89
(F)	.012 54	.011 41	.012 01	.004 23	.003 97	.002 41	.002 41	.002 91	.002 91
(G)	.063 25	.049 78	.057 60	.034 96	.032 74	.015 20	.014 54	.016 30	.016 30
(H)	.152 30	.134 85	.142 53	.029 32	.029 31	.016 78	.016 65	.011 94	.011 94
	Dimension $k = 4$								
(A)	.791 56	.785 32	.790 51	—	—	—	—	.283 65	.283 65
(B)	22.087	22.087	22.087	—	—	—	—	22.057	22.057
(C)	.166 91	.107 06	.137 28	—	—	—	—	.002 26	.002 26
(D)	.200 64	.170 60	.189 49	—	—	—	—	.000 00	.000 00
(E)	.048 84	.030 30	.038 32	—	—	—	—	.023 85	.023 85
(F)	.010 87	.010 06	.008 23	—	—	—	—	.000 92	.000 92
(G)	.064 75	.066 14	.053 88	—	—	—	—	.002 46	.002 46
(H)	.114 55	.118 67	.115 27	—	—	—	—	.001 72	.001 72

[a] Based on 1000 random samples for $k = 2$ and $k = 3$ and 100 random samples for $k = 4$ and each DGP.

[b] The following abbreviations are used: BRC for BRCENS, and TB$_{it.}$ for TA$_{iter}$–BRCENS.

[c] Only 50 000 iterations for $k = 2$.

Table 14.6 Average relative computation times

Const.	IPOL	BRCENS	TA$_5$	TA$_{20}$	TA$_{50}$	TA$_{100}$
1.0	9000	1.0	800	1300	2400	4100
0.5	8900	1.7	800	1300	2300	4000
0.0	8800	18.0	780	1300	2300	4000

and $C = 0.5$ are reported.[8]

Again, the results are clearly in favour of the threshold accepting implementation. Even with only 5000 iterations it achieves smaller average RMSD values of the parameter estimates for all but one DGP in dimension two. Increasing the number of iterations tends to further decrease these values. This finding can be interpreted as follows. Even if the threshold accepting implementation misses the global minimizer, it is likely to provide coefficient estimates very close to the true CQR estimate, while the other algorithms may get stuck in some local minimum quite far away from the correct CQR estimate, thus resulting in large values of the average RMSD.

14.4.4 Timing

The results reported in the previous sections clearly suggest a superior performance of the threshold accepting implementation for the CQR estimation problem compared to all other available algorithms — except for the exact enumeration algorithm, of course. Unfortunately, this advantage of the threshold accepting implementation does not come without any costs. Such costs are given by the increased computation time, in particular if threshold accepting is run with large numbers of iterations. Table 14.6 reports the average relative computation times required by the different algorithms in dimension $k = 3$.

The reported numbers are the ratios of average computation times across the data–generating processes (A)–(D) relative to BRCENS, with $C = 1$. For all algorithms, FORTRAN implementations were run on the same IBM RS 6000 workstation. The numbers are rounded up to two valid digits. In actual time, TA$_{100}$ required on average 15 seconds per sample.

Obviously, the computation time required by the threshold accepting algorithm is much larger than for the BRCENS algorithm, even if only 5000 iterations are used. This disadvantage becomes less pronounced if the degree of censoring is high (Const. = 1). Nevertheless, upon increasing the number of iterations to 100 000, the threshold accepting implementation needs almost the

---

[8] The results for the quantiles of absolute deviations can be found in Fitzenberger and Winker (1999).

computational resources required by the exact enumeration algorithm IPOL. Consequently, it might be difficult to argue in favour of the threshold accepting algorithm based on these timing results. However, it should be kept in mind that real applications of CQR typically use much larger values of k and T. Then, IPOL becomes infeasible, while threshold accepting can still produce high–quality results. In particular, it might be expected that the results provided by threshold accepting are still better — in terms of the objective function — and more robust — in terms of the coefficient estimates — than those provided by the standard algorithms. Therefore, the use of threshold accepting is supposed to become even more competitive when the size of the problem instances increases.

14.5 Conclusions

The application of threshold accepting presented in this chapter differs from the other applications for two reasons. First, the CQR estimation problem is a continuous global optimization problem. Thus, the methods discussed next in Chapter 15 could also be applied to this problem. However, the experience with other algorithms used to solve this problem indicate that results are improved when using the interpolation property of the estimation problem, which reduces it to an integer optimization problem. Secondly, the CQR estimation problem has been tackled by several highly specialized algorithms. The failure of these algorithms to produce high–quality results, in particular if the degree of censoring is high, provides a strong argument for the use of optimization heuristics. Furthermore, implementation details of these tailored algorithms can be used for the definition of local neighbourhoods.

Apart from the algorithms used for the solution of the CQR estimation problem in the literature, the interpolation property provides a mean for obtaining exact results by enumeration for very small problem instances. Using these results as a bench–mark in a simulation study for different data–generating processes in dimensions two to four exhibits a clear superiority of the heuristic optimization approach based on threshold accepting. This has advantages both in absolute and relative performances. Furthermore, even if the global optimum is not found, the coefficient estimators supplied by the threshold accepting algorithms are closer to the true CQR estimates than for the conventional algorithms.

These advantages of the threshold accepting implementations might justify the higher computational cost compared to standard algorithms, which becomes less pronounced for large–scale problems as they appear in practice.

15

Continuous Global Optimization

15.1 Introduction

The discussion of optimization problems arising in economics, statistics and econometrics in Chapters 2 and 3 mentioned several settings where an objective function f has to be minimized over a reel vector space $\mathbb{R}^d$. In particular, the maximum–likelihood estimation for complex models might lead to functions with several local optima which cannot be solved easily by standard algorithms. The maximum–likelihood estimation of Markov switching models, threshold autoregressive models and the censored quantile regression are examples of such problems. While for the latter problem it has been shown in Chapter 14 that it is possible to transform the problem into an integer optimization problem, such bypasses are not always available.

Despite of the seeming prevalence of continuous optimization problems in economics and econometrics, the threshold accepting heuristic was introduced solely for the case of a discrete search space in Part II of this book. Furthermore, all applications studied in Part III so far have covered only this case. Three arguments can be put forward for this concentration on discrete optimization problems. First, for a number of continuous optimization problems arising in economics and statistics classical optimization tools, e.g. gradient methods, offer an adequate solution strategy. These methods are well known and integrated in most econometric software packages. Consequently, the use of optimization heuristics is indicated solely in cases when these methods are expected to fail due to specific properties of the search space or objective function such as discontinuities or the existence of many local maxima, as shown earlier in Figure 1.4, or in Figure 15.1, an example taken from Beaumont and Bradshaw (1995, p. 162).

The function shown in Figure 15.1 is defined by

$$f(x_1, x_2) = -\sum_{i=1}^{2} \left((2\pi - |x_i|)|\sin(10|x_i|)| + \sqrt{|x_i|} \right) \text{ for } x_i \in [-\pi, \pi] \quad (15.1)$$

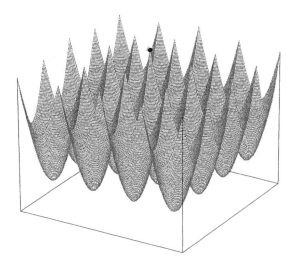

Figure 15.1 A continuous function with many local maxima.

and takes its global maximum in $x = (0,0)$, marked by the filled circle in the figure. It is evident that for optimization problems like this classical optimization methods fail. Thereby, a two–dimensional situation as depicted in Figure 15.1 is still much simpler than corresponding higher–dimensional problems as they arise as maximum–likelihood functions for some econometric models. It is for these kind of problems that optimization heuristics also have to be considered in a continuous setting.

A second argument for concentrating on discrete problems in the previous chapters is that some highly complex continuous optimization problems can be analysed in a discrete setting either by an approximation, which replaces the original search space by a fine grid, or by theoretical arguments, which imply that optimal solutions have to belong to a discrete subset of the search space. In fact, for the uniform design problem, the original problem was restricted to an integer optimization problem considering only designs on a finite grid, the optimal aggregation problem became an integer optimization problem due to the predefined 0–1 weights, and the censored quantile regression problem was transformed to an integer optimization problem by making use of a specific property of potential solutions. Thus, only the model selection problem was a discrete optimization problem from the beginning.

The third argument in favour of the concentration on integer optimization problems in the previous chapters is given by the observation that many optimization problems, which are tackled relying on ad hoc methods or not even considered as explicit optimization tasks, belong to the class of highly complex integer optimization problems.

Nevertheless, it cannot be expected that any complex optimization problem arising in econometrics or statistics allows for a transformation to an integer optimization problem. Therefore, the potential use of threshold accepting for continuous global optimization is discussed in this final chapter of Part III. Unfortunately, so far, there exists no implementation of the threshold accepting heuristic for continuous optimization problems. Consequently, the discussion of such an implementation in Section 15.2 has to remain at a quite general level. In particular, no detailed description of a threshold accepting implementation and no tuning details can be provided. Nevertheless, the findings of comparative implementation studies summarized in Section 7.4 allow for the conclusion that evidence from existing implementations of simulated annealing and genetic algorithms might permit a tentative conclusion about the potential of a future threshold accepting implementation for such problems. A summary of some contributions to this literature is provided in Section 15.3. However, before turning to these implementation and performance aspects, a reference is made to methods for detecting whether a continuous optimization problem exhibits some features which potentially let classical optimization methods fail.

Goffe (1997) proposes three tools for this purpose. The first tool comprises a graphical analysis, which shares some features with the empirical jump distributions used for discrete search spaces throughout Part III. For functions depending on up to two variables, a three–dimensional plot may contribute some first insights. However, often either a higher number of variables is involved or the three–dimensional plot does not provide enough information. In fact, its usefulness depends on the chosen part of the search space. In both cases, a plot of cross–sections may provide additional information. Therefore, the relevant part of the search space is included in a rectangular box. Then, at both ends of the box two points are randomly selected and the function value is plotted along a line connecting the two points in the search space. Such cross–section plots should be repeated several times for different points at the limit of the box. If some of these plots show several local minima, this might result from multiple minima of the function with regard to the Euclidian topography or, at least, from a curved valley, which also challenges standard continuous optimization algorithms. Figure 15.2 shows four cross–section plots for the example given in Figure 15.1.

Crossing the parameter space from one randomly selected point of the interval $[-\pi/5, \pi/5]$ for $x_2 = -\pi/5$ to another randomly selected point of this interval for $x_2 = \pi/5$ results in function values with several local maxima. Thus, the cross–section plots clearly indicate the presence of multiple local optima for this example.

A second tool is the use of radius plots. For these plots, the function is evaluated at many different points, which might be randomly generated, based on a rectangular grid or taken from a uniform point set. From the minimum point of this sample, the distances to all other points are recorded. The radius

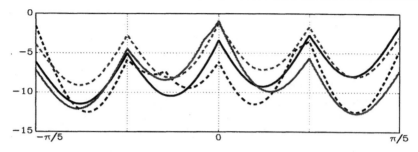

Figure 15.2   Cross–section plots.

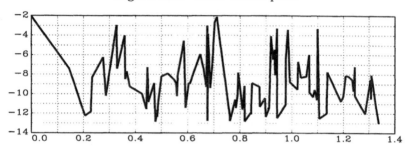

Figure 15.3   Radius plot.

plot shows the distance versus function value. For a flat surface, the radius plot
will be flat and increasing, while multiple minima, long valleys or spikes induce
a non–monotonic behaviour of the radius plot. For a maximization problem
such as that given by Figure 15.1, the construction is reversed. Figure 15.3
shows a radius plot for the example given in Figure 15.1.

For 100 randomly selected points (x_{i1}, x_{i2}) the value of the objective
function is calculated. Then, the point with the maximum function value
is identified. The Euclidean distances of all 100 points to this 'best' point
are calculated and plotted against the function value. For a well behaved
maximization problem, this plot should result in a monotonic decreasing line,
while the result plotted in Figure 15.3 strongly suggests the presence of many
local optima.

Finally, Goffe (1997) proposes a test for quadraticity. This test is
based on the observation that many optimization algorithms use quadratic
approximation schemes, i.e. it is assumed implicitly that the function to be
optimized can be approximated locally by a quadratic function. Conjugate
gradient algorithms theoretically converge in d steps if the function on $\mathbb{R}^d$
is quadratic. Thus, starting such an algorithm on different starting points
and measuring the distance between the points obtained after d steps of the
conjugate gradient algorithm provides a measure of non–quadraticity, which
works well on the examples provided by Goffe (1997, p. 11).

15.2   Implementation of Threshold Accepting

Provided that the results of a preliminary analysis of the objective function indicate that curved valleys, many local minima, or discontinuities hamper an efficient use of classical optimization methods, a threshold accepting implementation could be used. Following existing implementations of simulated annealing for global optimization over continuous variables as described, e.g. by Goffe et al. (1994), Romeijn and Smith (1994), and Jones and Forbes (1995), such an implementation operates quite similar to the discrete case described in Part II.

In general, the definition of neighbourhoods will use the available topology on $\mathbb{R}^d$. However, in order to generate a broad coverage of the objective function landscape in the beginning of optimization and a fine resolution at the end, a change of neighbourhood size is required as optimization proceeds. Goffe et al. (1994) propose the use of the following construction for this purpose. At each iteration, the situation is described by the current solution $\mathbf{x}^c$, a vector of step lengths $\mathbf{v} = (v_1, \ldots, v_d)$ and the current threshold T_i. Then, the neighbourhood of $\mathbf{x}^c$ is defined by

$$\mathcal{N}(\mathbf{x}^c) = \{\mathbf{x} | x_i \in [x_i^c - v_i, x_i^c + v_i] \text{ for all } i = 1, \ldots, d\},$$

i.e. each element x_i^c of $\mathbf{x}^c$ is varied randomly up to a predefined step length v_i. Reducing the v_i as optimization proceeds, a finer resolution is obtained for later iterations. The adjustment of v_i can be made data–driven. Goffe et al. (1994, p. 69) propose to choose it so that about 50% of all exchange steps are accepted. Given this definition of neighbourhoods, the acceptance rule for a randomly generated $\mathbf{x}^n \in \mathcal{N}(\mathbf{x}^c)$ remains the same as in the discrete threshold accepting implementation, i.e. $\mathbf{x}^n$ is accepted as the new current solution, if and only if $f(\mathbf{x}^n) - f(\mathbf{x}^c) \leq T_i$.

The threshold sequence can also be based on a data–driven method. Using the cross–section plots described in Section 15.1, typical step heights can be sampled depending on a given step length. Assuming a monotonically decreasing sequence of step lengths, a threshold sequence is obtained.

Using these ingredients, a straightforward threshold accepting implementation becomes possible. However, the continuous setting, in principle, allows for further refinements. For example, Goffe et al. (1994, p. 70) propose the use of the changes of best results obtained for each cooling temperature (corresponding to threshold values in the simulated annealing framework) as an indicator as to whether a global optimum is reached. Furthermore, if the objective function can be differentiated at least over parts of the parameter space, derivative information can also be used to define step lengths.

Obviously, a final assessment of useful and less useful features of such a threshold accepting implementation can be made solely based on a real implementation and using some of the bench–mark test functions discussed in the literature. This contribution is left for future research.

15.3 Performance of Optimization Heuristics for Continuous Global Optimization

Since a threshold accepting implementation for continuous optimization problems does not yet exist, the following remarks on performance are rather tentative. However, given the similar performance of simulated annealing and threshold accepting implementations for discrete optimization problems, the results obtained for simulated annealing and other optimization heuristics might provide a rough approximation to what might be expected from a well carried out threshold accepting implementation.

Goffe et al. (1994) compare a simulated annealing implementation with three classical optimization algorithms, namely a simplex algorithm, a conjugate gradient algorithm, and a quasi–Newton algorithm, with the last two algorithms using numerical derivatives. 100 randomly selected different starting values are employed to account for the sensitivity of conventional algorithms to the choice of starting values. The test functions cover an artificial problem with two local minima, a rational expectations model derived from the standard stock–adjustment money demand equations, a frontier cost function, and a function from the neural network literature. For the first test case, simulated annealing always comes up with the global optimum, while the frequencies for the other algorithms to obtain the global optimum range between 40 and 48%. However, simulated annealing takes much more time as might have been expected from the comparison provided in Table 14.6 in the previous chapter. However, a further tuning of the simulated annealing implementation reduces the required number of function evaluations dramatically. For the rational expectations model, all algorithms faced difficulties in obtaining a minimum. However, after eliminating some regions of the parameter space, simulated annealing was much more consistent in finding the minimum. For the translog cost frontier model, only the simulated annealing algorithm provided reasonable estimates of the global minimum, while all other algorithms either failed for numerical reasons or provided estimates far away from the optimum. For the neural network problem, the results obtained by simulated annealing were even an order of magnitude smaller than the best results obtained by the conventional algorithms (22.8 versus 228.21). From these findings, it might be concluded that many continuous optimization problems appearing in econometrics are difficult to solve by using conventional algorithms. In fact, the solutions provided by these algorithms may be only local minima, results of numerical problems, and far away from globally optimal values (McCullough and Vinod, 1999).

Romeijn and Smith (1994) use a slightly different simulated annealing implementation. In particular, the generation of neighbouring points differs from the one described in Section 15.2 for a potential threshold accepting implementation. Given a current solution $\mathbf{x}^c$, each point of the search space can be the next candidate point, however with a probability decreasing at an exponential rate with the distance from x^c. This so–called 'hide–and–seek' version of simulated annealing exhibits nice theoretical convergence properties (Romeijn and Smith, 1994, pp. 107ff.). The empirical implementation uses eight different test problems from the literature, including unconstrained and linearly constrained problems. However, all test problems are artificial and not derived from econometric problems. Summarizing the findings, the proposed simulated annealing implementation seems to work well on the test cases, including problems having non convex, or even disconnected, feasible regions. Furthermore, time complexity seems to grow only linearly with the dimension of the problem.

The adaptive simulated annealing algorithm presented by Jones and Forbes (1995) uses heuristic methods for the construction of the cooling schedule and the generation of neighbours, which resemble the data–driven concept proposed for threshold accepting in Chapter 8 and Section 15.1. Again, the test cases are artificial with no relation to econometric or statistic problems. The results indicate that the adaptive cooling schedule and neighbour generation lead to high performance. Thus, it seems possible to use this simulated annealing implementation without extensive tests of parameter constellations as required for a standard simulated annealing algorithm.

In contrast to the last two mentioned papers, Dorsey and Mayer (1995) consider not only artificial problems, but also test cases taken from the econometric literature. These authors analyse the performance of a genetic algorithm implementation for estimation problems which may exhibit multiple local maxima. One application covers a canonical disequilibrium model for the USA commercial loan market described in Mayer (1989), which consists of two equations for loan demand and supply:

$$D_t = \beta_{10} + \beta_{11}(RL_t - RA_t) + \beta_{12}IP_{t-1} + \varepsilon_{1t} \qquad (15.2)$$

$$S_t = \beta_{20} + \beta_{21}(RL_t - RT_t) + \beta_{22}TD_t + \varepsilon_{2t} \qquad (15.3)$$

Thereby, demand for bank loans D_t is supposed to depend on the cost of financing RL, measured by the average prime rate, relative to alternative means of financing RA, given by the Aaa corporate bond rate. The expected effect of this cost spread is negative, i.e. $\beta_{11} < 0$. The industrial production index IP measures a firm's expectations about future economic activity. The supply of loans depends on relative returns, where RT is the three–month Treasury bill rate and β_{21} is expected to be positive. Total deposits TD measure the available funds. For the example studied in Dorsey and Mayer (1995), it is assumed that changes in the loan rate do not convey any information about rationing of one side of the market. The original maximum–likelihood estimates are provided in column (1) of Table 15.1. This was obtained by an SAS program using some quadratic hill–climbing algorithm. Mayer (1989, p. 314) indicates that this problem may exhibit several local

Table 15.1   ML estimation results for a disequilibrium model of the USA loan market

	Original solution	GA solution
Parameter	(1)	(2)
Demand equation		
β_{10}	−1.347	61.115
β_{11}	−4.163	27.587
β_{12}	3.234	2.567
Supply equation		
β_{20}	−83.257	−89.249
β_{21}	3.129	0.439
β_{22}	0.339	0.354
$\sigma_{\varepsilon_1}^2$	1031.530	79.429
$\sigma_{\varepsilon_2}^2$	104.132	54.069
Log–likelihood	−313.596	−296.555

maxima, as different starting values resulted in at least two different parameter estimates. The one presented in Table 15.1 corresponds to the higher value of the likelihood function.

The results obtained by Dorsey and Mayer (1995) using a genetic algorithm (GA) are reported in column (2) of Table 15.1. The higher value of the log–likelihood indicates that the solution reported in (1) is not the maximum-likelihood estimator of the problem. Furthermore, the improved result does not change only with regard to the likelihood function, but also with regard to the coefficient estimates. While the impact of the volume indicators IP and TD, respectively, is similar for both estimates, the impact of relative prices is much smaller for the improved estimates. In fact, β_{11} has the 'wrong' sign in column (2), compared to a priori expectations and β_{21}, while still having the 'right' sign, becomes much smaller. This example highlights the potential consequences of relying on inadequate standard optimization tools for the complex estimation problems of modern econometrics.

Pham and Karaboga (2000, pp. 27ff.) compare the performance of different optimization heuristics, including genetic algorithms and simulated annealing on the following four artificial bench–mark test functions:

$$F_1(\mathbf{x}) = \sum_{i=1}^{3} x_i^2$$

$$F_2(\mathbf{x}) = 100(x_1^2 - x_2)^2 + (1 - x_2)^2$$

$$F_3(\mathbf{x}) = \sum_{i=1}^{5} [x_i]$$

$$F_4(\mathbf{x}) = [0.002 + \sum_{j=1}^{25} \frac{1}{j + \sum_{i=1}^{2} (x_i - a_{ij})^6}]^{-1},$$

where $[x_i]$ denotes the greatest integer less than or equal to x_i, and the a_{ij} are given by

$$\{(a_{1j}, a_{2j})\}_{j=1}^{25} = \{(-32, -32), (-16, 32), (0, -32), (16, -32), (32, -32),$$
$$(-32, 16), (-16, -16), (0, -16), (16, -16), (32, -16), \ldots,$$
$$(-32, 32), (-16, 32), (0, 32), (16, 32), (32, 32)\}.$$

It turns out that the ranking of the algorithms differs quite substantially depending on which objective function is considered. A general finding is that during the early phase of optimization — corresponding to a low number of iterations — tabu search and simulated annealing seem superior to genetic algorithms and neural networks. The latter may eventually result in better solution quality, when the number of function evaluations increases. While for the test functions F_1 and F_2, neural networks outperform the other tested algorithms for a high number of iterations, for the test functions F_3 and F_4 simulated annealing is competitive for all number of iterations.

Finally, Křivý, Tvrdík and Krpec (2000) present an application of stochastic algorithms to nonlinear regression problem including a standard local search implementation and a genetic algorithm. These authors use small nonlinear regression problems for which the standard algorithms implemented in statistical software often fails to provide reasonable results, while the optimization heuristics show good performance. Since the problems depend on a small number of parameters (up to seven), it is surprising that standard algorithms fail so often. On the other hand, it is not clear whether the good performance of the presented algorithms will persist if larger problem instances are considered.

15.4   Conclusions

Despite the fact that so far no threshold accepting implementation for global continuous optimization is available, the material presented in this chapter allows for the following main conclusions.

First, not all continuous optimization problems require the use of global continuous optimization heuristics, as a part of these problems can be solved by using the conventional tools. A second part can be transformed to discrete optimization problems by making use of specific features of optimal solutions. Only for the third part of these problems, which are too complex to be solved by conventional algorithms, and for which no adequate transformation to a discrete optimization problem can be found, should a continuous optimization heuristic be considered. For these problems, the efficiency of a continuous optimization heuristic should be compared with the efficiency of a discrete optimization heuristic working on a discrete approximation of the continuous optimization problem. Since a discrete approximation will lead to some approximation error, a direct approach on the continuous problem is superior if computing resources are not a binding constraint. However, if only a small number of iterations can be performed, a discrete version of threshold accepting working on the discrete approximation may be more efficient.

Secondly, existing results on the performance of simulated annealing implementations and genetic algorithms allow for the conclusion that for this third class of problems heuristics may improve results considerably as compared to conventional algorithms. Although, the use of optimization heuristics may require more computer resources, these costs are compensated for by achieving results which are more likely to be close to the global optimum, while using standard algorithms may lead to erroneous conclusions about economic issues of importance.

Thirdly, following existing implementations of simulated annealing for global continuous optimization and including the knowledge about threshold accepting implementations for discrete optimization problems presented in the previous chapters, it seems possible to obtain a high–performance threshold accepting implementation for the purpose of solving complex continuous optimization problems as they appear in econometrics and statistics. However, a real test of such an implementation has to be left for future research.

Part IV

Conclusion and Outlook

16

Conclusion

This book has presented some material on the classical and heuristic optimization paradigms in economics, statistics, and econometrics, introduced the optimization heuristic threshold accepting in more detail, and presented several applications where threshold accepting exhibited a good performance. While this description of the contents of the book might serve as a condensed summary, it is not enough to summarize the intention of writing this book and the conclusions for optimization applications in statistics and econometrics. It is the purpose of this short chapter to provide this information.

16.1 The Optimization Paradigms

The first part of the book was devoted to an overview on the importance of optimization techniques in economics, statistics, and econometrics. First, the role of optimization in economic theory was found to be a crucial one. However, the optimization paradigm commonly used in theory differs markedly from the one employed by economic subjects in reality. In particular, in economic theory the writing down of an optimization problem is treated equivalently to its exact solution. In reality, however, optimization problems are often not dealt with explicitly. Even if an optimization problem is described explicitly as such, often ad hoc heuristics are used instead of some exact solution strategy. It is argued that this behaviour is rational due to the high inherent complexity of many optimization problems met in economic practice. Economic theory has recently started to take this feature into account and to develop models which require less strong assumptions about the solution of optimization problems.

A slightly different situation is found in statistics and econometrics, where optimization problems have to be solved practically. Therefore, a distinction is made between the optimization problem per se and its solution. However, not all optimization problems arising in statistics and econometrics are tackled by using explicit optimization techniques. Instead, often ad hoc heuristics are employed or the optimization aspect of a problem is completely neglected, e.g. when choosing the variables used in an econometric model to mimic some theoretic counterparts.

A closer look at the different approaches to optimization problems in statistics and econometrics indicates that the probability that an optimization problem is treated explicitly is high within the classical optimization paradigm, i.e. if analytical solutions are available or standard numerical methods provide high–quality solutions with certainty or high probability. For example, a linear regression model is not 'solved' by simply assuming some reasonable values of the regression parameters after regarding some scatter plots, but by explicitly solving the underlying system of linear equations.

On the other side, those problems which do not fit easily within this framework are much less likely to be treated explicitly as optimization problems. It is argued that it is basically the high inherent complexity of such problems which hinders the successful application of standard optimization techniques. In these cases, either explicit optimization is replaced by some ad hoc approaches, or the optimization problem is treated with inadequate conventional optimization tools. Ad hoc approaches often require a substantial amount of subjective input, thus making the results difficult to interpret. On the other side, applying conventional algorithms to these kinds of problems is not a solution either. In fact, disregarding the obstacle imposed by high inherent complexity and applying standard routines might lead to erroneous results, as the examples provided by McCullough and Vinod (1999) and Dorsey and Mayer (1995) indicate.[1]

To sum up the findings of Part I, it can be stated that optimization is ubiquitous and essential in economics, statistics and econometrics. The scope of optimization is even larger than usually considered when the highly complex problems are taken into account, which so far are often tackled by using ad hoc approaches. Furthermore, a few examples show that optimization heuristics are used by economic agents and that some methods used in statistics and econometrics can also be interpreted as optimization heuristics. Therefore, their use might also be indicated for the just mentioned highly complex problems when standard approaches fail to be effective.

16.2   The Art of Implementing Threshold Accepting

The second part of the book aimed at introducing a specific optimization heuristic — threshold accepting — as one tool to overcome the binding constraints of the standard optimization paradigm. First, an overview on different optimization techniques, both from the standard and the heuristic optimization paradigm, presented in Chapter 5 shows that there exist substantial similarities between standard algorithm such as Newton's method and local search heuristics. However, the differences are also outlined, which allow the more refined heuristics to work efficiently even in settings of

---

[1] The comparison of different standard and nonstandard algorithms for the censored quantile regression problem in Chapter 14 gives a further example.

nonconvex objective functions with several local minima, when standard methods often fail. Besides threshold accepting, a few other optimization heuristics are considered which are more broadly used in applications in engineering or operational research, such as the simulated annealing algorithm or genetic algorithms.

The implementation of threshold accepting is described in detail in Chapter 6, which also includes a simple example to demonstrate, how the threshold accepting step allows the algorithm to escape 'bad' local minima in order to converge eventually to the global minimum. This heuristic motivation is substantiated by the global convergence results for simulated annealing and threshold accepting reported in Chapter 7, which show that in contrast to standard algorithms, these search heuristics will also provide high–quality solutions for highly complex problems. However, since these convergence results are not constructive, it is difficult to draw conclusions for real life implementations. To this end, available evidence on the relative performance of different optimization heuristics is gathered in the final section of Chapter 7. This evidence supports the claim that threshold accepting is a reasonable choice for highly complex optimization problems. In particular, if the size of the problem limits the number of iterations as compared to the instance size, simulated annealing and threshold accepting seem to outperform genetic algorithms.

The performance of optimization heuristics depends on some tuning parameters. Chapter 8 is devoted to a simulation analysis of the impact of such tuning parameters on the performance of threshold accepting. A mainly data–driven method for generating these parameters is proposed and tested, which appears to be very efficient both for the travelling salesman example analysed in Chapter 8 and the applications to problems in statistics and econometrics covered in Part III. Consequently, the expense for the implementation of threshold accepting to a new problem can be reduced considerably when using this data–driven method, which can be motivated from some theoretical arguments of the convergence analysis given in Chapter 7. Based on the results of Chapter 7 and Chapter 8, it can be concluded that a threshold accepting algorithm can be considered as a useful choice for tackling highly complex optimization problems. At least, without extensive implementation costs, it provides a bench–mark which might be hard to better even by more problem–specific methods, and which provides a lower bound for the deviation of ad hoc approaches from the global optimum.

Finally, the arguments gathered on the implementation of threshold accepting are summarized in a practical guideline in Chapter 9, which should provide the necessary information for implementing the threshold accepting heuristic to complex optimization problems as they arise in econometrics and statistics.

16.3   Using Threshold Accepting in Statistics and Econometrics

The results of four applications of threshold accepting to highly complex optimization problems in statistics and econometrics are summarized in Part III. These results substantiate the conclusion that threshold accepting is a reasonable choice for such problems. For the uniform design problem covered in Chapter 11 and the censored quantile regression problem treated in Chapter 14 other algorithms have been used before. Consequently, the results obtained by these algorithms may serve as a bench–mark for the threshold accepting implementation. It turns out that by using the threshold accepting heuristic non–trivial improvements can be obtained. Often these improvements in solution quality come at the price of higher computational costs. Nevertheless, since an improvement cannot be obtained by increasing computing time for the alternative methods, this potential drawback of threshold accepting does not seem too important in times of dramatically decreasing costs of computation. This conclusion holds true at least as long as the improvements in solution quality can be obtained by increasing the number of iterations of the threshold accepting algorithm with a low–degree polynomial rate with respect to solution quality. For the other two examples covered in Chapters 12 and 13, no similar bench–marks are available, because only ad hoc approaches have been used so far. The threshold accepting implementations easily better these ad hoc approaches in terms of the objective function, as should have been expected.

Thus, drawing a first general conclusion on the real applications studied in Part III, it can be stated that threshold accepting is able to provide high–quality results even for problems of high inherent complexity. In particular, it improves the results obtained by existing methods and ad hoc approaches. Furthermore, the obtained results have some importance for the specific applications, which are summarized in the following paragraphs.

In Chapter 11, it has been shown that by using a threshold accepting implementation it is possible to generate highly uniform point sets. Due to the flexibility of the implementation, it can also be used to generate point sets for specific regression models. The implication of the possibility to obtain such improved point sets are far reaching in different areas of statistics and econometrics. In particular, in experimental economics or response surface analysis either the number of necessary evaluations (experiments) can be reduced holding the error bound constant, or by using the same number of evaluations (experiments), the approximation quality can be significantly improved. A further field of application is given by all kinds of simulation methods. However, so far only results for a comparatively small number of points have been generated by the threshold accepting implementation. In order to be useful in general simulation problems under the heading of quasi–Monte Carlo methods, this has to be extended to larger instances.

For the model selection problem discussed in Chapter 12, there exists a strong theoretical background. Nevertheless, practical implementations, e.g. in the context of VAR or VEC models, often rely on ad hoc methods for this purpose, due to the inherent complexity of a complete evaluation of all alternatives. The application of threshold accepting allows us to replace these ad hoc methods. Consequently, the subjectivity inherent in ad hoc model selection can be avoided. Of course, a correct model specification is of great interest in its own right. However, often further steps of the analysis depend on a suitable model selection in prior steps. This is the case, e.g. for Johansen's cointegration test, thus making a good model specification even more important.

The problem of optimal aggregation discussed in Chapter 13 is even more complex than the modelling problem. In fact, the appendix presented in this chapter shows that it belongs to the class of NP–complete problems. Nevertheless, aggregation is ubiquitous in econometric modelling and can hardly be avoided. The results of the threshold accepting implementation indicate that aggregation matters, as ad hoc aggregation and optimized aggregation lead to quite different economic interpretations of the model of international price spillovers which was taken as an example. Furthermore, the results indicate that threshold accepting is able to provide optimized results even for this most complex problem studied in Part III, which are much better than the modes of aggregation provided by statistical agencies. However, it cannot be expected that the global optimum is reached for all instances, and the generation of complete hierarchical classification systems is still beyond the scope of this implementation.

The application studied in Chapter 14 differs from the other applications for two reasons. First, the censored quantile regression problem is a continuous global optimization problem. Consequently, it could be tackled by using the methods outlined in Chapter 15. However, a theoretical result, the interpolation property, allows it to be reduced to a discrete optimization problem, which appears to be easier to solve. Secondly, for this problem several other algorithms have been presented in the literature which serve as a bench–mark for threshold accepting. A further bench–mark can be obtained by enumeration for very small problem instances. Using these bench–marks in a simulation study for different data–generating processes in dimensions two to four exhibits a clear superiority of the threshold accepting implementation. This has advantages both in absolute and relative performances. Furthermore, even if the global optimum is not found, the coefficient estimators supplied by the threshold accepting algorithms are closer to the true estimates than for the conventional algorithms.

Part III concludes in Chapter 15 with an introduction to the application of optimization heuristics such as threshold accepting to continuous global optimization problems. Given the similarity between simulated annealing and threshold accepting, it is argued that threshold accepting implementations are also possible for these kinds of problems. However, no actual implementation is presented. This is left for further research along with several other aspects which will be mentioned in Chapter 17.

17

Outlook for Further Research

Although this book tries to provide a comprehensive introduction to the application of threshold accepting in statistics and econometrics, it shares the fate of any documentation on ongoing research in applied statistics or econometrics — it is never completed! Like the introduction of new estimation methods, new data sets or a different statistical approach, the introduction of optimization heuristics such as threshold accepting helps to answer a few questions and to improve knowledge for specific implementations. At the same time, however, the use of new methods tends to raise new questions. These questions may concern the new results obtained by threshold accepting for some optimization problem. Then, the questions raised have to be answered within the context of these applications and are not a primary concern of this present book. However, the results of threshold accepting implementations do not only challenge existing solutions obtained by other methods, but also raise questions about the performance of the implementation itself and about further implementations to similar problems. Thus, it is not surprising that despite of the extensive coverage of different aspects of optimization heuristics in statistics and econometrics, in particular considering the threshold accepting heuristic, this book rather seeks to mark the beginning of a research field, as opposed to its completion.

Given the broad coverage of this volume, an outlook for further research can hardly be given in a few sentences. In fact, a detailed description of further research required within this context would require a second volume. Therefore, instead of providing a precise catalogue of the next steps to be undertaken in this field, a few comments may outline the areas, where I personally expect further developments to take place in the near future. These future developments may be summarized in four categories. First, given the advances in available optimization techniques and the increase in computing resources, the change in the optimization paradigm outlined in the first part of this book may reach further areas. Secondly, a better understanding of the mathematics behind local search heuristics may contribute to a different concept of convergence, covering at the same time asymptotic convergence of statistics of interest and the asymptotic convergence of the optimization algorithm. Thirdly, such a better understanding of asymptotic behaviour may also contribute to further improvements of the performance for a finite number of iterations. To the same end, further tuning experiments have to be conducted and combinations of the threshold accepting algorithm with other meta–heuristics should be considered. Finally, the application of threshold accepting is not constrained to the problems analysed within this present book. The following sections will outline these four aspects in a little bit more detail.

17.1   The Change in the Optimization Paradigm

The examples of optimization problems presented in Chapters 2 and 3 are selected as typical optimization problems in economics, statistics and econometrics, respectively. Nevertheless, the enumeration of a few examples does not yet allow for a clear–cut conclusion about the relative importance of different categories of optimization problems. For the purposes of this book, it was sufficient to show that there exist optimization problems in these areas which cannot be tackled relying on the standard optimization paradigm.

The gathering of information on a larger number of typical optimization problems met in these areas will help to assess the relative importance of the different categories. Thereby, special attention should be given to new developments in economic and econometric modelling. While available theory and statistical methodology is based to a large extent on the standard optimization paradigm, it might be expected that current and future developments are more likely to fall outside the range of this standard paradigm.

Transferring findings from the mathematical theory of complexity and the operational research literature, it seems likely that a closer look at all optimization problems in economics and econometrics — including those which so far have not been treated as explicit optimization problems — will indicate that it is rather the exception than the rule that such a problem can be solved within the classical paradigm, unless very strong additional assumptions are imposed.

Of course, those problems which allow a high–quality solution by classical optimization methods will always be solved relying on these fast and well known algorithms. However, for the other class of problems, which is expected to increase and to become eventually dominant, the use of optimization heuristics such as threshold accepting should be taken into account.

17.2   Asymptotic Considerations

Chapter 7 provided some strong convergence results for the threshold accepting heuristic. Nevertheless, the stochastic nature of optimization heuristics such as threshold accepting might be regarded as a shortcoming, as for a finite number of iterations the final solution might be different from the global optimum. Of course, such an argument does not allow for the conclusion that one should rather rely on standard algorithms or ad hoc solutions for complex problems, as these methods will fail even asymptotically. Consequently, either one restrains from a further analysis of the problem, or one has to living with the stochastic nature of the results provided by optimization heuristics.

Fortunately, statisticians and econometricians are used to live with stochastic outcomes. In fact, the usual asymptotic analysis in statistics raises a similar problem. While it is possible to obtain strong convergence results as the number of observations goes to infinity, much less is known about finite samples. In particular, any statistic which is calculated based on such a finite sample is stochastic and will converge only asymptotically to some well defined distribution if all assumptions are met.

Given this similarity, it seems a valuable topic of further research to attempt to gain a deeper understanding of the mathematics of refined local search algorithms and their convergence to the global optimum. By using this information, one might strive for a concept of convergence which combines asymptotic aspects of statistical theory and asymptotic aspects of the optimization routines. So far, no uniform treatment of both aspects is available, and the nonconstructive arguments of the convergence results for simulated annealing and threshold accepting do not lend themselves easily to such an extension. However, using a broader, probabilistic concept of convergence covering both aspects might be feasible. Such a concept should provide some uniform convergence of the statistical method and the optimization heuristic, i.e. with the number of observations going to infinity, the number of iterations used within the optimization heuristics may also be allowed to tend to infinity. If the number of observations has only a small–order polynomial influence on the instance size of the optimization problem, one might be able to derive a result that asymptotically, not only will the statistic converge to its true population value, but also the optimization heuristic will provide this true value with a probability tending to one.

17.3 Improvements of Performance

A better understanding of the asymptotic behaviour of threshold accepting might also provide further insights in the optimal choice of parameters for implementations when using only a finite number of iterations. Of course, this performance is crucial for real applications, even if it becomes possible to derive some generalized convergence results along the lines outlined in the previous section.

Apart from asymptotic arguments, further tuning experiments, such as the one presented in Chapter 8, might provide additional information. Therefore, the experimental set–up has to be chosen carefully and a large number of different problems should be considered. Besides tuning experiments, very small problem instances could be used to generate optimal threshold sequences (Althöfer and Koschnik, 1991, pp. 194f.), i.e. threshold sequences maximizing the probability of ending with the global optimum after a small number of iterations.

Finally, one can experiment with adaptive schemes for neighbourhood structures and the threshold sequence. In fact, even heuristic optimization methods can be used to obtain improved threshold sequences or neighbourhood structures (Bölte and Thonemann, 1996). In addition, hybrid algorithms including, e.g. arguments from tabu search are worth considering. As a result, automatic data–based methods similar to those employed for the applications in Part III should be developed which will eventually allow for their inclusion in standard software tools.

17.4 Further Applications

While a change in the dominant optimization paradigm in economics, statistics and econometrics cannot be expected to take place within a decade, further research on the performance of threshold accepting will produce results within a shorter time–scale. Furthermore, the application of threshold accepting to new complex optimization problems can take place almost immediately, while the application of existing implementations to new problem instances is possible instantaneously. Hence, in the short run most further research results will stem from such applications and extensions of existing implementations.

Starting with the application of threshold accepting to the problem of optimal uniform designs covered in Chapter 11, further research might cover three aspects. First, optimized designs can be obtained for larger problem instances by using different concepts for measuring the uniformity. Secondly, a closer analysis of the link between uniform and orthogonal designs might result in further new orthogonal designs obtained by the application of threshold accepting. Thirdly, the optimized designs can be used in applications to assess the improvements in solution quality. If the application of threshold accepting also becomes possible for a larger number of points, it can be used to base quasi–Monte Carlo methods on point sets generated by the threshold accepting implementation instead of some number theoretic method.

For the implementation to multivariate lag order selection, some straightforward generalizations can be considered. The available evidence can be enlarged by using a larger set of different models for Monte Carlo simulations. Furthermore, the implementation is easily extended to cover exogenous variables in the vector autoregressive model or even error correction terms. Then, Monte Carlo simulations can be used to assess the impact of optimized lag order selection on further steps of the econometric analysis, e.g. cointegration testing. Finally, the method will be applied to real data sets, when either the dynamics are of interest on their own, or further analysis requires an optimal choice in this first modelling step.

The results of the optimal aggregation implementation presented in Chapter 13 are already meaningful in their own right, although further economic analysis is required to exploit this information. In particular, the results will provide improved understanding of the sectoral structure of international trade and price spillovers. Apart from this analysis of the results obtained by the presented threshold accepting implementation, the robustness of the results can be analysed in more detail by using, e.g. slightly different samples or subsets of the commodities. However, the largest scope for further developments linked to this application is seen in an improvement of the evaluation of the objective function on the one side, and applications to further data sets on the other.

For the final application presented in Part III, i.e. the censored quantile estimation problem, the available Monte Carlo evidence clearly documents the superiority of the threshold accepting implementation, in particular, if censoring is a relevant phenomenon in the analysed data set. However, for the small instances covered in the Monte Carlo simulation this improved performance came at the cost of much higher computational cost. Typically, the performance of optimization heuristics in terms of computational costs is relatively better when the problem instances become larger and standard approaches also require a much higher computational input. Consequently, the relative performance of threshold accepting should be analysed for larger problem instances, including real data sets.

The previous sections have highlighted some areas for future research linked to the applications of threshold accepting presented in Part III. However, a much larger number of optimization problems has to be solved in statistics and econometrics where standard approaches have failed to provide reasonable results. Hence, future research will extend the application of the heuristic optimization paradigm to a large number of such instances, including, e.g. continuous optimization problems as briefly outlined in Chapter 15. Any optimization problem withstanding standard techniques is a natural candidate for a threshold accepting implementation.

References

Aarts, E. H. L., Korst, J. H. M. and van Laarhoven, P. J. M. (1997). Simulated annealing, *in* E. Aarts and J. K. Lenstra (eds), *Local Search in Combinatorial Optimization*, Wiley, Chichester, pp. 91–120.

Aarts, E. H. L. and Lenstra, J. K. (1997). Local search in combinatorial optimization: Introduction, *in* E. Aarts and J. K. Lenstra (eds), *Local Search in Combinatorial Optimization*, Wiley, Chichester, pp. 1–17.

Aarts, E. H. L., Lenstra, J. K., van Laarhoven, P. J. M. and Ulder, N. L. J. (1994). A computational study of local search algorithms for job shop scheduling, *ORSA Journal on Computing* **6**: 118–125.

Abadir, K. M., Hadri, K. and Tzavalis, E. (1999). The influence of VAR dimensions on estimator biases, *Econometrica* **67**: 163–181.

Acworth, P., Broadie, M. and Glasserman, P. (1998). A comparison of some Monte Carlo and quasi–Monte Carlo techniques for option pricing, *in* H. Niederreiter, P. Hellekalek, G. Larcher and P. Zinterhof (eds), *Monte Carlo and Quasi–Monte Carlo Methods 1996*, Vol. 127 of *Lecture Notes in Statistics*, pp. 1–18.

Aho, A. V., Hopcroft, J. E. and Ullman, J. D. (1974). *The Design and Analysis of Computer Algorithms*, Addison–Wesley, Reading, MA.

Akaike, H. (1969). Fitting autoregressive models for prediction, *Annals of the Institute of Statistical Mathematics* **21**: 243–247.

Akaike, H. (1971). Autoregressive model fitting for control, *Annals of the Institute of Statistical Mathematics* **23**: 163–180.

Akaike, H. (1973). Information theory and an extension of the maximum likelihood principle, *in* B. N. Petrov and F. Csaki (eds), *Second International Symposium on Information Theory*, Budapest, pp. 267–281.

Akerlof, G. A. and Yellen, J. L. (1985). Can small deviations from rationality make significant differences to economic equilibria?, *American Economic Review* **75**: 708–721.

Althöfer, I. and Koschnik, K.-U. (1991). On the convergence of threshold accepting, *Applied Mathematics and Optimization* **24**: 183–195.

Arifovic, J. (1994). Genetic algorithm learning and the cobweb model, *Journal of Economics Dynamics and Control* **18**: 3–28.

Atkinson, A. C. and Donev, A. N. (1992). *Optimum Experimental Designs*, Clarendon Press, Oxford.

Ausiello, G., Crescenzi, P., Gambosi, G., Kann, V., Marchetti-Spaccamela, A. and Protasi, M. (1999). *Complexity and Approximation*, Springer, Berlin.

Ausiello, G. and Protasi, M. (1995). Local search, reducibility and approximability of NP–optimization problems, *Information Processing Letters* **54**: 73–79.

Aznar, A. (1989). *Econometric Model Selection: A New Approach*, Kluwer, Dordrecht.

Bagnoli, M., Salant, S. and Swierzbinski, J. (1989). Durable–goods monopoly with discrete demand, *Journal of Political Economy* **97**: 1459–1478.

Balk, B. M. (1983). Does there exist a relation between inflation and relative price–change variability?, *Economics Letters* **13**: 173–180.

Barr, R. S., Golden, B. L., Kelly, J. P., Resende, M. G. C. and Stewart, W. R. (1995). Designing and reporting on computational experiments with heuristic methods, *Journal of Heuristics* **1**: 9–32.

Basar, T. and Vallée, T. (1999). Computation of the closed–loop stackelberg solution using the genetic algorithm, *in* H.-F. Chen, D.-Z. Cheng and J.-F. Zhang (eds), *Proceedings of the 14th World Congress of the International Federation of Automatic Control*, Vol. M, Elsevier, Oxford, pp. 69–74.

Beaumont, P. M. and Bradshaw, P. T. (1995). A distributed parallel genetic algorithm for solving optimal growth models, *Computational Economics* **8**: 159–179.

Beck, J. and Chen., W. W. L. (1987). *Irregularities of Distribution*, Cambridge University Press, Cambridge.

Beck, J. and Sós, V. T. (1995). Discrepancy theory, *in* R. Graham and M. G. L.Lovász (eds), *Handbook of Combinatorics*, Elsevier, Amsterdam, pp. 1405–1446.

Biethahn, J. and Nissen, V. (1995). *Evolutionary Algorithms in Management Applications*, Springer, Berlin / Heidelberg.

Birchenhall, C. (1995). Modular technical change and genetic algorithms, *Computational Economics* **8**: 233–253.

Bölte, A. and Thonemann, U. W. (1996). Optimizing simulated annealing schedules with genetic programming, *European Journal of Operational Research* **92**: 402–416.

Bochachevsky, I. O., Johnson, M. E. and Stein, M. L. (1986). Generalized simulated annealing for function optimization, *Technometrics* **28**: 209–217.

Bose, R. C. and Manvel, B. (1984). *Introduction to Combinatorial Theory*, Wiley, New York.

Brüggemann, I. and Wolters, J. (1998). Money and prices in Germany: Empirical results for 1962 to 1996, *in* R. Galata and H. Küchenhoff (eds), *Econometrics in Theory and Practice*, Physica, Heidelberg, pp. 205–225.

Brind, C., Muller, C. and Prosser, P. (1995). Stochastic techniques for resource management, *BT Technology Journal* **13**: 55–63.

Brooks, C. and Burke, S. P. (1998). Forecasting exchange rate volatility using conditional variance models selected by information criteria, *Economics Letters* **61**: 273–278.

Brown, S. R. and Melamed, L. E. (1990). Experimental design and analysis, *Technical Report 07–074*, Sage University, Newbury Park, CA.

Brown, S. R. and Melamed, L. E. (1993). Experimental design and analysis, *in* M. S. Lewis-Beck (ed.), *Experimental Design and Methods*, Sage, London, pp. 75–159.

Buchinsky, M. (1994). Changes in the U.S. wage structure 1963–1987: Applications of quantile regression, *Econometrica* **62**: 405–458.

Bundschuh, P. and Zhu, Y. C. (1993). A method for exact calculation of the discrepancy of low–dimensional finite point set (I), *Abhandlungen aus dem Mathematischen Seminar der Univeristät Hamburg* **63**: 115–133.

Burnham, K. P. and Anderson, D. R. (1995). *Model Selection and Inference*, Springer, New York.

Caflisch, R. E. and Morokoff, W. (1996). Valuation of mortage–backed securities using the quasi-Monte Carlo method, *Technical Report 96–16*, Department of Mathematics UCLA, Los Angeles, CA.

Canner, N., Mankiw, N. G. and Weil, D. N. (1997). An asset allocation puzzle, *American Economic Review* **87**: 181–191.

Canova, F. (1995). Vector autoregressive models: Specification, estimation, inference and forecasting, *in* M. H. Pesaran and M. R. Wickens (eds), *Handbook of Applied Econometrics: Macroeconomics*, Blackwell, Oxford, pp. 73–138.

Cerny, V. (1985). A thermodynamical approach to the travelling salesman problem: An efficient simulated annealing algorithm, *Journal of Optimization Theory and Applications* **45**: 41–51.

Chan, L.-Y., Fang, K.-T. and Winker, P. (1998). An equivalence theorem for orthogonality and D–optimality, *Technical Report 186*, Hong Kong Baptist University, Hong Kong.

Chao, J. C. and Phillips, P. C. B. (1999). Model selection in partially nonstationary vector autoregressive processes with reduced rank structure, *Journal of Econometrics* **91**: 227–271.

Charon, I. and Hudry, O. (1993). The noising method: A new method for combinatorial optimization, *Operations Research Letters* **14**: 133–137.

Chipman, J. S. (1975). Optimal aggregation in large–scale econometric models, *The Indian Journal of Statistics* **37**: 121–159.

Chipman, J. S. (1976). Estimation and aggregation in econometrics: An application of the theory of generalized inverses, *in* M. Z. Nashed (ed.), *Generalized Inverses and Applications*, Academic Press, New York, pp. 553–773.

Chipman, J. S. (1985). Testing for reduction of mean–square error by aggregation in dynamic econometric models, *in* P. R. Krishnaiah (ed.), *Multivariate Analysis — VI*, North–Holland, Amsterdam, pp. 97–119.

Chipman, J. S. and Winker, P. (1995a). Optimal industrial classification, *New Trends in Probability and Statistics*, Vol. 3 of *Multivariate Statistics and Matrices in Statistics, Proceedings of the 5th Tartu Conference*, TEV, Vilnius, pp. 229–234.

Chipman, J. S. and Winker, P. (1995b). Optimal industrial classification by threshold accepting, *Control and Cybernetics* **24**: 477–494.

Clements, M. P. and Krolzig, H.-M. (1998). A comparison of the forecast performance of Markov–switching and threshold autoregressive models of US GNP, *Econometrics Journal* **1**: C47–C75.

Cook, S. (1971). The complexity of theory proving procedures, *Proceedings of the 3rd Annual ACM Symposium on Theory of Computing*, Association for Computing Machinery, New York, pp. 151–158.

Cook, W. J., Cunningham, W. H., Pulleyblank, W. R. and Schrijver, A. (1998). *Combinatorial Optimization*, Wiley, New York.

Cotterman, R. and Peracchi, F. (1992). Classification and aggregation: An application to industrial classification in CPS data, *Journal of Applied Econometrics* **7**: 31–51.

Davidson, R. and MacKinnon, J. G. (1993). *The Practice of Econometrics – Classic and Contemporary*, Oxford University Press, New York.

Dean, A. and Voss, D. (1999). *Design and Analysis of Experiments*, Springer, New York.

Dennis, J. E. and Schnabel, R. B. (1983). *Numerical Methods for Unconstrained Optimization and Nonlinear Equations*, Prentice–Hall, Englewood Cliffs, NJ.

Dennis, J. E. and Schnabel, R. B. (1989). A view of unconstrained optimization, *in*

G. L. Nemhauser, A. H. G. R. Kan and M. J. Todd (eds), *Handbooks in Operations Research and Management Science*, Vol. 1, North–Holland, Amsterdam, pp. 1–72.

Donev, A. N. and Atkinson, A. C. (1988). An adjustment algorithm for the construction of exact D–optimum experimental designs, *Technometrics* **30**: 429–433.

Dorsey, B. and Mayer, W. J. (1995). Genetic algorithms for estimation problems with multiple optima, nondifferentiability and other irregular features, *Journal of Business and Economic Statistics* **13**: 53–66.

Dueck, G. (1993). New optimization heuristics: The great–deluge algorithm and the record–to–record–travel, *Journal of Computational Physics* **104**: 86–92.

Dueck, G. and Scheuer, T. (1990). Threshold accepting: A general purpose algorithm appearing superior to simulated annealing, *Journal of Computational Physics* **90**: 161–175.

Dueck, G. and Winker, P. (1992). New concepts and algorithms for portfolio choice, *Applied Stochastic Models and Data Analysis* **8**: 159–178.

Dueck, G. and Wirsching, J. (1991). Threshold accepting algorithms for 0–1 knapsack problems, *in* H. Wacker and W. Zulehner (eds), *Proceedings of the Fourth European Conference on Mathematics in Industry*, B. G. Teubner, Stuttgart, pp. 225–262.

Ericsson, N. R. and Marquez, J. (1998). A framework for economic forecasting, *Econometrics Journal* **1**: C228–C266.

Fackler, J. S. (1985). An empirical analysis of the markets for goods, money and credit, *Journal of Money, Credit and Banking* **17**: 28–42.

Fang, K.-T. and Hickernell, F. J. (1995). The uniform design and its applications, *Bulletin of the International Statistical Institute* **50th Session, Book 1**(11): 49–65.

Fang, K.-T., Hickernell, F. J. and Winker, P. (1996). Some global optimization algorithms in statistics, *in* D.-Z. Du, X.-S. Zhang and W. Wang (eds), *Operations Research and Its Applications*, Lecture Notes in Operations Research 2, World Publishing Corporation, Beijing, pp. 14–26.

Fang, K.-T., Lin, D. K. J., Winker, P., and Zhang, Y. (2000). Uniform design: Theory and application, *Technometrics* **42**: 237–248.

Fang, K.-T., Ma, C.-X. and Winker, P. (2000). Centered L_2–discrepancy of random sampling and latin hypercube design, and construction of uniform designs, *Mathematics of Computation* : forthcoming.

Fang, K.-T. and Wang, Y. (1994). *Number–theoretic Methods in Statistics*, Chapman & Hall, London.

Farley, A. M. and Jones, S. (1994). Using a genetic algorithm to determine an index of leading economic indicators, *Computational Economics* **7**: 163–173.

Fase, M. M. G. (1994). In search for stability: An empirical appraisal of the demand for money in the G7 and EC countries, *De Economist* **142**: 421–454.

Ferrante, J. and Rackoff, C. (1979). The computational complexity of logical theories, *Technical Report 718*, Springer, Berlin.

Fisher, W. D. (1962). Optimal aggregation in multi–equation prediction models, *Econometrica* **30**: 769–774.

Fitzenberger, B. (1997a). A guide to censored quantile regressions, *in* G. S. Maddala and C. R. Rao (eds), *Handbook of Statistics*, Vol. 15: Robust Inference, North–Holland, Amsterdam, pp. 405–437.

Fitzenberger, B. (1997b). Computational aspects of censored quantile regression, *in* Y. Dodge (ed.), *Proceedings of the 3rd International Conference on Statistical*

Data Analysis based on the L_1 - Norm and Related Methods, IMS Lecture Note Series 31, Hayword, CA, pp. 171–186.

Fitzenberger, B. and Franz, W. (1999). Industry–level wage bargaining: A partial rehabilitation — the German experience, *Technical Report 99-33*, ZEW, Mannheim.

Fitzenberger, B., Koenker, R. and Machado, J. A. F. (2000). Economic applications of quantile regressions, *Empirical Economics* : forthcoming.

Fitzenberger, B. and Winker, P. (1998). Using threshold accepting to improve the computation of censored quantile regression, *in* R. Payne and P. Green (eds), *COMPSTAT 1998: Proceedings in Computational Statistics*, Physica, Heidelberg, pp. 311–316.

Fitzenberger, B. and Winker, P. (1999). Improving the computation of censored quantile regressions, *Technical Report 568-99*, Institute of Economics and Statistics, Mannheim.

Floyd, R. W. and Rivest, R. L. (1975). Expected time bounds for selection, *Communications of the ACM* **18**: 165–173.

Fox, B. L. (1994). Random restarting versus simulated annealing, *Computers, Mathematics and Applications* **27**(6): 33–35.

Fox, B. L. (1995). Faster simulated annealing, *SIAM Journal on Optimization* **5**: 405–488.

Franz, W., Göggelmann, K., Schellhorn, M. and Winker, P. (2000). Quasi–Monte Carlo methods in stochastic simulations, *Empirical Economics* **25**: 247–259.

Freedman, D. (1983). *Markov Chains*, 2nd Edn, Springer, New York.

Friedman, D. and Sunder, S. (1994). *Experimental Methods*, Cambridge University Press, Cambridge.

Frigon, N. L. and Mathews, D. (1997). *Practical Guide to Experimental Design*, Wiley, New York.

Gallant, R. and Tauchen, G. (1996). Which moments to match?, *Econometric Theory* **12**: 657–681.

Garcia-Diaz, A. and Phillips, D. T. (1995). *Principles of Experimental Design and Analysis*, Chapman & Hall, London.

Garey, M. R. and Johnson, D. S. (1979). *Computers and Intractability*, Freeman, New York.

Gelfand, S. B. and Mitter, S. K. (1991). Weak convergence of Markov Chain sampling methods and annealing algorithms to diffusions, *Journal of Optimization Theory and Applications* **68**: 483–498.

Geweke, J. (1985). Macroeconometric modeling and the theory of the representative agent, *American Economic Association Papers and Proceedings* **75**: 206–210.

Gidas, B. (1985). Nonstationary Markov Chains and covergence of the annealing algorithm, *Journal of Statistical Physics* **39**: 73–131.

Gill, P. E., Murray, W. and Wright, M. H. (1991). *Numerical Linear Algebra and Optimization*, Vol. 1, Addison-Wesley, Redwood, CA.

Gilli, M. and Këllezi, E. (2000). Portfolio selection, tail risk and heuristic optimization, *Technical report*, Department of Econometrics, University of Geneva.

Glass, C. A. and Potts, C. N. (1996). A comparison of local search methods for flow shop scheduling, *Annals of Operations Research* **63**: 489–509.

Glass, C. A., Potts, C. N. and Shade, P. (1994). Unrelated parallel machine scheduling using local search, *Mathematical and Computer Modelling* **20**: 41–52.

Goel, R. K. and Ram, R. (1993). Inflation and relative–price variability: the effect of commodity aggregation, *Applied Economics* **25**: 703–709.

Goffe, W. L. (1997). A toolkit for optimizing functions in economics, *Technical Report USM–97–001*, University of Southern Mississippi, Hattiesburg, MS.

Goffe, W. L., Ferrier, G. and Rogers, J. (1994). Global optimization of statistical functions with simulated annealing, *Journal of Econometrics* **60**: 65–99.

Gonzalo, J. and Pitarakis, J.-Y. (1998). Specification via model selection in vector error correction models, *Economics Letters* **60**: 321–328.

Gourieroux, C. A., Monfort, A. and Renault, E. (1993). Indirect inference, *Journal of Applied Econometrics* **8**: 85–118.

Greenwood, G. W. and Hu, X. (1998). On the use of random walks to estimate correlation in fitness landscapes, *Computational Statistics and Data Analysis* **28**: 131–137.

Greville, T. N. E. (1969). Introduction to spline functions, *in* T. N. E. Greville (ed.), *Theory and Application of Spline Functions*, Academic Press, New York, pp. 1–35.

Haines, L. M. (1987). The application of the annealing algorithm to the construction of exact optimal designs for linear-regression models, *Technometrics* **29**: 439–447.

Hajek, B. (1988). Cooling schedules for optimal annealing, *Mathematics of Operations Research* **13**: 311–329.

Hamming, R. W. (1950). Error detecting and error correcting codes, *Bell System Technical Journal* **29**: 147–160.

Hanafi, S., Freville, A. and Abdellaoui, A. E. (1996). Comparison of heuristics for the 0–1 multidimensional knapsack problem, *in* I. H. Osman and J. P. Kelly (eds), *Meta-Heuristics: Theory and Applications*, Kluwer, Boston, MA, pp. 449–465.

Hannan, E. J. and Quinn, B. G. (1979). The determination of the order of an autoregression, *Journal of the Royal Statistical Society* B **41**: 190–195.

Hansen, B. E. (2000). Sample splitting and threshold estimation, *Econometrica* **68**: 575–603.

Hatanaka, M. (1952). Note on consolidation within a Leontief system, *Econometrica* **20**: 301–303.

Hawkins, D. M. and Olive, D. (1999). Applications and algorithms for least trimmed sum of absolute deviations regression, *Computational Statistics and Data Analysis* **32**: 119–134.

Hedayat, A. S., Sloane, N. J. A. and Stufken, J. (1999). *Orthogonal Arrays*, Springer, New York.

Heinrich, S. (1995). Efficient algorithms for computing the L_2–discrepancy, *Technical Report 267/95*, Department of Computing, University of Kaiserslautern.

Hickernell, F. J. (1998). A generalized discrepancy and quadrature error bound, *Mathematics of Computation* **67**: 299–322.

Hildenbrand, W. (1994). *Market Demand*, Princeton University Press, Princeton, NJ.

Hildenbrand, W. (1998). How relevant are specifications of behavioral relations on the micro-level for modelling the time path of population aggregates?, *European Economic Review* **42**: 437–458.

Hocking, R. R. and Leslie, R. N. (1967). Selection of the best subset in regression analysis, *Technometrics* **9**: 531–540.

Holland, O. A. (1987). *Schnittebenenverfahren für Travelling–Salesman und verwandte Probleme*, Rheinische Friedrich–Wilhelm–Universität, Bonn.

Hopcroft, J. E. and Ullman, J. D. (1979). *Introduction to Automata Theory, Languages and Computation*, Addison–Wesley, Reading, MA.

Horowitz, J. and Neumann, G. (1987). Semiparametric estimation of employment duration models, *Econometric Reviews* **6**: 5–40.

Horst, R. and Tuy, H. (1996). *Global Optimization*, 3rd Edn, Springer, Berlin.

Hu, T. C., Kahng, A. B. and Tsao, C.-W. A. (1995). Old bachelor acceptance: A new class of non–monotone threshold accepting methods, *ORSA Journal on Computing* **7**: 417–425.

i Silvestre, J. L. C., i Rosselló, A. S. and Ortuno, M. A. (1999). Response surfaces estimates for the Dickey–Fuller unit root test with structural breaks, *Economics Letters* **63**: 279–283.

Isaacson, D. L. and Madsen, R. W. (1976). *Markov Chains: Theory and Applications*, Wiley, New York.

Jacobsen, T. (1995). On the determination of lag order in vector autoregressions of cointegrated systems, *Computational Statistics* **10**: 177–192.

Johansen, S. (1988). Statistical analysis of cointegration vectors, *Journal of Economic Dynamics and Control* **12**: 231–254.

Johansen, S. (1991). Estimation and hypothesis testing of cointegration vectors in gaussian vector autoregressive models, *Econometrica* **59**: 1551–1580.

Johansen, S. (1992). Determination of cointegration rank in the presence of a linear trend, *Oxford Bulletin of Economics and Statistic* **54**: 383–397.

Johansen, S. (1995). *Likelihood–based Inference in Cointegrated Vector Autoregressive Models*, Oxford University Press, Oxford.

Johansen, S. and Juselius, K. (1990). Maximum likelihood estimation and inference on cointegration – with applications to the demand for money, *Oxford Bulletin of Economics and Statistic* **52**: 169–211.

Johnson, D. S. and McGeoch, L. A. (1997). The traveling salesman problem: A case study, *in* E. Aarts and J. K. Lenstra (eds), *Local Search in Combinatorial Optmization*, Wiley, Chichester, pp. 215–310.

Johnston, J. (1972). *Econometric Methods*, 2nd Edn, McGraw–Hill Kogakusha, Tokyo.

Jones, A. E. W. and Forbes, G. W. (1995). An adaptive simulated annealing algorithm for global optimization over continuous variables, *Journal of Global Optimization* **6**: 1–37.

Judd, K. D. (1998). *Numerical Methods in Economics*, MIT Press, Cambridge, MA.

Kagel, J. H. and Roth, A. E. (eds) (1995). *Handbook of Experimental Economics*, Princeton University Press, Princeton, NJ.

Kan, A. H. G. R. and Timmer, G. T. (1989). Global optimization, *in* G. L. Nemhauser, A. H. G. R. Kan and M. J. Todd (eds), *Handbooks in Operations Research and Management Science*, Vol. 1, North–Holland, Amsterdam, pp. 631–662.

Kang, C.-S. A., Bedworth, D. D. and Rollier, D. A. (1982). Automatic identification of autoregressive integrated moving average time series, *IEEE Transactions* **14**: 156–166.

Karp, R. M. (1972). Reducibility among combinatorial problems, *in* R. E. Miller and J. W. Thatcher (eds), *Complexity of Computer Computations*, Plenum Press, New York, pp. 85–103.

Kemeny, J. G. and Snell, J. L. (1976). *Finite Markov Chains*, Springer, New York.

Kirkpatrick, S., Gelatt, C. and Vecchi, M. (1983). Optimization by simulated annealing, *Science* **220**: 671–680.

Knuth, D. E. (1981). *The Art of Computer Programming*, 2nd Edn, Addison–Wesley, Reading, MA.

Koenker, R. (1997). L_1 computation: An interior monologue, *in* Y. Dodge (ed.),

L_1-Statistical Procedures and Related Topics, Lecture Notes–Monograph Series, Vol. 31, Institute of Mathematical Statistics, Hayward, CA, pp. 15–32.

Koenker, R. and Bassett, G. (1978). Regression quantiles, Econometrica 46: 33–50.

Koenker, R. and Park, B. J. (1996). An interior point algorithm for nonlinear quantile regression, Journal of Econometrics 71: 265–283.

Křivý, I., Tvrdík, J. and Krpec, R. (2000). Stochastic algorithms in nonlinear regression, Computational Statistics and Data Analysis 33: 277–290.

Leamer, E. E. (1990). Optimal aggregation of linear net export systems, in T. Barker and M. H. Pesaran (eds), Disaggregation in Econometric Modelling, Routledge, London, pp. 150–170.

L'Ecuyer, P. and Hellekalek, P. (1998). Random number generators: Selection criteria and testing, in P. Hellekalek and G. Larcher (eds), Random and Quasi-Random Point Sets, Lecture Notes in Statistics 138, Springer, New York, pp. 223–265.

Lee, K., Pesaran, M. H. and Pierse, R. G. (1990). Aggregation bias in labour demand equations for the UK economy, in T. Barker and M. H. Pesaran (eds), Disaggregation in Econometric Modelling, Routledge, London, pp. 113–149.

Li, J. S. and Winker, P. (2000). Time series simulation with quasi–Monte Carlo methods, Technical Report forthcoming, Department of Economics, Pennsylvania State University.

Li, W. and Fang, K.-T. (1995). A global optimum algorithm on two factor uniform design, in K.-T. Fang and F. J. Hickernell (eds), Proceedings of the Workshop on Quasi–Monte Carlo Methods and their Applications., Hong Kong Baptist University, Hong Kong, pp. 189–201.

Lin, C. K. Y., Haley, K. B. and Sparks, C. (1995). A comparative study of both standard and adaptive versions of threshold accepting and simulated annealing algorithms in three scheduling problems, European Journal of Operational Research 83: 330–347.

Lütkepohl, H. (1985). Comparison of criteria for estimating the order of a vector autoregressive process, Journal of Time Series Analysis 6: 35–52.

Lütkepohl, H. (1991). Introduction to Multiple Time Series Analysis, Springer, Berlin.

Lütkepohl, H. and Wolters, J. (1998). A money demand system for German M3, Empirical Economics 23: 371–386.

Lütkepohl, H. and Wolters, J. (eds) (1999). Money Demand in Europe, Physica, Heidelberg.

Ma, C. and Fang, K.-T. (2000). A new approach in constructing orthogonal and nearly orthogonal arrays, Metrica : forthcoming.

Malinvaud, E. (1956). L'agrégation dans les modéles économiques, Cahiers du Séminaire d'Econométrie 4: 69–146.

Markowitz, H. (1952). Portfolio selection, Journal of Finance 7: 77–91.

Martin, V. L. and Wilkins, N. P. (1997). Indirect estimation of ARFIMA and VARFIMA models, Technical Report 547, University of Melbourne.

Mayer, W. J. (1989). Estimating disequilibrium models with limited a priori price-adjustment information, Journal of Econometrics 41: 303–320.

Mayr, E. W. and Meyer, A. R. (1982). The complexity of the word problems for computative semigroups and polynomial ideals, Advances in Mathematics 46: 305–329.

McClave, J. (1975). Subset autoregression, Technometrics 17: 213–220.

McCullough, B. D. and Vinod, H. D. (1999). The numerical reliability of econometric software, *Journal of Economic Literature* **38**: 633–665.

McFadden, D. F. (1999). Rationality for economists, *Journal of Risk and Uncertainty* **19**: 73–105.

Metropolis, W., Rosenbluth, A., Rosenbluth, M., Teller, A. and Teller, E. (1953). Equation of the state calculations by fast computing machines, *Journal of Chemical Physics* **21**: 1087–1092.

Meyn, S. P. and Tweedie, R. L. (1993). *Markov Chains and Stochastic Stability*, Springer, London.

Mitchell, M. (1996). *An Introduction to Genetic Algorithms*, MIT Press, Cambridge, MA.

Morokoff, W. J. and Caflisch, R. E. (1998). Quasi–Monte Carlo simulation of random walks in finance, *in* H. Niederreiter, P. Zinterhof and P. Hellekalek (eds), *Monte Carlo and Quasi–Monte Carlo Methods 1996*, Lecture Notes in Statistics 127, Springer, New York, pp. 340–352.

Neck, R. and Karbuz, S. (1995). Optimal budgetary and monetary policies under uncertainty. A stochastic control approach, *Annals of Operations Research* **58**: 379–402.

Niederreiter, H. (1992). *Random Number Generation and Quasi–Monte Carlo Methods*, Regional Conference Series in Applied Mathematics, SIAM CBMS–NSF, Philadelphia, PE.

Nissen, V. and Paul, H. (1995). A modification of threshold accepting and its application to the quadratic assignement problem, *OR Spektrum* **17**: 205–210.

Norris, J. R. (1997). *Markov Chains*, Cambridge University Press, Cambridge.

Nurmela, K. J. (1993). Constructing combinatorial designs by local search, *Research Report 27*, Helsinki University of Technology.

Oke, T. and Lyhagen, J. (1999). Small-sample properties of some test for unit root with data–based choice of the degree of augmentation, *Computational Statistics and Data Analysis* **30**: 457–469.

Osman, I. H. and Christofides, N. (1994). Capacitated clustering problems by hybrid simulated annealing and tabu search, *International Transactions in Operational Research* **1**: 317–336.

Osman, I. H. and Kelly, J. P. (1996). Metaheuristics: An overview, *in* I. H. Osman and J. P. Kelly (eds), *Meta–Heuristics: Theory and Applications*, Kluwer, Boston, Ma, pp. 1–21.

Osman, I. H. and Laporte, G. (1996). Metaheuristics: A bibliography, *Annals of Operations Research* **63**: 513–623.

Pagano, M. (1972). An algorithm for fitting autoregressive systems, *Applied Statistics* **21**: 274–281.

Papadimitriou, C. H. and Steiglitz, K. (1982). *Combinatorial Optimization*, Prentice–Hall, Englewood Cliffs, NJ.

Paul, W. J. (1978). *Komplexitätstheorie*, B. G. Teubner, Stuttgart.

Pesaran, M. H., Pierse, R. G. and Kumar, M. S. (1989). Econometric analysis of aggregation in the context of linear prediction models, *Econometrica* **57**: 861–888.

Pesaran, M. H. and Timmermann, A. (1995). Predictability of stock returns: Robustness and economic significance, *The Journal of Finance* **50**: 1201–1228.

Pham, D. T. and Karaboga, D. (2000). *Intelligent Optimisation Techniques*, Springer, London.

Phillips, P. C. B. (1995). Bayesian model selection and prediction with empirical applications, *Journal of Econometrics* **69**: 289–331.

Pierce, D. A. and Haugh, L. D. (1977). Causality in temporal systems, *Journal of*

Econometrics **5**: 265–293.

Pinske, C. A. P. (1993). On the computation of semiparametric estimates in limted dependent variable models, *Journal of Econometrics* **58**: 185–205.

Pirlot, M. (1992). General local search heuristics in combinatorial optimization: A tutorial, *Belgian Journal of Operations Research, Statistics and Computer Science* **32**: 7–67.

Polak, E. (1997). *Optimization*, Springer, New York.

Pötscher, B. M. (1991). Effects of model selection on inference, *Econometric Theory* **7**: 163–185.

Potter, S. (1995). A nonlinear approach to U.S. GNP, *Journal of Applied Econometrics* **10**: 109–125.

Potvin, J.-Y. (1996). Genetic algorithms for the traveling salesman problem, *Annals of Operations Research* **63**: 339–370.

Powell, J. L. (1984). Least absolute deviations estimation for the censored regression model, *Journal of Econometrics* **25**: 303–325.

Press, W. H., Teukolsky, S. A., Vetterling, W. T. and Flannery, B. P. (1992). *Numerical Recipes in Fortran 77*, 2nd Edn, Cambridge University Press, New York.

Pukelsheim, F. (1993). *Optimal Design of Experiments*, Wiley, New York.

Quinn, B. G. (1980). Order determination for a multivariate autoregression, *Journal of the Royal Statistical Society* B **42**: 182–185.

Raghavarao, D. (1971). *Constructions and Combinatorial Problems in Designs of Experiments*, Wiley, New York.

Reischuk, K. J. (1990). *Einführung in die Komplexitätstheorie*, B. G. Teubner, Stuttgart.

Revuz, D. (1984). *Markov-Chains*, 2nd Edn, North-Holland, Amsterdam.

Rodepeter, R. and Winter, J. K. (1999). Rules of thumb in life–cycle savings models, *Discussion Paper 99-81*, University of Mannheim.

Romeijn, H. E. and Smith, R. L. (1994). Simulated annealing for constrained global optimization, *Journal of Global Optimization* **5**: 101–126.

Rossier, Y., Troyon, M. and Liebling, T. M. (1986). Probabilistic exchange algorithm and euklidean traveling salesman problems, *OR Spectrum* **8**: 151–164.

Ryan, J. (1995). The depth and width of local minima in discrete solution spaces, *Discrete Applied Mathematics* **56**: 75–82.

Sacks, J., Welch, W. J., Mitchell, T. J. and Wynn, H. P. (1989). Design and analysis of computer experiments, *Statistical Science* **4**: 409–435.

Samuelson, P. A. (1953). Prices of factors and goods in general equilibrium, *Review of Economic Studies* **21**: 1–20.

Schrimpf, G., Schneider, J., Stamm-Wilbrandt, H. and Dueck, G. (1998). *Record Breaking Optimization Results Using the Ruine and Recreate Principle*, IBM Scientific Center Heidelberg, Heidelberg.

Schwarz, G. (1978). Estimating the dimension of a model, *The Annals of Statistics* **6**: 461–464.

Selten, R. (1998). Features of experimentally observed bounded rationality, *European Economic Review* **42**: 413–436.

Seneta, E. (1981). *Non-negative Matrices and Markov-Chains*, Springer, New York.

Shepard, R. W. (1981). *Cost and Production Functions*, Princeton University Press, reprinted by Springer, Berlin.

Shibata, R. (1976). Selection of the order of an autoregressive model by Akaike's information criterion, *Biometrika* **63**: 117–126.

Siedentopf, J. (1995). An efficient scheduling algorithm based upon threshold accepting, *Technical Report No. 16*, Institute of Production Management and Industrial Information Management at the University of Leipzig, Leipzig.

Sims, C. A. (1972). Money, income and causality, *American Economic Review* **62**: 540–552.

Sims, C. A. (1988). Bayesian scepticism on unit root econometrics, *Journal of Economic Dynamics and Control* **12**: 463–474.

Sinclair, M. (1993). Comparison of the performance of modern heuristics for combinatorial optimization on real data, *Computers Operations Research* **20**: 687–695.

Stewart, B. S., Liaw, C.-F. and White, C. (1994). A bibliography of heuristic search research through 1992, *IEEE Transactions on Systems, Man and Cybernetics* **24**: 268–293.

Strassen, V. (1969). Gaussian elimination is not optimal, *Numerische Mathematik* **13**: 354–356.

Tajuddin, W. A. and Abdullah, W. (1994). Seeking global minima, *Journal of Computational Physics* **110**: 320–326.

Tezuka, S. (1998). Financial applications of Monte Carlo and quasi–Monte Carlo methods, *in* P. Hellekalek and G. Larcher (eds), *Random and Quasi-Random Point Sets*, Lecture Notes in Statistics 138, Springer, New York, pp. 303–332.

Theil, H. (1954). *Linear Aggregation of Economic Relations*, North–Holland, Amsterdam.

Thompson, G. D. and Lyon, C. C. (1992). A generalized test for perfect aggregation, *Economics Letters* **40**: 389–396.

Thornton, D. L. and Battan, D. S. (1985). Lag–length selection and tests of Granger causality between money and income, *Journal of Money, Credit and Banking* **27**: 164–178.

Today, H. Y. and Phillips, P. C. B. (1994). Vector autoregression and causality: A theoretical overview and simulation study, *Econometric Reviews* **13**: 259–285.

Toutenburg, H. (1995). *Experimental Design and Model Choice*, Physica, Heidelberg.

Trinca, L. A. and Gilmour, S. G. (2000). An algorithm for arranging response surface designs in small blocks, *Computational Statistics and Data Analysis* **33**: 25–43.

Tsurumi, H. and Wago, H. (1991). Mean squared errors of forecast for selecting nonnested linear models and comparison with other criteria, *Journal of Econometrics* **48**: 215–240.

Ulder, N. L. J., Aarts, E. H. L., Bandelt, H.-J., van Laarhoven, P. J. M. and Pesch, E. (1990). Genetic local search algorithms for the traveling salesman problem, *Lecture Notes in Computer Sciences*, Vol. 496, pp. 109–116.

Velupillai, K. (2000). *Computable Economics*, Oxford University Press, Oxford.

Warnock, T. T. (1972). Computational investigations of low discrepancy point sets, *in* S. K. Zaremba (ed.), *Applications of Number Theory to Numerical Analysis*, Academic Press, New York, pp. 319–343.

Webb, R. H. (1985). Toward more accurate macroeconomic forecasts from vector autoregressions, *Economic Review of the Federal Reserve Bank of Richmond*, pp. 3–11.

Weyl, H. (1916). Über die Gleichverteilung der Zahlen mod Eins, *Mathematische Annalen* **77**: 313–357.

Wilf, H. S. (1986). *Algorithms and Complexity*, Prentice–Hall, Englewood Cliffs, NJ.

Williams, H. P. (1978). *Model Building in Mathematical Programming*, Wiley, Chichester.

Winker, P. (1991). Die Platzkomplexität des Wortproblems für kommutative Halbgruppenpräsentationen und des Zugehörigkeitsproblems für rationale Polynomideale, *Master Thesis*, Universität Konstanz, Konstanz. http://www.mathe.uni-konstanz.de/~strassen/lehre_und_forschung.html.

Winker, P. (1995). Identification of multivariate AR–models by threshold accepting, *Computational Statistics and Data Analysis* **20**: 295–307.

Winker, P. (1999a). Sluggish adjustment of interest rates and credit rationing, *Applied Economics* **31**: 267–277.

Winker, P. (1999b). Quasi–Monte Carlo policy simulations in a macroeconometric disequilibrium model, *Proceedings of the 14th World Congress International Federation of Automatic Control*, Elsevier, Oxford, pp. 45–50.

Winker, P. (2000a). Automatic multivariate lag structure identification, *Advances in Artificial Intelligence in Economics, Finance and Management*, Vol. , forthcoming.

Winker, P. (2000b). Optimized multivariate lag structure selection, *Computational Economics* : forthcoming.

Winker, P. and Fang, K.-T. (1997). Application of threshold accepting to the evaluation of the discrepancy of a set of points, *SIAM Journal on Numerical Analysis* **34**: 2028–2042.

Winker, P. and Fang, K.-T. (1998). Optimal U–type designs, *in* H. Niederreiter, P. Zinterhof and P. Hellekalek (eds), *Monte Carlo and Quasi–Monte Carlo Methods 1996*, Lecture Notes in Statistics 127, Springer, New York, pp. 436–448.

Womersley, R. S. (1986). Censored discrete l_1 approximation, *SIAM Journal of Scientific Statistical Computations* **7**: 105–122.

Zaman, A. (1984). Avoiding model selection by the use of shrinkage techniques, *Journal of Econometrics* **25**: 73–85.

Zellner, A. (1962). An efficient method of estimating seemingly unrelated regressions, and tests for aggregation bias, *Journal of the American Statistical Association* **57**: 348–368.

Zhang, H.-M. and Wang, P. (1994). A new way to estimate orders in time series, *Journal of Time Series Analysis* **15**: 545–559.

List of Symbols

Author Index

Subject Index

WILEY SERIES IN PROBABILITY AND STATISTICS

ESTABLISHED BY WALTER A. SHEWHART AND SAMUEL S. WILKS

Editors

*Vic Barnett, Noel A. C. Cressie, Nicholas I. Fisher, Iain M. Johnstone,
J. B. Kadane, David W. Scott, Bernard W. Silverman,
Adrian F. M. Smith, Jozef L. Teugels, Ralph A. Bradley, Emeritus,
J. Stuart Hunter, Emeritus, David G. Kendall, Emeritus*

Probability and Statistics Section

*ANDERSON · The Statistical Analysis of Time Series
ARNOLD, BALAKRISHNAN, and NAGARAJA · A First Course in Order Statistics
ARNOLD, BALAKRISHNAN, and NAGARAJA · Records
BACCELLI, COHEN, OLSDER, and QUADRAT · Synchronization and Linearity: An
    Algebra for Discrete Event Systems
BARNETT · Comparative Statistical Inference, *Third Edition*
BASILEVSKY · Statistical Factor Analysis and Related Methods: Theory and
    Applications
BERNARDO and SMITH · Bayesian Theory
BILLINGSLEY · Convergence of Probability Measures, *Second Edition*
BOROVKOV · Asymptotic Methods in Queuing Theory
BOROVKOV · Ergodicity and Stability of Stochastic Processes
BRANDT, FRANKEN, and LISEK · Stationary Stochastic Models
CAINES · Linear Stochastic Systems
CAIROLI and DALANG · Sequential Stochastic Optimization
CONSTANTINE · Combinatorial Theory and Statistical Design
COOK · Regression Graphics
COVER and THOMAS · Elements of Information Theory
CSÖRGŐ and HORVÁTH · Weighted Approximations in Probability Statistics
CSÖRGŐ and HORVÁTH · Limit Theorems in Change Point Analysis
*DANIEL · Fitting Equations to Data: Computer Analysis of Multifactor Data, *Second
    Edition*
DETTE and STUDDEN · The Theory of Canonical Moments with Applications in
    Statistics, Probability, and Analysis
DEY and MUKERJEE · Fractional Factional Plans
*DOOB · Stochastic Processes
DRYDEN and MARDIA · Statistical Shape Analysis
DUPUIS and ELLIS · A Weak Convergence Approach to the Theory of Large
    Deviations
ETHIER and KURTZ · Markov Processes: Characterization and Convergence
FELLER · An Introduction to Probability Theory and Its Applications, Volume 1, *Third
    Edition*, Revised; Volume II, *Second Edition*
FULLER · Introduction to Statistical Time Series, *Second Edition*
FULLER · Measurement Error Models

*Now available in a lower priced paperback edition in the Wiley Classics Library.

*Now available in a lower priced paperback edition in the Wiley Classics Library.

*Now available in a lower priced paperback edition in the Wiley Classics Library.

Applied Probability and Statistics (Continued)

DANIEL · Biostatistics: A Foundation for Analysis in the Health Sciences, *Sixth Edition*

DAVID · Order Statistics, *Second Edition*

*DEGROOT, FIENBERG, and KADANE · Statistics and the Law

DODGE · Alternative Methods of Regression

DOWDY and WEARDEN · Statistics for Research, *Second Edition*

DUNN and CLARK · Applied Statistics: Analysis of Variance and Regression, *Second Edition*

*ELANDT-JOHNSON and JOHNSON · Survival Models and Data Analysis

*FLEISS · The Design and Analysis of Clinical Experiments

FLEISS · Statistical Methods for Rates and Proportions, *Second Edition*

FLEMING and HARRINGTON · Counting Processes and Survival Analysis

GALLANT · Nonlinear Statistical Models

GLASSERMAN and YAO · Monotone Structure in Discrete-Event Systems

GNANADESIKAN · Methods for Statistical Data Analysis of Multivariate Observations. *Second Edition*

GOLDSTEIN and LEWIS · Assessment: Problems, Development, and Statistical Issues

GREENWOOD and NIKULIN · A Guide to Chi-squared Testing

*HAHN · Statistical Models in Engineering

HAHN and MEEKER · Statistical Intervals: A Guide for Practitioners

HAND · Construction and Assessment of Classification Rules

HAND · Discrimination and Classification

HEDAYAT and SINHA · Design and Inference in Finite Population Sampling

HEIBERGER · Computation for the Analysis of Designed Experiments

HINKELMAN and KEMPTHORNE · Design and Analysis of Experiments, Volume 1: Introduction to Experimental Design

HOAGLIN, MOSTELLER, and TUKEY · Exploratory Approach to Analysis of Variance

HOAGLIN, MOSTELLER, and TUKEY · Exploring Data Tables, Trends and Shapes

HOAGLIN, MOSTELLER, and TUKEY · Understanding Robust and Exploratory Data Analysis

HOCHBERG and TAMHANE · Multiple Comparison Procedures

HOCKING · Methods and Applications of Linear Models: Regression and the Analysis of Variables

HOGG and KLUGMAN · Loss Distributions

HOSMER and LEMESHOW · Applied Logistic Regression

HØYLAND and RAUSAND · System Reliability Theory: Models and Statistical Methods

HUBERTY · Applied Discriminant Analysis

HUNT and KENNEDY · Financial Derivatives in Theory and Practice

JACKSON · A User's Guide to Principal Components

JOHN · Statistical Methods in Engineering and Quality Assurance

JOHNSON · Multivariate Statistical Simulation

JOHNSON & KOTZ · Distributions in Statistics

JOHNSON, KOTZ, and BALAKRISHNAN · Continuous Univariate Distributions, Volume 1, *Second Edition*

JOHNSON, KOTZ, and BALAKRISHNAN · Continuous Univariate Distributions, Volume 2, *Second Edition*

JOHNSON, KOTZ, and BALAKRISHNAN · Discrete Multivariate Distributions

JOHNSON, KOTZ, and KEMP · Univariate Discrete Distributions, *Second Edition*

JUREČKOVÁ and SEN · Robust Statistical Procedures: Asymptotics and Interrelations

KADANE · Bayesian Methods and Ethics in a Clinical Trial Design

KADANE and SCHUM · A Probabilistic Analysis of the Sacco and Vanzetti Evidence

---

*Now available in a lower priced paperback edition in the Wiley Classics Library.

---

*Now available in a lower priced paperback edition in the Wiley Classics Library.

Applied Probability and Statistics (Continued)

ROLSKI, SCHMIDLI, SCHMIDT, and TEUGELS · Stochastic Processes for Insurance and Finance

ROUSSEEUW and LEROY · Robust Regression and Outlier Detection

RUBIN · Multiple Imputation for Nonresponse in Surveys

RUBINSTEIN · Simulation and the Monte Carlo Method

RUBINSTEIN and MELAMED · Modern Simulation and Modeling

RYAN · Statistical Methods for Quality Improvement, *Second Edition*

SCHUSS · Theory and Applications of Stochastic Differential Equations

SCOTT · Multivariate Density Estimation: Theory, Practice, and Visualization

*SEARLE · Linear Models

SEARLE · Linear Models for Unbalanced Data

SEARLE, CASELLA, and McCULLOCH · Variance Components

STOYAN, KENDALL, and MECKE · Stochastic Geometry and Its Applications, *Second Edition*

STOYAN and STOYAN · Fractals, Random Shapes, and Point Fields: Methods of Geometrical Statistics

SUTTON, ABRAMS, JONES, SHELDON and SONG · Methods for Meta-analysis in Medical Research

THOMPSON · Empirical Model Building

THOMPSON · Sampling

THOMPSON · Simulation: A Modeler's Approach

TIJMS · Stochastic Modeling and Analysis: A Computational Approach

TIJMS · Stochastic Models: An Algorithmic Approach

TITTERINGTON, SMITH, and MARKOV · Statistical Analysis of Finite Mixture Distributions

UPTON and FINGLETON · Spatial Data Analysis by Example, Volume 1: Point Pattern and Quantitative Data

UPTON and FINGLETON · Spatial Data Analysis by Example, Volume II: Categorical and Directional Data

VAN RIJCKEVORSEL and DE LEEUW · Component and Correspondence Analysis

VIDAKOVIC · Statistical Modeling by Wavelets

WEISBERG · Applied Linear Regression, *Second Edition*

WESTFALL and YOUNG · Resampling-Based Multiple Testing: Examples and Methods for p-Value Adjustment

WHITTLE · Systems in Stochastic Equilibrium

WINKER · Optimization Heuristics in Econometrics

WOODING · Planning Pharmaceutical Clinical Trials: Basic Statistical Principles

WOOLSON · Statistical Methods for the Analysis of Biomedical Data

*ZELLNER · An Introduction to Bayesian Inference in Econometrics

Texts and References Section

AGRESTI · An Introduction to Categorical Data Analysis

ANDERSON · An Introduction to Multivariate Statistical Analysis, *Second Edition*

ANDERSON and LOYNES · The Teaching of Practical Statistics

ARMITAGE and COLTON · Encyclopedia of Biostatistics. 6 Volume set

BARTOSZYNSKI and NIEWIADOMSKA-BUGAJ · Probability and Statistical Inference

BENDAT and PIERSOL · Random Data: Analysis and Measurement Procedures, *Third Edition*

*Now available in a lower priced paperback edition in the Wiley Classics Library.

Texts and References Section (Continued)

BERRY, CHALONER, and GEWEKE · Bayesian Analysis in Statistics and Econometrics: Essays in Honor of Arnold Zellner

BHATTACHARYA and JOHNSON · Statistical Concepts and Methods

BILLINGSLEY · Probability and Measure, *Second Edition*

BOX · R. A. Fisher, the Life of a Scientist

BOX, HUNTER, and HUNTER · Statistics for Experimenters: An Introduction to Design, Data Analysis, and Model Building

BOX and LUCEÑO · Statistical Control by Monitoring and Feedback Adjustment

BROWN and HOLLANDER · Statistics: A Biomedical Introduction

CHATTERJEE and PRICE · Regression Analysis by Example, *Third Edition*

COOK and WEISBERG · An Introduction to Regression Graphics

COOK and WEISBERG · Applied Regression Including Computing and Graphics

COX · A Handbook of Introductory Statistical Methods

DILLON and GOLDSTEIN · Multivariate Analysis: Methods and Applications

*DODGE and ROMIG · Sampling Inspection Tables, *Second Edition*

DRAPER and SMITH · Applied Regression Analysis, *Third Edition*

DUDEWICZ and MISHRA · Modern Mathematical Statistics

DUNN · Basic Statistics: A Primer for the Biomedical Sciences, *Second Edition*

EVANS, HASTINGS and PEACOCK · Statistical Distributions, *Third Edition*

FISHER and VAN BELLE · Biostatistics: A Methodology for the Health Sciences

FREEMAN and SMITH · Aspects of Uncertainty: A Tribute to D. V. Lindley

GROSS and HARRIS · Fundamentals of Queueing Theory, *Third Edition*

HALD · A History of Probability and Statistics and their Applications Before 1750

HALD · A History of Mathematical Statistics from 1750 to 1930

HELLER · MACSYMA for Statisticians

HOEL · Introduction to Mathematical Statistics, *Fifth Edition*

HOLLANDER and WOLFE · Nonparametric Statistical Methods, *Second Edition*

HOSMER and LEMESHOW · Applied Logistic Recession, *Second Edition*

HOSMER and LEMESHOW · Applied Survival Analysis: Regression Modeling of Time to Event Data

JOHNSON and BALAKRISHNAN · Advances in the Theory and Practice of Statistics: A Volume in Honor of Samuel Kotz

JOHNSON and KOTZ (editors) · Leading Personalities in Statistical Sciences: From the Seventeenth Century to the Present

JUDGE, GRIFFITHS, HILL, LÜTKEPOHL, and LEE · The Theory and Practice of Econometrics, *Second Edition*

KHURI · Advanced Calculus with Applications in Statistics

KOTZ and JOHNSON (editors) · Encyclopedia of Statistical Sciences. Volumes 1 to 9 with Index

KOTZ and JOHNSON (editors) · Encyclopedia of Statistical Sciences: Supplement Volume

KOTZ, REED, and BANKS (editors) · Encyclopedia of Statistical Sciences: Update Volume 1

KOTZ, REED, and BANKS (editors) · Encyclopedia of Statistical Sciences: Update Volume 2

LAMPERTI · Probability: A Survey of the Mathematical Theory, *Second Edition*

LARSON · Introduction to Probability Theory and Statistical Inference, *Third Edition*

LE · Applied Categorical Data Analysis

LE · Applied Survival Analysis

MALLOWS · Design, Data, and Analysis by Some Friends of Cuthbert Daniel

MARDIA · The Art of Statistical Science: A Tribute to G. S. Watson

---

*Now available in a lower priced paperback edition in the Wiley Classics Library.

WILEY SERIES IN PROBABILITY AND STATISTICS

ESTABLISHED BY WALTER A. SHEWHART AND SAMUEL S. WILKS

Editors
Robert M. Groves, Graham Kalton, J. N. K. Rao, Norbert Schwarz, Christopher Skinner

Survey Methodology Section

*Now available in a lower priced paperback edition in the Wiley Classics Library.